NELSON

SENIOR GRAPHICS

FOR QUEENSLAND SCHOOLS

KRISTEN GUTHRIE
Consultant: CHRIS RALPH

Australia • Brazil • Japan • Korea • Mexico • Singapore • Spain • United Kingdom • United States

Nelson Senior Graphics for Queensland Schools
1st Edition
Kristen Guthrie
Consultant: Chris Ralph

Publishing editor: Tanya Wasylewski
Project editor: Nadine Anderson-Conklin
Editor: Maria Bascombe
Proofreader: Nadine Anderson-Conklin
Indexer: Bruce Gillespie
Permissions researcher: Debbie Gallagher
Art direction: Luana Keays
Text design: Sarah Anderson
Cover design: Kristen Guthrie
Cover image: Kristen Guthrie/Ben Jennings/Mark Wilken/Tom Grech
Typesetter: Q2AMedia
Reprint: Jess Lovell

Any URLs contained in this publication were checked for currency during the production process. Note, however, that the publisher cannot vouch for the ongoing currency of URLs.

For product information and technology assistance,
in Australia call **1300 790 853**;
in New Zealand call **0800 449 725**

For permission to use material from this text or product, please email **aust.permissions@cengage.com**

National Library of Australia Cataloguing-in-Publication Data
Guthrie, Kristen, author.
Nelson senior graphics : for Queensland / Kristen Guthrie ;
Chris Ralph, consultant.

9780170349994 (paperback)

Graphic arts--Study and teaching (Secondary)
Graphic arts--Textbooks.
Design--Study and teaching (Secondary)
Design--Textbooks.

741.6

Cengage Learning Australia
Level 7, 80 Dorcas Street
South Melbourne, Victoria Australia 3205

Cengage Learning New Zealand
Unit 4B Rosedale Office Park
331 Rosedale Road, Albany, North Shore 0632, NZ

For learning solutions, visit **cengage.com.au**

Printed in China by China Translation & Printing Services.
4 5 6 7 8 9 10 20 19 18 17 16

CONTENTS

ISBN 9780170349994

C

Design factors 81

D

Graphical representations 185

ISBN 9780170349994

INTRODUCTION TO SENIOR GRAPHICS

'Design is thinking made visual.'
Saul Bass

Design is embedded in most aspects of our daily lives. Many of the spaces in which we live, the products we use and the images we see originated from some form of design process.

As a means of presenting products, identifying environments and of expressing ideas and information, we rely on design to innovate, streamline, beautify, engage and enhance our lives. Design can influence how we feel within a space, react to an issue, form an opinion or purchase a product.

Professional designers working in many design areas, including graphic design, industrial design and built environment design (architecture, interior design and landscape architecture) understand the power of the visual, and they apply various techniques to express ideas, develop concepts and make a visual connection with an audience.

The universal applications of digital technologies have meant that the communication of visual ideas and the presentation of graphical products now occur in rapid and sophisticated forms. Developments in technology and production methods, as well as social changes, serve to redefine the boundaries of visual language. Understanding developments in design areas and having a grasp of their origins helps us to understand shifts and changes in the visual world.

It is important for Senior Graphics students to develop an informed and analytical understanding of the scope and potential of design. The study of Senior Graphics will help you, as a student, to understand the important role that design plays in our lives.

Throughout your studies in Senior Graphics you will be asked to convey concepts and devise solutions to design problems that rely on visual means for clarification. This book is designed to guide you through the core knowledge and assist you in the development of many key skills required for successful production of design outcomes.

One of the central elements of Senior Graphics is the design process. Facilitating a creative progression from the products, the design process provides a defined yet flexible space for imaginative responses and experimentation where concepts are developed and refined. You will be given many opportunities to develop creative and innovative solutions to design problems, challenging you to extend your ideas, skills and knowledge within the framework of the design process.

Senior Graphics covers a vast amount of information and introduces many practical techniques and theoretical knowledge. All aspects of the subject are intrinsically linked, and a comprehensive grasp of all areas fosters an ability to successfully develop, evaluate and produce effective designs.

This book is divided into four parts, to reflect the core subject matter of the Senior Graphics syllabus. You will find explanations and visual examples of content directly related to these areas. This book provides a framework for your critical and creative thinking, and provides you with the tools to understand, develop and apply effective design tools to your own work. The websites referred to throughout the book can be accessed directly via the free, unprotected weblinks site at http://nsg.nelsonnet.com.au. A secure NelsonNet teacher website is also provided.*

Disclaimer

*Please note that complimentary access to NelsonNet is only available to teachers who use the accompanying student textbook as a core educational resource in their classroom. Contact your sales representative for information about access codes and conditions.

ISBN 9780170349994

WHERE CAN SENIOR GRAPHICS TAKE YOU?

Many students choose to study Senior Graphics to prepare or build a folio for entry to a tertiary institution. Many visual art and design courses at TAFE and university accept students after viewing a visual art folio during a preliminary interview. Keep in mind that many universities insist on good academic results as well as an impressive folio.

Senior Graphics can lead to many different study and career paths, some of which are listed below.

+ Advertising
+ Animation
+ Architectural drafting
+ Architecture
+ Cartography
+ Cartooning
+ Costume design
+ Education
+ Engineering
+ Fashion design
+ Furniture design
+ Graphic design/ Communication design
+ Graphic prepress, printing and production
+ Illustration
+ Industrial design
+ Interior decoration
+ Interior design
+ Landscape architecture
+ Multimedia development
+ Set/Theatre design
+ Signwriting
+ Visual merchandising
+ Web design

It does not matter whether you choose to study Graphics to move towards a career in design, because it offers stimulation and challenges not found elsewhere in the curriculum, or simply because you like to draw. You have made your first move into a study that will change the way you view the world around you!

HOW TO USE THIS BOOK

To assist you in using this book, icons are placed throughout to indicate the following:

FYI **For your information**

Information to read that may expand your interest in the topic

Tip

Helpful information to assist in developing your skills

Tech tip

Helpful information to assist in developing your skills with digital media

Weblink

Websites that contain information that may assist your learning; access weblinks directly at http://nsg.nelsonnet.com.au.

Activity

Practical exercises to help develop your skills

abc **Key word**

Fundamental vocabulary

Key point

Essential information

AUTHOR ACKNOWLEDGEMENTS

Thank you to my students, past and present, whose talents and skills never fail to inspire and surprise me!

A huge thanks must go to designer Mark Wilken of Studio Workshops for his great illustrations and advice. Thanks also to the professional designers who were happy to have their work featured throughout this book.

Many thanks to the Cengage team, especially the publisher, Tanya Wasylewski, and the project editor, Nadine Anderson-Conklin, for their support and hard work. Special thanks also to editor Marcia Bascombe and permissions researcher Debbie Gallagher for their patience, conscientiousness and expertise.

–Kristen Guthrie

ISBN 9780170349994

PART A

DESIGN AREAS

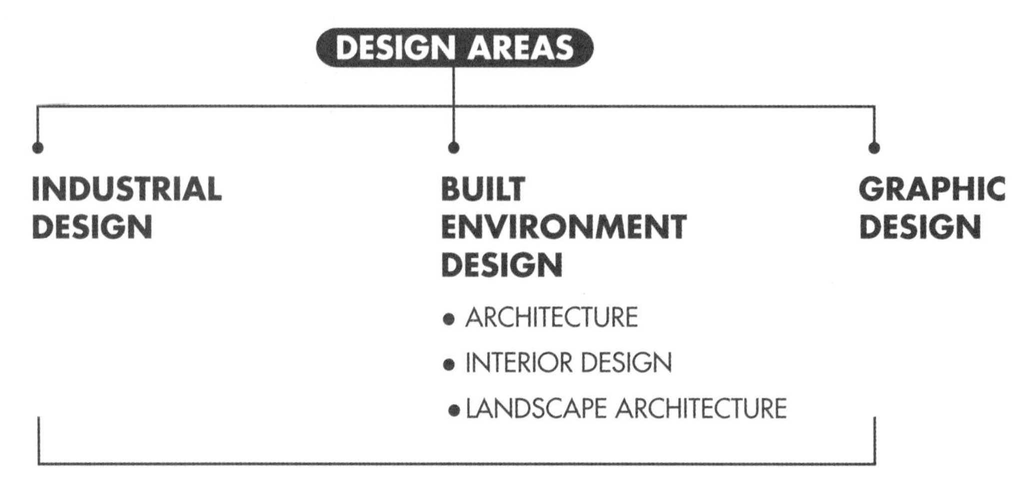

Designers work in a variety of professional contexts and environments.

In all design areas, designers apply a design process that involves research,exploration and development of ideas, collaboration, evaluation and production.

Three main design areas are identified in the Senior Graphics syllabus: industrial design, built environment design and graphic design. Each design area has its own language, traditions, origins and influences that distinguish it from other professional fields. Designers in all areas follow an established process to make the most efficient use of time, skills and resources. The following chapters go some way towards explaining the skills, responsibilities and design processes applied by design professionals.

Design is a fluid professional field and designers may find themselves working independently or within a team at different times in their career. Similarly, many designers work across disciplines, combining product and interior design, architectural design and landscape design, graphic design and multimedia design. Design is a profession that is dynamic and ever changing.

However, designers work in a range of professional configurations; three of the most common contexts are as freelance designers, studio-based designers and in-house designers.

PROFESSIONAL CONTEXTS

FREELANCE DESIGNERS

Freelance designers are individuals who work independently rather than as an employee of a design business or other organisation. Often, freelancers operate as a one-person firm, the smallest of small businesses. They may work from a home-based office or in a shared studio with other professionals. Freelance designers, illustrators, photographers, animators and other creative professionals acquire work from direct contact with clients or as an external contractor, employed by an advertising agency or design firm on a short-term basis as part of a larger team.

A freelance designer is usually responsible not only for their creative output but also for administrative aspects of the business, such as finance, tax and all communications. For many, the appeal of freelance work lies in its autonomy and independence but, as with many small and single-person businesses, the working conditions are dependent on the amount of work available.

ISBN 9780170349994

DESIGN STUDIOS

Many designers work for studios, which can vary in size from small partnerships to large organisations. Within a design studio, the designer rarely works alone on a project, and is often part of a large team. The team may include a creative director or project manager whose role is to manage both the project itself and the professionals involved in the project's development and production. A design team may consist of designers, administrative staff and support staff, as well as external contractors and consultants. Contact with the client may be restricted to the creative director, project manager or partner of the studio, who will then pass on the client needs and details of the brief to the larger team, usually at team meetings.

A team structure provides a creative network for the flow of ideas and possible concepts, and facilitates feedback and evaluation. Unlike freelancers, designers who work for design studios are often provided with administrative support, thereby enabling them to focus entirely on the creative aspects of the design process.

IN-HOUSE DESIGNERS

Large organisations often employ designers to manage design tasks in-house. Many government organisations and private companies have full-time design teams working as an integral part of the company. Companies, such as automotive manufacturers or film production companies for example, require the services of designers as an essential part of their product design and development.

Although the main focus of a company or organisation, such as a hospital or airport, may be fundamentally unrelated to design, there is often a requirement for the design and production of materials for promotion, employee training, shareholder information and the like. Some organisations may need designers to produce annual reports, signage, newsletters and training materials.

Although many companies and organisations outsource work to independent design firms, for large projects, such as corporate identity design, they often employ in-house designers to manage the ongoing application of that identity and to produce company- or organisation-specific materials.

Irrespective of the variety of approaches and professional circumstances, there are many similarities in the way professional designers approach key aspects of the design process.

THE DESIGN BRIEF

The design brief is the starting point of the design process for professional designers in all design areas, and the nature and detail of that brief can vary widely. A client may sometimes approach a designer with a detailed description of a design need, but more often, the client simply has an idea and is depending on the designer to assist in expanding and clarifying the feasibility of that idea. Designers may also gain project work by submitting an application to a client, known as a tender proposal. A client may receive numerous proposals from different design firms; within each proposal there is information on design ideas, costs and a timeline. To determine the most appropriate designer for the task, a client may request a 'pitch' where the designer presents their ideas and proposal to the client directly.

In the early stages of the design process a designer will spend time with a client to gain a clear and detailed understanding of the client's needs and to ascertain whether a suitable working relationship can be established. Designers should listen carefully to the needs of the client, and it is at this stage that a mutual understanding of both the nature of the work and the fee is negotiated.

It is in the designer's interest to ensure that the brief is detailed, contains all the necessary information and provides a clear direction for the designer. Key questions that designers ask at this stage include the following: What does the client want to achieve? What are the expectations? What are the client's criteria for success? What is the budget? What is the preferred timeline? Is it realistic? What existing material – such as imagery, content and detailed market research analysis – is available?

ISBN 9780170349994

Client contact continues throughout the design process. Many designers require that clients accept each stage of the process by signing off (that is, approving) the work carried out so far. Formal client approval is particularly important on large projects, where the cost of the client changing their mind can be extremely high. Conditions covering changes to the contract or brief are usually established at the beginning of the process.

The scale of the task is often influenced by constraints, such as the cost, time and the context of the final visual communication, as well as the specific needs of the client. Constraints are usually clarified at the beginning of the design process but they can arise at any time. Consequently, the design process is never static and remains flexible.

Some designers use a return brief at this stage of the design process. This involves the designer writing a brief based on information gathered from the client at initial meetings. The return brief is then sent to the client as the designer's interpretation of the client's need. With further discussion, a final design brief can be established that suits both the client and the designer.

In general terms, the brief is about definition and clarification. In defining the actual needs of the client, the target audience and the purpose of the final design, a designer can gain the clear and concise direction they require.

RESEARCH

Once the brief has been clarified, the research phase is undertaken. Research can take several forms, and may include client research and audience research.

Client research

Designers may need to research the client background to establish a clearer understanding of their current needs in the light of their previous history. In many instances, the client may be a large organisation with a well-established corporate or organisational 'culture'. Just like any other social structure, a company is made up of history, hierarchy and traditions. These affect not only the way the company works, but also the way people respond to and recognise it and its reputation within the wider community. Research into the background of a company is often an important aspect of a design brief. Understanding the company can lead to a clearer understanding of what the company or client really needs. If a client wishes to alter the culture of the organisation through changes to its brand, processes, location, manufacturing, services or products, a designer needs to understand not only the changes that need to occur but also why those changes need to occur. An understanding of the client can be just as important as an understanding of the purpose, target audience and context of the finished visual communication.

Key questions for client research

+ Who is the client?
+ What does the client do?
+ What is the size of the company or organisation?
+ What size is it perceived to be by the public?
+ What values is it perceived to hold?
+ What is the corporate culture perceived to be?

Designers will look at the design history of a company and establish the background of previous design work.

+ What are the existing graphical products used by the client?
+ Is there a corporate style?
+ What is the existing style or aesthetic?
+ How does the client feel about previous designs?
+ What other designs does the client like, both in and outside their field of interest?

Market research

In some instances, clients who have the resources to do so will undertake extensive research into their target audience. Specialist market research companies and social or trend forecasters act as consultants, gathering and analysing information about a specific market and/or general social trends.

Key questions for market research

+ Who is the audience?
+ Who falls within the company or organisation's existing market?

ISBN 9780170349994

+ Who does the company perceive as the target market for the new visual communication?
+ Is this a different or new market for the company? If so, why?
+ What are the company's primary and secondary target markets? (For a product aimed at teenagers, for example, the teenagers themselves are a primary market and the parents who will pay for the item are a secondary market.)
+ What research has been done to establish this market?
+ What are the details of the market? (These include age, income, background, interests, purchasing patterns, ethnicity, location, familiarity with product, technological knowledge, etc.)
+ What other products appeal to this market?

Other research

In researching the design brief, designers look to a range of resources and material for information and inspiration. These sources include market research, current trends and the work of other designers in the same or similar fields. Research is important throughout the design process and covers not only the appearance or form of the final design but also materials, methods of production and advances in technology.

Ongoing self-education is important for designers, as changes in fashion, technology, social attitudes and world events all impact on design. Books, magazines, journals and the Internet provide imagery and information that help to maintain a designer's awareness of change. Some design magazines produce an annual edition giving an overview of the year. These publications offer an insight into trends and the application of materials and technology.

Conferences, trade events and seminars give designers the opportunity to observe and discuss innovation and trends in design. Worldwide conferences may be specific to a particular area of design or may offer an overview of current and proposed methodology. Conferences offer networking opportunities for designers and provide a forum for discussion and the sharing of opinions and ideas. Online forums and articles also provide the opportunity to communicate with others working in the same or similar fields.

DESIGN AND PRODUCTION

Although many designers are multiskilled and competent in a range of creative areas, they sometimes call upon the skills of specialists within their organisation or external contractors. Throughout the design process, therefore, the designer interacts with any skilled professionals whose expertise may be required in the design and production stages.

The skills of a photographer or illustrator, for example, may be called upon during the development of a graphic design, or the expertise of a printer or web designer may be required during the final stages of the design process.

Specialist professionals

+ 3D modeller
+ Artist
+ Builder
+ Engineer
+ Flash animator
+ Illustrator
+ Model maker
+ Photographer
+ Photoshop artist
+ Printer
+ Project manager
+ Standards tester (for Australian and international standards)
+ Stylist
+ Tradesperson
+ Web designer

Sales representatives and agents who communicate with designers provide information about developments in materials, processes and technologies. Such information may have an impact on the outcome of a design process.

TEAMWORK

The degree of interaction between professionals may vary according to the demands of the design brief. Designers working within teams will often collaborate on key aspects of the task. This is particularly valuable for sharing ideas and discussing the various directions a design concept might take.

ISBN 9780170349994

Freelance designers, although working independently, rarely work in a vacuum. They will often be involved as an external provider to an existing team, offering a particular area of expertise, or they may answer to an art director or project manager.

ETHICAL AND LEGAL ISSUES IN DESIGN

As in any professional area, designers need to consider the legal and ethical issues that affect their field. As professionals they have responsibilities towards their clients, users and the wider community (see Chapter 11 for more information about legal responsibilities in design).

These issues include the following:

+ attribution
+ copyright
+ cultural sensitivities
+ image manipulation
+ plagiarism
+ safety
+ sustainability.

THE DESIGN PROCESS

Methods of developing ideas and concepts vary greatly and are dictated by the individual skills of the design professional. Approaches vary from designer to designer and from design team to design team. Some designers begin their design process with thumbnail sketches and rough drawings, while others feel more comfortable using a computer as an initial tool. Word lists, discussions and models are used widely in the planning of a design strategy.

Client involvement in the design process also varies from project to project and designer to designer.

In the application of design elements and principles, many designers use team feedback to judge the effectiveness of their concepts. Methods of evaluation vary but peer approval is often a powerful influence on decision making. Initial concepts may also be tested on small groups picked from the target market in order to test the effectiveness of the design direction. Feedback is analysed and fed into the continuing design process.

The materials used by designers are dictated by many factors and are as varied as the design briefs. The constraints of the brief – such as time, materials, costs and location – will affect the application of materials, as will less tangible factors. These include the preferences of the client and external influences such as planning restrictions or public sensibilities.

Manufacturing capabilities, access to materials and the availability of technology are other factors that will determine production methods and materials.

EVALUATION

In evaluating the final concept, designers might use a range of methods. In some instances, formal criteria will be used to assess the success of a design. Criteria may have been set within the brief or – in the case of many manufactured products – may exist as part of an established industry standard. Evaluation is often based on subjective factors such as the aesthetic appeal of a design and its appeal in the target marketplace. Tangible factors such as sales figures and financial turnover provide more concrete indicators of the success of a design.

At the conclusion of the design process, some designers may remain involved in a management role. As part of ongoing project management, many designers produce special documentation, style guides and manuals that detail the function, use and application of the finished design. Such guides enable the original integrity of the design to be maintained even without the continued presence or involvement of the designer.

ISBN 9780170349994

PROFESSIONAL PRACTICES BY DESIGN AREA

Design area		Research methods	Skills	Specialist practitioners	Decision making	Evaluation techniques	Legal and ethical considerations
Graphic design	Graphic designer	Observation of environment Books Internet Site visits Photographs Past experiences Market research Seasonal trends Client history Market history	Communication skills Drawing skills Computer skills (esp. Adobe Photoshop, Illustrator and InDesign) Visualisation skills Skill in the selection of appropriate materials and media Organisational skills	Photographer Printer Web designer Copywriter Illustrator Animator Photoshop retouch artist Typographer	(All designers) Reading the brief assists in making appropriate and correct decisions. Understanding and applying the constraints of the task as outlined in the design brief. The testing of materials and media to ascertain the most appropriate outcome. Testing and experimentation with colour options and thematic issues. Decision making based on issues such as: • suitability • durability • ergonomics • environmental impact • cost • sustainability. Decisions are made based on the appropriateness to the target market. Whether or not a design is positioned correctly can be established through market testing.	Through the use of a mock-up or rough draft. The success of the design based on sales or statistics before and after the completed design. The use of post-design analysis to determine client satisfaction and usability of the product/design.	Attribution Copyright Cultural sensitivities Image manipulation Plagiarism Sustainability
Industrial design	Industrial designer	Sample materials Similar products Observation and analysis of competitor products Environmental factors Safety and manufacturing standards Ergonomics	Practical modelling skills Computer skills (esp. 3D) Skill in the use of appropriate technical language/terminology Organisational and planning skills	Model maker 3D computer modelling artist Engineers (e.g. electrical, systems) Manufacturer Toolmaker		Referral to other jobs due to success. Use of user questionnaires and market research.	Safety Standards & regulations Sustainability Plagiarism Contractual legalities
Built environment design	Architect	Books Magazines Internet Site visits and evaluation (Site analysis) Observation of location Environmental and historical area research	Drawing 3D model construction Project management (e.g. planning, budgeting) Understanding of materials Ability to read and understand plans and technical drawings	Model maker Builder and construction professionals Computer 3D artist Interior designer Engineers (e.g. structural) Draftsperson Planning professionals			Safety Building regulations Sustainability Plagiarism Contractual legalities

ISBN 9780170349994

Design area		Research methods	Skills	Specialist practitioners	Decision making	Evaluation techniques	Legal and ethical considerations
Industrial design	Interior designer	Analysis of client's needs Existing and historical buildings Observation Books Magazines Internet The space/environment Existing style of the building/space Trends Analysis of client's needs	Skill in the use of appropriate terminology Understanding of building standards Visualisation skills Ability to create themes and integrate varied design elements Interpretation of trends Drawing skills Management of varied materials and elements together	Architect Tradespeople (e.g. painters, cabinet makers, builders) Lighting designer Textile designer		Client satisfaction The use of post-design analysis to determine client satisfaction and usability of the space/design. Referral to other jobs due to success.	Safety Building regulations Sustainability Plagiarism Contractual legalities Attribution
Built environment design	Landscape designer	Climate Land forms Soil quality History/background of the site Analysis of client's needs	Horticultural knowledge Environmental knowledge Drawing skills Skill in interpretation of plans	Architect Interior designer Horticulturalist Nursery		Success can be visibly seen in successful planting and survival Client satisfaction Feedback from users	

ISBN 9780170349994

CHAPTER 1
INDUSTRIAL DESIGN

'Products are a form of communication – they demonstrate your value system, what you care about.'

Jony Ive

Industrial design is an area of design established in the mid-19th century during the latter part of the Industrial Revolution. As a design discipline, the origins of industrial designs can be traced back to the influential design movements of the early 20th century in Europe including the Deutscher Werkbund and the Bauhaus. Members of both movements recognised that a formal visual language of function was overtaking the decorative designs of previous art and design movements (such as Art Nouveau). In the chaos of the First World War and its aftermath, designers identified a need for accessible, standardised, simple forms and eschewed the highly decorative and hand built in favour of a streamlined, 'machine aesthetic'.

The Bauhaus was a highly influential design school established in 1919 that promoted a functional aesthetic in all areas of design. Core studies at the Bauhaus focused on the logical analysis of form and function. The use of materials such as steel, Plexiglass®, rayon and even cellophane in design were radical departures from the traditional visual arts training that had gone before. Students were taught to use instruments in their drawings; items such as the compass and the straightedge ruler, which had previously been the tools of engineers and draftspeople, became part of the creative process within the Bauhaus.

▲ 'Barcelona Chair' by Ludwig Mies Van Der Rohe. Bauhaus 1929.

ISBN 9780170349994

The Bauhaus director, Walter Gropius, believed that the making of objects and constructions was an important social and intellectual pursuit, and he encouraged students to follow a functional aesthetic. Studies at the Bauhaus included graphic design, typography, furniture design, architecture, textiles and metal.

The alumni of the Bauhaus influenced design around the world and led to the development of the highly influential International Style, seen as most typical of the clean lines of Modernist design. The characteristics of modernism still influence many contemporary industrial designs; the focus on function, the application of clean, unambiguous forms and the innovative use of materials.

▲ Le Corbusier (designer) France 1887–1965
Charlotte Perriand (designer) France 1903–99
Cassina, Milan (manufacturer) Italy est. 1927
LC/4 chaise longue 1928 (designed), c.1970 (manufactured)
horse skin, leather, chromium plated steel, painted steel, rubber, cotton, metal, dacron
(a–b) 70.9 × 160.5 × 57.3 cm (overall)
National Gallery of Victoria, Melbourne Purchased, 1971
© Le Corbusier, Charlotte Perriand, Pierre Jeanneret/ADAGP. Licensed by Viscopy, 2012.

Architect and designer, Dieter Rams, as chief of design at German company Braun, was highly influential in the design of consumer products. His modernist designs for shavers, audio visual equipment and small domestic appliances were highly influential and many are still in production today. Most famously, Rams defined ten principles of good design, which are celebrated as defining effective modern industrial design.

Good design . . .

Is innovative. The possibilities for progression are not, by any means, exhausted. Technological development is always offering new opportunities for original designs. But imaginative design always develops in tandem with improving technology, and can never be an end in itself.

Makes a product useful. A product is bought to be used. It has to satisfy not only functional, but also psychological and aesthetic criteria. Good design emphasises the usefulness of a product while disregarding anything that could detract from it.

Is aesthetic. The aesthetic quality of a product is integral to its usefulness because products are used every day and have an effect on people and their wellbeing. Only well-executed objects can be beautiful.

Makes a product understandable. It clarifies the product's structure. Better still, it can make the product clearly express its function by making use of the user's intuition. At best, it is self-explanatory.

Is unobtrusive. Products fulfilling a purpose are like tools. They are neither decorative objects nor works of art. Their design should therefore be both neutral and restrained, to leave room for the user's self-expression.

Is honest. It does not make a product appear more innovative, powerful or valuable than it really is. It does not attempt to manipulate the consumer with promises that cannot be kept.

Is long-lasting. It avoids being fashionable and therefore never appears antiquated. Unlike fashionable design, it lasts many years – even in today's throwaway society. Is thorough down to the last detail – Nothing must be arbitrary or left to chance. Care and accuracy in the design process show respect towards the consumer.

Is environmentally friendly. Design makes an important contribution to the preservation of the environment. It conserves resources and minimises physical and visual pollution throughout the lifecycle of the product.

Is as little design as possible. Less, but better – because it concentrates on the essential aspects, and the products are not burdened with non-essentials. Back to purity, back to simplicity.

ISBN 9780170349994

1.1 INDUSTRIAL DESIGN PROFESSIONALS

WHAT DO THEY DO?

Industrial designers consider form and function to create consumer or industrial products both large and small, including but not limited to motor vehicles, consumer electronics, lighting, furniture, medical equipment, toys, recreational products, industrial machinery and water craft. Most industrial designers strive to create products that are sustainable, efficient and effective by using innovative technologies and materials, principles of design and appealing aesthetics.

Industrial designers are usually conceptual thinkers who are trained to respond to design problems in practical yet creative ways. They are required to confidently work to a design brief in the development of effective, attractive and marketable products, from initial research through to final prototype.

WHAT SPECIALIST SKILLS DO THEY HAVE?

In training at university level, industrial designers develop skills in specialist materials and manufacturing, ergonomics and engineering. They undertake three-dimensional design and drawing studies to assist them in experimenting with ideas and documenting their design process. Industrial designers often use design sketching to visualise their early design ideas and this is a key part of their design process. Using pictorial drawing methods, such as isometric or perspective drawing, designers sketch and render their ideas before venturing into digital media.

Graduate industrial designers may specialise in an area of interest or offer general skills. Specialist designers may include automotive designers, furniture designers and lighting designers. Generalist industrial designers may be required to design a wider variety of products from hair dryers

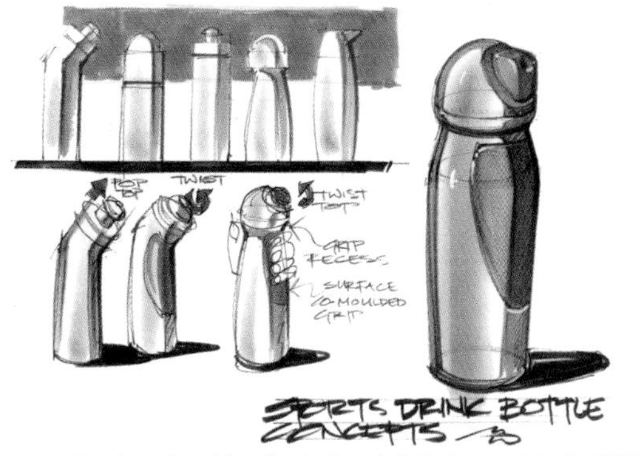

▲ Design sketching by industrial designer, Mark Wilken

Mark Wilken

and kettles to gymnasium equipment and water bottles. Manufacturers often approach industrial designers with an idea for a product for which the designer will be engaged to visualise and make real. Ability to communicate with clients and interact with other specialists are additional skills that an industrial designer needs to acquire.

WHO DO THEY WORK WITH?

Industrial designers work with a range of specialists. Although industrial designers often have a comprehensive understanding of how a product works, they also rely on the expertise of others to ensure that needs such as safety, functionality, durability and reliability are met. In working on a small domestic appliance, such as a blender, an industrial designer may need to consult with an electrical or systems engineer whose role it is to create an interactive interface for the product. Engineers often work with industrial designers, bringing their specific technical expertise to projects; the automotive industry, in particular sees engineers and industrial designers work in close partnership.

Although most industrial designers are required to produce models during their studies, professional projects may see them utilise the skills of a model maker to create a prototype of a design. Very often, prototypes are used for testing and evaluation before the expensive manufacturing process begins; for example, all new motor vehicle designs are created in clay as part of the design process. Likewise, a specialist 3D digital modeller may be involved in creating digital representations of the final design.

WHAT RESOURCES DO THEY USE?

Industrial designers are required to maintain wide knowledge of innovations in materials and technologies. Advancements in materials and manufacturing processes can increase the durability, sustainability and functionality of products so it is essential that designers remain up to date. Designers might access sample materials, attend conferences and seminars, use online resources and magazines to enhance their knowledge of changes and trends in design.

Product designs are invariably bound by manufacturing standards and regulations. Familiarity with Australian and international standards is essential to ensure that new designs are compliant. Industrial designers need to access information about user-related factors in their designs including ergonomics, interface design and aesthetic preferences. Research is a key component in building knowledge about these factors.

WHAT DESIGN TECHNOLOGIES DO THEY USE?

Industrial designers use a wide range of design technologies from pencils, pens and markers for design sketching to sophisticated CADD software such as Revit, SolidWorks and AutoCAD. They use 2D and 3D printing technologies to create plans and prototypes of product designs and a range of modelling techniques including moulding, forming and constructing to create scale models.

1.2 SIGNIFICANT INDUSTRIAL DESIGNERS

+ Dieter Rams
+ Raymond Loewy
+ Marc Newson
+ Philippe Starck
+ Karim Rashid
+ IDEO
+ Smart Design
+ Hilary Cottam

HELPFUL LINKS FOR INDUSTRIAL DESIGN

Core77: An essential site for all things industrial design

Industrial design served: Portfolios of the latest industrial and products design from around the world

Design Institute Australia: Professional association for designers in all areas. Includes helpful definitions of design specialisations

Industrial design community and resources

MADE Quarterly: Innovative Australian magazine dedicated to industrial design, interior design and architecture, photography and food

Dexigner: Design hub for news and resources relevant to all areas of design

Yanko Design: Webzine focused on contemporary international industrial design, technology, interior design, architecture, exhibition design and fashion

Access all weblinks directly at http://nsg.nelsonnet.com.au.

1.3 INDUSTRIAL DESIGN CASE STUDY: F+ TASK CHAIR

INDUSTRIAL DESIGNER: NICHOLAS FOUND

A client who required a comfortable and portable chair approached industrial designer Nicholas Found. The client, an accountant whose job

ISBN 9780170349994

involved visiting many city-based companies, had sustained a serious back injury. Due to the nature of her work, the client spent many hours seated at a desk in different office environments and found sitting in different chairs caused considerable pain and discomfort. The client approached Nicholas for a seating solution that accommodated her injury and allowed for portability.

Nicholas Found responded with the design of the F+ Task Chair; a flexible seating solution that uses ergonomic principles to support all areas of the body, provides movement for the lower back and legs, and folds for easy transport and storage.

THE DESIGN BRIEF

The design brief outlined the requirements of the client. It was a written document that outlined the need for the chair design as well as suggestions for materials and a neutral colour scheme. Nick was also provided with a budget and a timeline of eight months.

RESEARCH

Nicholas' research focused on a number of areas. He first investigated the nature of the client's back injury, speaking to health professionals including the client's doctor and physiotherapist. Following these conversations, he was able to build an understanding of the nature of the injury and the postures that would be most comfortable for the client. Nicholas also looked into both the mechanics and aesthetics of chair design. His investigations took in the long tradition of chair design over the past 100 years and allowed him to analyse the aesthetic qualities that have endured.

GENERATION AND DEVELOPMENT OF DESIGN CONCEPTS

Early in the design process, and using the information gathered from the health professionals, Nicholas identified that the use of a small inflatable ball as a structural feature under the chair was a suitable option to address the needs outlined in the brief.

In experimenting with the Swiss ball concept, Nicholas created a scale prototype (1:4) using a 3D printer, to test the viability of the concept. The model enabled him to address connecting the ball to the seating unit and test its functionality

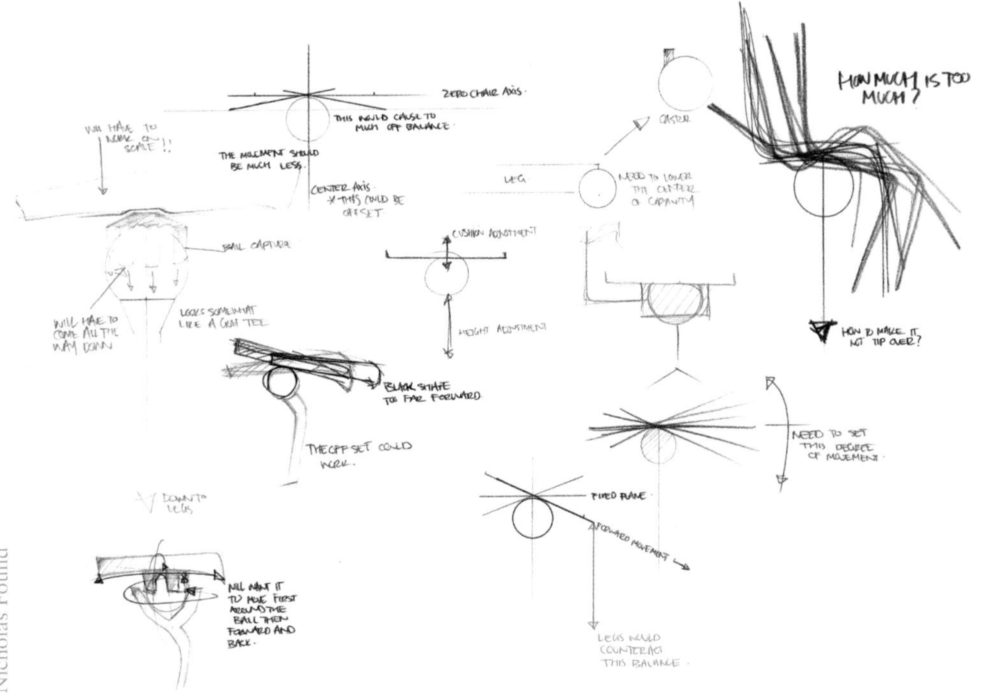

Initial sketches for the 'Swiss ball' mechanism on the chair

Nicholas Found

ISBN 9780170349994

in supporting the weight of a user. The prototype successfully showed that with the weight of the user on the chair the ball mechanism was released into a small supporting frame allowing for flex and comfort. Without weight on the chair seat, the ball sprang into a fixed position at the top of the frame to prevent the seat from falling over.

Nicholas Found

▲ Using a 3D printer, the prototype was created for testing. This image shows the ball mechanism in the fixed (upper) position.

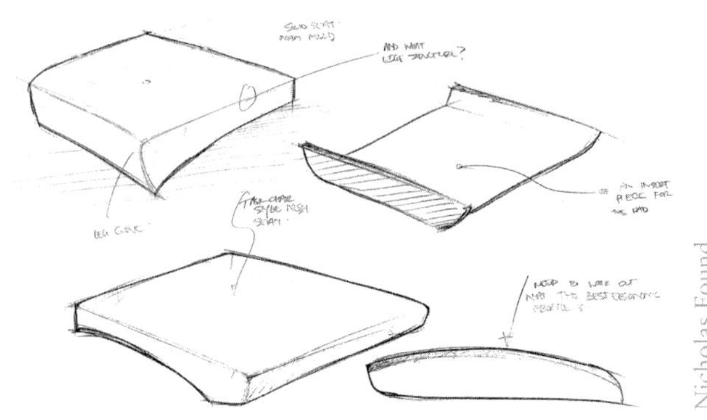

Nicholas Found

▲ As well as offering portability, the chair needed to offer a comfortable seating position for the user. Ergonomic principles were taken into account throughout the design process.

Factors such as ergonomics to address issues of comfort were considered by Nicholas, as were the mechanistic aspects of the design including adjustable height and a collapsible structure.

Three-dimensional modelling of the final design using the Lightworks software program allowed the designer to see the visual appearance of the design concept and to identify areas that required further development.

This method enabled a good understanding of the product's structure and appearance without the expense of production.

▶ Nick used drawing throughout the process to visualise ideas; note the use of annotation to critically analyse each concept.

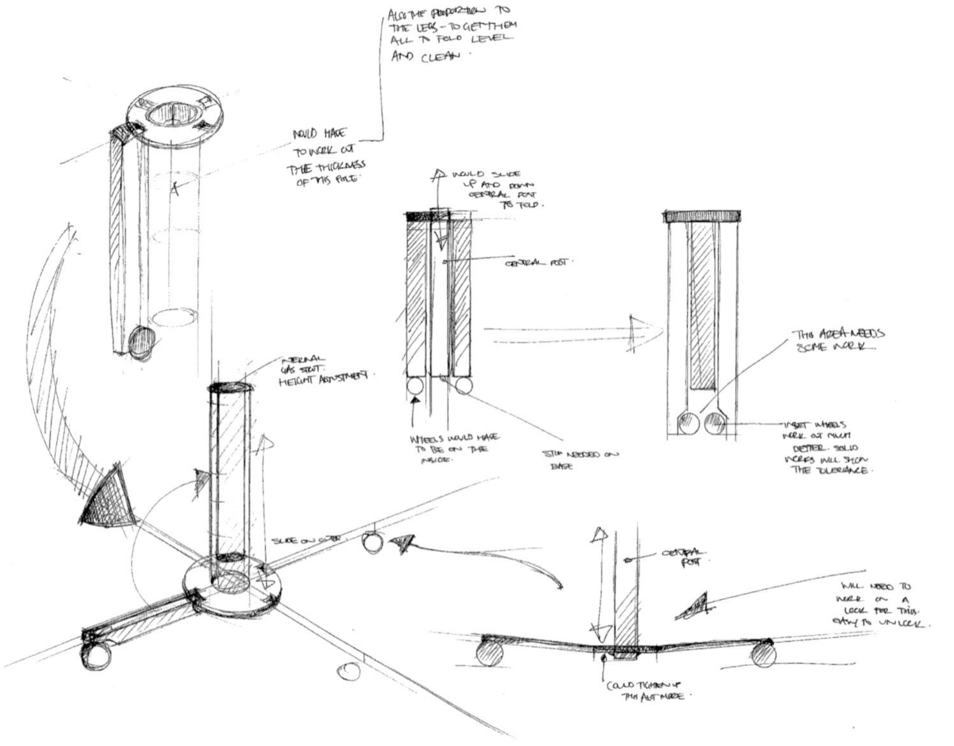

Nicholas Found

ISBN 9780170349994

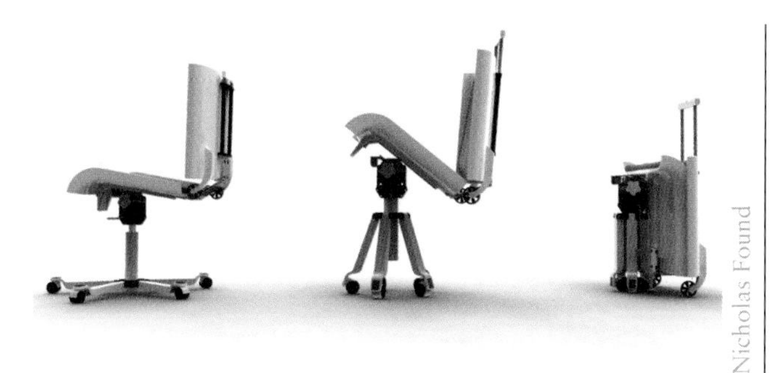

▲ Modelled image of the chair design indicating function and appearance

Using the modelled images, Nicholas was able to identify, evaluate and annotate areas to be refined.

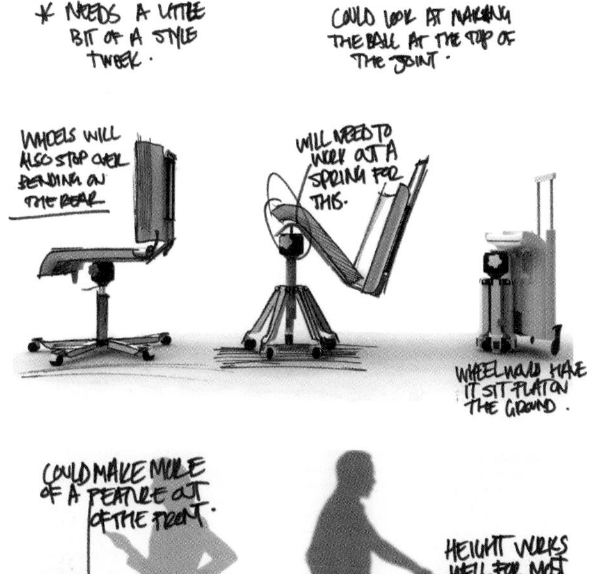

▲ Annotated images identifying concerns and possible changes to the design

REFINED DESIGN AND PRODUCTION

Once changes were identified and adjusted to suit both the needs of the brief and the limitations of manufacturing capability, a final three-dimensional rendering was created. This final design tool allowed for a last evaluation of the product prior to manufacture. Once satisfied with the product, Nick created detailed technical drawings using CADD. These drawings included third-angle orthogonal drawings of parts, section views and an exploded isometric view of the chair. The drawings and written specifications of materials were sent to the manufacturer so that tooling (preparation of machinery for manufacture) and production could begin.

▲ The finished F+ Task Chair. Note the addition of a logo and the changes to the colour palette from the previous designs.

WORKING WITH SPECIALISTS

Nicholas briefly consulted a mechanical engineer in the early stages of the process to ensure that the fabrication of his chair design was viable. Together, they created a number of modifications to proposed design concepts to ensure the maximum functionality.

EVALUATION

Although his original brief was the design of a single, purpose-built chair, Nicholas identified that such a product may suit a broader audience. The original client now has her comfortable chair and Nicholas is in discussions to manufacture a version of the F+ Task Chair on a wider scale. Ultimately, Nicholas may analyse consumer feedback and sales figures to determine the commercial success of the venture.

ISBN 9780170349994

INDUSTRIAL DESIGNER ~ DR RAFAEL GOMEZ

Dr Rafael Gomez has more than 15 years industry experience as a designer for small, medium and large enterprises and has completed projects for the aviation, construction, medical, government and consumer electronics industries. He is design director for Propaganda Mill, a multidisciplinary design company working across product design, branding, graphics, high-end visualisation and projection graphics. Dr Gomez is a design lecturer at Queensland University of Technology and has lectured in a variety of design subjects. He is currently second-year industrial design coordinator at QUT, and has lectured and written curriculum for design visualisation, design history and design usability. Dr Gomez is founder and president of the Design and Emotion Society Australia and a member of the Design Institute of Australia.

What do you see as some of the key skills offered by professional industrial designers?

Industrial design is a multifaceted profession whose services can be used and applied in various capacities. In the traditional sense, the skill sets that professionals need to possess include: (**i**) people knowledge (**ii**) manufacturing knowledge (**iii**) business knowledge and (**iv**) strong communication skills.

(**i**) People knowledge: Perhaps the most important expertise offered by industrial designers is knowledge of the people (users) they design for. Designers are the main advocates of the end users and acquire knowledge about the physical needs, cognitive requirements and emotional experience through research and experience.

(**ii**) Manufacturing knowledge: Traditionally, industrial design involves the production of mass manufactured products. As a result, professionals need to possess strong knowledge in established manufacturing processes, methods and techniques as well as new techniques like rapid prototyping and 3D printing.

(**iii**) Business knowledge: Equally, industrial designers acquire knowledge of business needs. Designing successful products involves understanding how a business functions effectively. Designers need to align business needs with manufacturing knowledge and information about the people they are designing for.

(**iv**) Communication skills: Finally, industrial designers display strong communication skills. These include strong sketching and illustration, an inclination to model-making and computer-aided design skills. Solid interpersonal and verbal skills are also an important part of an industrial designer's skill set.

These are the established key skills and knowledge areas that industrial designers offer. Nevertheless, more and more, designers are moving into new areas of knowledge, including designing services and proposing new ways to visualise complex systems. These new areas of application involve novel expertise, including an eye for detail as well as being able to see the big picture, and a greater demand for creativity.

Dr Rafael Gomez

Thinking of a recent brief, what were some of the constraints that had to be considered in the design?

A recent project that I was involved with was the design of a medical device. In many countries around the world, including Australia, the medical field is highly regulated and because of this, many constraints are placed on the designer about what can and cannot be done with a project. In Australia the governing body that defines these guidelines is called the Therapeutic Goods Administration (TGA). Under TGA rules, medical devices fall into specific classes that define the product's complexity and categories, and impose different guidelines for the design and regulation. This particular device was classified as Class I,

ISBN 9780170349994

meaning it was a simple device with minor levels of regulation. Nevertheless, there were specific standards, guidelines and rules that needed to be followed or the device would be prohibited in Australia.

Further, the user group imposed specific constraints on the project. The typical user of the device was a child or an elderly person and so there were certain constraints because of these users. These constraints included limitations on font size, colour, device size and shape for holding as well as limitations on the forces required to use any part of the device. Usability of the device was also important and consideration had to be paid to any sharp or irregular surfaces on the device so that it did not pose a hazard for users to hold.

How do you research ideas for a design brief?

One of the first places that I obtain information from is the client themselves. It is common that clients already have some level of research on their idea prior to engaging a designer. Sometimes it is some simple research on their potential users or customers, while other times (mainly with larger businesses) they have a marketing department that can provide in-depth analysis on users, customers, technology, competitor products and many other types of data that can be useful.

Another source of information during the research phases is the users themselves. Often as a designer you are required to conduct what is referred to as user testing, which involves talking to potential end users of the product and performing some basic experiments. This could be performing interviews or surveys about the product in question, or once a prototype is developed it could also involve testing users with the prototype.

Another source of information is to explore and analyse existing products, or what is referred to as product benchmarking. This can be a useful technique to find out what has already been done in the market, avoid the failures and improve on the successes of existing designs. Likewise, inspiration can be drawn from graphics, photographs, logos, branding and other collateral existing around the market.

Finally, the traditional resources, including books and the Internet, are easily accessible sources of information for any project. This could include looking at scholarly journals on a special or specific topic, as well as international and Australian standards that need to be adhered to for the device type.

What part does drawing play in your design process?

Drawing is an essential part of the process. Nothing can replace an effective drawing and it is an increasingly critical part of the design development process. The reason is that many designers think visually and drawing acts as a very rapid way of capturing that thinking process so that it can be recorded and communicated to others quickly. Drawing is relevant during all stages of the design process especially at the early stages, but serves a purpose during the middle and later stages of product development. The key aspect of drawing is not its precision (it doesn't have to be perfect) but rather it has to be rapid and effective – not even the latest software or visualisation tools can compare to the speed and freedom allowed by simple pen and paper.

Ellaspede Motorcycles and Tammy Law Photography

ISBN 9780170349994

When working for a client, what input do they have in the design process?

The client's input varies depending on the project. Nevertheless, they always have some involvement into the process because they often determine time frames and overall project duration. Some clients like to work closely with the designer throughout the project and this is usually the case with small- to medium-sized businesses where the client has a small number of projects on the go. With larger businesses there is usually more of a distant relationship between client and designer yet there is always a manager who is the key person liaising between the client and designer during the process.

How are decisions made about design concepts?

Decisions are usually made between designer and client during concept presentation stages. This is often based on how closely the various concepts adhere to the design criteria set out in the design brief. However, it is common practice to have an internal evaluation prior to this stage to pick the number of concepts to be presented to the client. This internal decision-making process is different for every design firm but usually involves presentation of the various concepts to a senior design manager who assists in making the final choice about the concepts to present.

What factors inform the selection of materials in your designs?

The primary factors that determine material selection have to do with cost of material that meets performance criteria and sustainability requirements. For instance, if it is a device to be used underwater or near water then materials that do not rust or malfunction in water need to be considered while maintaining cost considerations and environment considerations. There are other important factors that influence what material to use, including function, mechanical, chemical, thermal and electrical properties and manufacturability. Finally, there are additional factors that can affect material selection, including features like colour, finish or how well the material wears out. For example, many designers will prefer to use wood in some circumstances because it wears attractively in a natural way rather than plastic that scratches and becomes damaged easily.

What other professionals do you often work with on a design?

There are many other professions that industrial designers work with due to the wide range of industries that designers work in. Nevertheless, some common professions include mechanical and electrical engineering, marketing, business, graphic designers, web developers, interaction designers, interior designers and architects.

What technologies are used in the execution of your designs?

There are many technologies used in the execution of design including computer-aided drafting and design (CADD) as well as rapid prototype techniques like additive manufacturing and

▲ Laser cutting

▲ Model making

ISBN 9780170349994

3D printing techniques. CADD is often used during the middle and end phases of the design process, usually when the design is more resolved or ready for manufacturing. Rapid prototyping techniques can be used at all times of the design process because they permit designers to quickly evaluate physical models of varying degrees for concept selection.

How is a final decision about the design direction arrived at?

The final decision is usually agreed upon by consultation between the designers and the client. The decision is dependent on many factors but usually takes into consideration the criteria determined in the design brief and which design fulfils the criteria best.

Design ideas are usually presented in drawings, videos and computer and physical models or any combination of these. Drawings and basic physical models are usually used in the early phases of the project to give a sense of the design still at concept stage. Computer and high-end physical models are usually reserved for later stages of the process once the design is more resolved. Movies and other presentation techniques are usually displayed to communicate the story or narrative of the design in context.

What legal and ethical issues are considered in the design work you do?

Most products are governed by some international or Australian standard. These are legal regulations that must be adhered to for health and safety reasons. Further, if the designer does not adhere to these standards the design may be prohibited from being sold or used in Australia or the country where it is being exported. Further, there are environmental regulations that control the process, including manufacturing techniques, chemicals used in and during manufacture and material selection.

What are some of your design inspirations? Where do you seek ideas and inspiration? How do you stay up to date with innovations in technology and materials?

Design inspiration can come from many different sources. Personally, I get inspiration from movies, music, photography and other designs and designers. I usually collect these during the early stages of my design process and present a carefully chosen selection of these to the client as inspiration images for the design. Clients usually respond well to this because they can see the rationale for the design look and feel rather than having to verbally explain it to the clients.

Staying up to date with innovations in technologies and materials is often difficult but the main source of information is through technology databases, websites and magazines. Having contacts in technology and materials engineering fields is also important but these take time and experience to develop.

What advice would you give students seeking tertiary studies and a career in industrial design?

The most important criterion for a student to have is passion for design. Industrial design can be a challenging and demanding profession, but if you have a love for what you do then your career can be an enjoyable one.

In a more practical sense, building skills in drawing, model-making and computer-aided design are important skills to have early on. These skills will serve you well during your studies in industrial design but will also assist in gaining a job fresh out of university.

Finally, consideration for improving the quality of life for humans and the environment that we exist in is also an important characteristic of future designers. We can no longer ignore the negative consequences that past human inventions have had on the environment, and designers are in the unique position and have the distinct opportunity to positively influence the impacts of designs for now and in the future.

CHAPTER 2
BUILT ENVIRONMENT DESIGN

'Once you enter the world of architecture you get hooked – that's the problem – and then you can't go back.'

Shelley McNamara

The design of environments for human shelter and comfort is evident in much of history. Built environment design, including architecture, interior design and landscape architecture, has a long tradition that links back to ancient civilisations. It is possible to identify cultural and historical change through the appearance and function of architectural structures, interiors, parks and gardens.

Structural and aesthetic developments in architecture, interior design and landscape design are often a reflection of the times in which they were built. The most magnificent churches and temples were constructed when religious institutions were at their most influential. Castles and palaces were designed to reflect the status and power of those who commissioned them. In more recent times, corporate wealth has often been celebrated by the building of bigger, taller and more imposing office towers. However, not all built environment design is about imposing structures; domestic architecture has evolved to embrace secure and comfortable dwellings that reflect the aesthetic of their time and location.

Today, we still use many classical architectural conventions that can be traced back to ancient Greece and Rome. Domes, vaults and arches are structural innovations that first appeared more than 2000 years ago, designed by Romans, such as Pollio, then rediscovered and refined during the Renaissance by great architects such as Brunelleschi and Alberti.

Designers who have responded to needs identified by client, location and climate have formed our modern environment over many years. Changing technologies and materials have defined

ISBN 9780170349994

▲ Glasshouse, 1949 by Philip Johnson. The design of the landscape was by David Whitney. The design of Johnson's Glasshouse was inspired by the Farnsworth House by leading architect and former Bauhaus member Ludwig Mies Van der Rohe.

the form, height and appearance of structures. For example, skyscrapers are a 19th–20th century development that evolved from advancements in engineering. In Australia our built environment has often been defined by climate and lifestyle; iconic Australian designs of domestic homes, government buildings and parks reflected in Colonial, Federation, Queenslander and tropical architecture, for example.

Design of the built environment is an area where innovation and experimentation are publicly debated, challenged and celebrated. Developments in materials and construction, shifts in aesthetics and taste as well as changes in urban planning priorities combine to stimulate innovative and confronting spatial designs. Given its highly visible and public nature, this area of design can, at times, be the most controversial with many designers required to address the response of the public in addition to clients and stakeholders.

Generational change usually sees cutting-edge built environment practices absorbed and adapted over time. For example, Philip Johnson's 1949 steel-framed Glasshouse eschewed solid walls for sheets of glass that embraced the surrounding landscape. At the time such a design challenged the architectural norms, while the use of glass walls and open plan spaces are common aspects of contemporary home design today.

2.1 BUILT ENVIRONMENT PROFESSIONALS

WHAT DO ARCHITECTS DO?

Architects are concerned with creating, enhancing and defining the built environment. Working with space and form, architects work closely with their clients to create environments that meet needs and solve design problems. They work on domestic, public, cultural and private commercial buildings on a small and large scale. Architects generally work as part of a team with assistance from junior designers and project managers. They work with builders, engineers, site managers and construction industries to see the original design take shape. Their work demonstrates a balance between creative ideas and construction technology. Architects design plans and elevations, using a range of design technologies to create two- and three-dimensional representations of their concepts.

ISBN 9780170349994

WHAT DO INTERIOR DESIGNERS DO?

Interior designers work with interior space. They explore how people engage with their environment and using this knowledge, as well as an understanding of building technologies they create spaces that address functional needs and communicate themes and ideas. Interior designers work on the spaces inside domestic, commercial and cultural buildings using many of the design technologies and drawing methods applied by architects. As well as structural changes within an environment, interior designers devise solutions for highly varied needs that might include the design of customised furniture or the identification of the most efficient movement through spatial zones. Interior designers also work on exhibition and display design and theatrical productions.

WHAT DO LANDSCAPE ARCHITECTS DO?

Landscape architects work within the natural environment to create outdoor spaces including parks, recreational spaces, gardens and landscapes associated with major infrastructure systems such as roads. Often working in partnership with architects, landscape architects have deep knowledge of environmental factors such as climate, horticulture and geography. The work of landscape architects is often collaborative and focused on creating designs that are sympathetic and appropriate for the environment, surrounding structures, urban landscape and climate. Like other built environment professionals, landscape architects use a range of design technologies to express and develop their design ideas.

WHAT SPECIALIST SKILLS DO THEY HAVE?

As in all design areas, built environment designers are highly creative problem solvers who have significant knowledge about the technical aspects of their specialised area. They have deep understanding of standards and safety practices and apply these to their projects. Given that they often work in multidisciplinary teams, designers often must be skilled managers and communicators to ensure that projects stay on track and on budget.

Designers in these areas have skill in visualising three-dimensional forms and creating these using both traditional and digital technologies. They are skilled in the presentation of concepts allowing non-professionals to comprehend the appearance, scale and proportion of a proposed design.

WHO DO THEY WORK WITH?

Built environment designers work with a wide range of professionals that varies according to the context and nature of the design projects. Specialists might include model makers, engineers, surveyors, building professionals, tradespeople, drafting professionals, horticulturalists, and lighting and building automation specialists among others. A shared language and use of appropriate terminology is important in built environment design because it facilitates communication of ideas between key parties involved in the design and construction of a project.

WHAT RESOURCES DO THEY USE?

Built environment designers use a range of technologies including the traditional drawing techniques. Although fewer designers use drawing boards to hand draw plans, it is not unheard of. Drawing, as in other design areas, is seen as the best method to quickly 'ideate' or visualise design ideas. Sketches may be done by hand or by tablet onto a screen.

ISBN 9780170349994

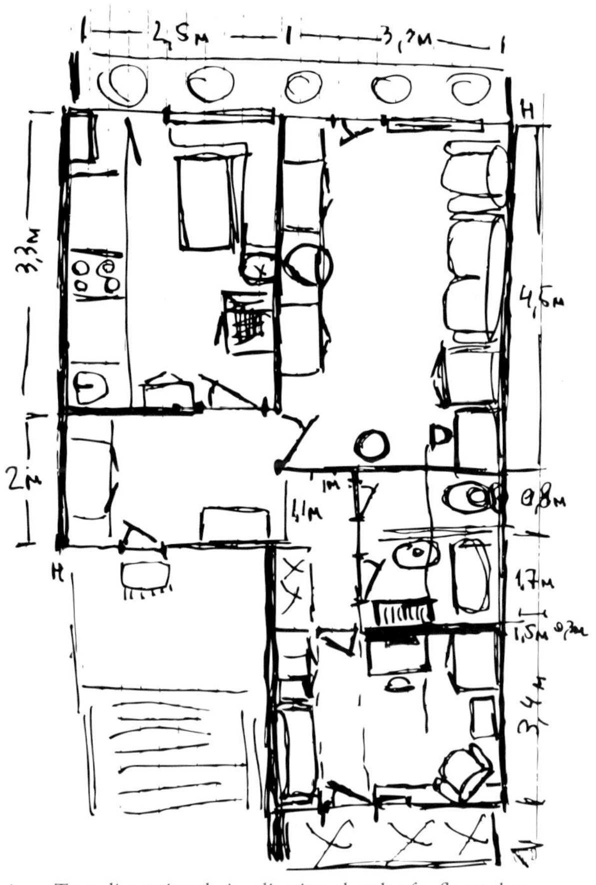

Mark Wilken

▲ Two-dimensional visualisation sketch of a floor plan

Digital technologies are commonly applied in all areas of built environment design. CADD software enables designs to be explored and analysed in detail before undertaking an expensive construction process. Technical specifications, lists of materials and fittings, finishes and fixtures are outlined in the planning stages of the design process and contained within plans and specification documentation.

Research is an important aspect of built environment design. All designers will undertake a site survey to ascertain the qualities and characteristics of the location of the construction, landscape or interior. Site analysis can identify issues such as privacy, challenging landforms and planning restrictions. Aerial photography may also be ordered to allow for a comprehensive visual understanding of the site. An investigation of regulatory requirements (ResCode, the Building Code, planning regulations, building regulations, etc.) is undertaken to determine restrictions and requirements set by local authorities.

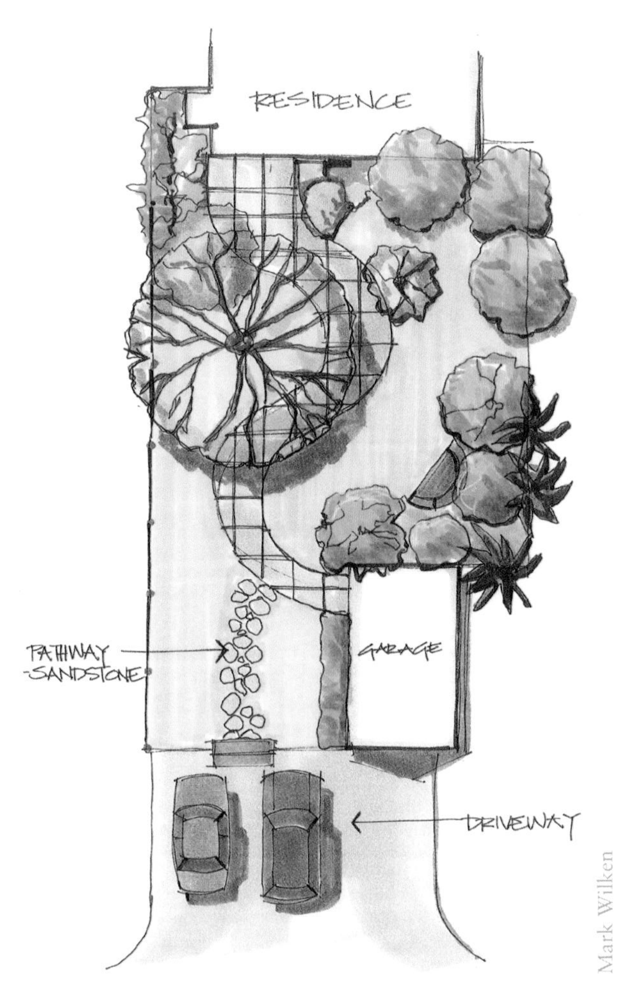

Mark Wilken

▲ Two-dimensional visualisation sketch of a site plan

2.2 SIGNIFICANT BUILT ENVIRONMENT DESIGNERS

+ Frank Lloyd Wright
+ Harry Seidler
+ Walter Burley-Griffin
+ Zaha Hadid
+ Frank Gehry
+ Sir Norman Foster
+ Patricia Urquiola
+ Sean Godsell
+ Arkhefield
+ Conrad Gargett Riddel
+ Thomas Heatherwick

ISBN 9780170349994

HELPFUL LINKS FOR BUILT ENVIRONMENT DESIGN

RAIA Royal Australian Institute of Architects: The home of the Australian professional association for built environment designers including all current and past award winners

Australian Institute of Landscape Architects: Resources and links about the landscape architecture profession in Australia

Design Institute Australia: Professional association for designers in all areas. Includes helpful definitions of design specialisations

Yellowtrace: Blog covering the latest in contemporary interior design and architecture as well as photography, art and furniture design

Butterpaper: Architecture and urban design links and articles

Access all weblinks directly at http://nsg.nelsonnet.com.au.

2.3 BUILT ENVIRONMENT DESIGN CASE STUDY: HILL HOUSE

ARCHITECT: ANDREW MAYNARD

Andrew Maynard Architects (AMA) were approached to modify and expand an existing home to meet the needs of a growing family. The existing dwelling was situated on a small

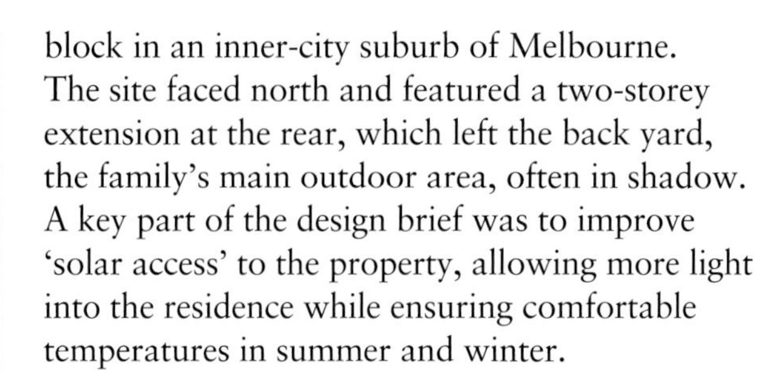

block in an inner-city suburb of Melbourne. The site faced north and featured a two-storey extension at the rear, which left the back yard, the family's main outdoor area, often in shadow. A key part of the design brief was to improve 'solar access' to the property, allowing more light into the residence while ensuring comfortable temperatures in summer and winter.

RESEARCH

An understanding of client needs is critical in environmental design and of particular concern in residential architecture. Client consultation, site visits and a deep understanding of the most appropriate solution to suit the client's lifestyle is key. The investigation of building regulations, to determine any restrictions and requirements set by the local authorities, is also important.

In this case, photography was used to document the site and existing structures. In analysing the existing dwelling, it was decided that the demolition of the older, rear extension was required to maximise the possibilities for an effective design solution.

DESIGN

In investigating the location and position of the site, AMA established that rear access to the property via a laneway offered opportunity to reorient the focus of the home. After considerable experimentation through the use of drawings and diagrams, ideas for alternative uses of the site were developed.

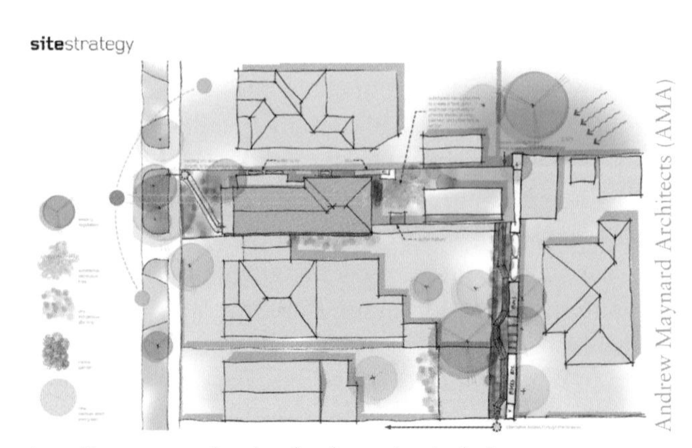

▲ Site strategy planning for the project including consideration of landscape design

 ISBN 9780170349994

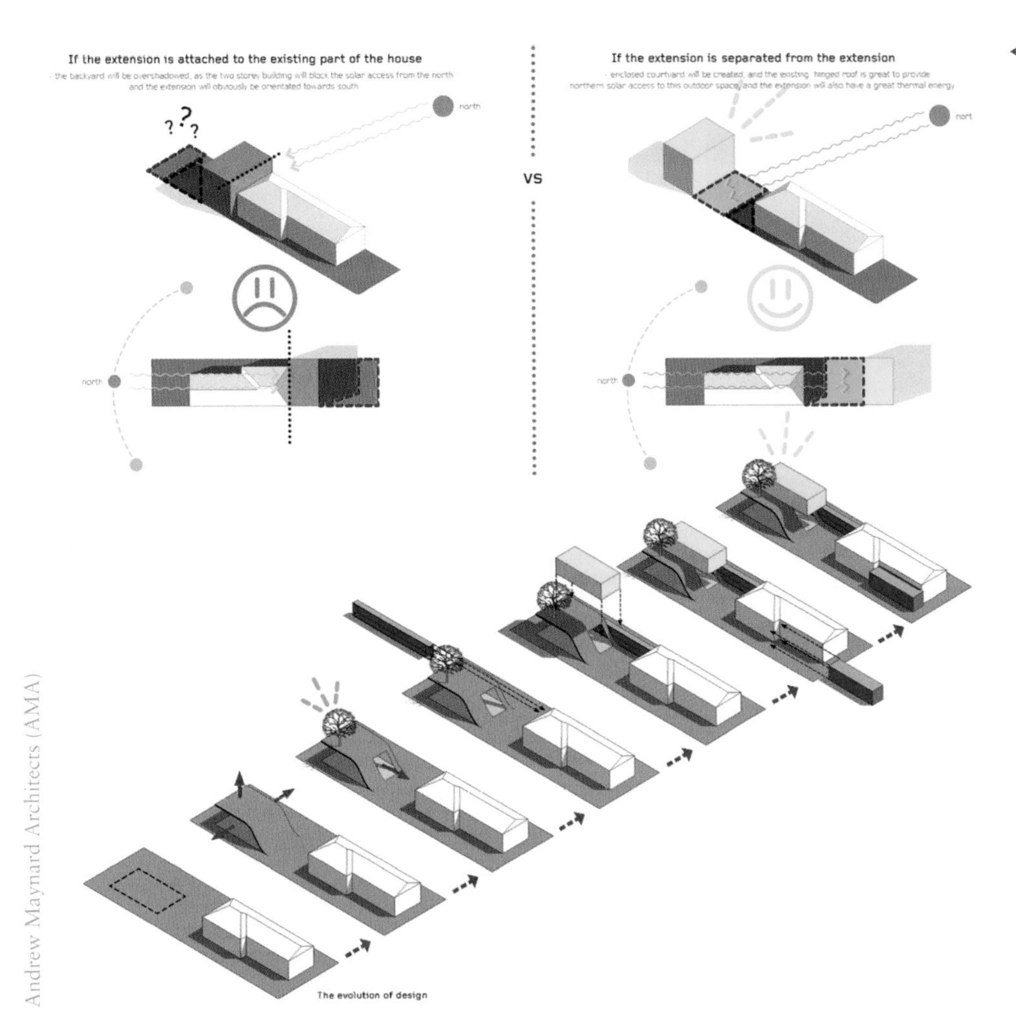

This illustration shows the impact that different structural forms have on the amount of sunlight in the property. This visual analysis communicates design options to the client and the larger building team.

Following extensive site analysis, the main design proposal was to relocate the main living areas of the home to the rear of the block. Although the original home would be retained as accommodation, it would no longer be the focus of the home. There would instead be a garden area in the centre of two connected dwellings.

Great consideration was given to the form of the construction. Many contemporary extensions use rectangular forms as their basis; however, Andrew Maynard remained focused on the maximisation of sunlight and developed the concept of a raised, cantilevered form sitting upon a visually striking 'hill'. The elevated structure was designed to provide shade to the outdoor areas below while facing the sun and 'employing passive solar gain', which would saturate the interior and garden with sunlight.

Architects are constrained by residential building codes and consideration must be given to the requirements set down. In the design of Hill House, issues around 'overlooking' needed to be addressed. In the design, windows had to be placed above head height to avoid looking directly at the neighbours. AMA used this to advantage by placing windows in such a way as to allow views of leafy neighbourhood gardens and nearby trees instead.

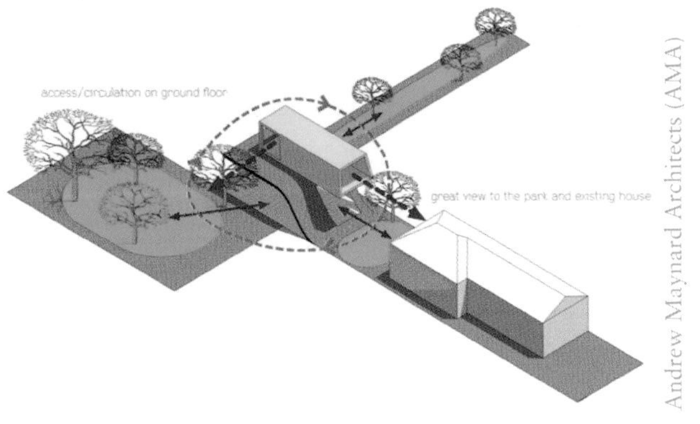

▲ This diagram illustrates the position of windows and their views from the new dwelling.

▲ Kitchen and living areas looking out towards the garden and existing dwelling

The selection of materials is often affected by their application, suitability and cost. Steel was used extensively in the construction of this project for its strength. Good support for the cantilevered structure was essential. Steel was also applied in the kitchen area for its aesthetic appearance, designed to develop a visually intriguing patina as it ages.

Sustainability was integral to the buildings' design; all windows were double glazed and the interior wood finishes used were VOC-free (Volatile Organic Compound). The use of synthetic grass (which also has insulation properties) in most areas of the garden reduces the need for water. A white roof increases solar resistance and reduces heat gain, and the use of mechanically operated louvres on the windows allows for the control of air flow.

WORKING WITH SPECIALISTS

Planning approval was sought for the project prior to the construction phase. The expertise of building inspectors and structural engineers ensured that the construction met safety standards and building regulations. Architects work closely with builders and it is through the provision of detailed plans and elevations that much of this communication occurs. Additionally, specifications for materials, fixtures and finishes in the project are outlined in documentation

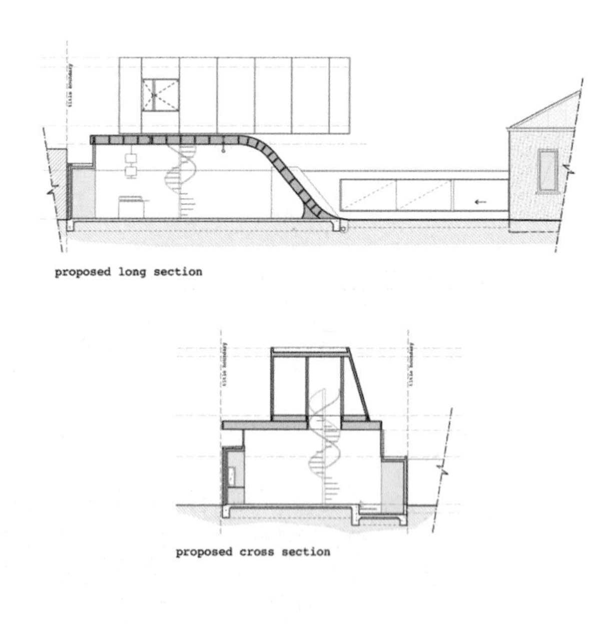

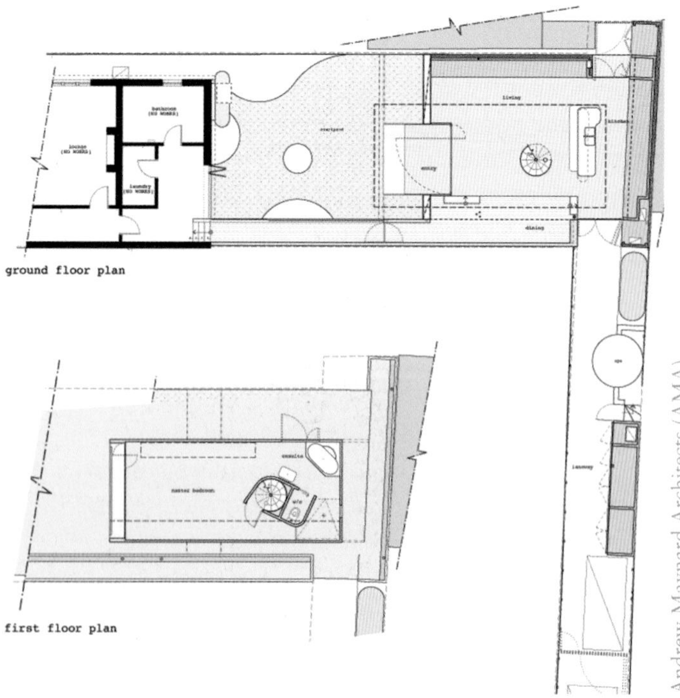

▲ Plans and section views

ISBN 9780170349994

that is provided to trade specialists for electrical, plumbing and interior finishes. AMA developed comprehensive plans, sections and elevations, and all relevant documentation required for construction.

EVALUATION

As with most residential architectural projects, the major test of Hill House is client satisfaction. The conclusion of a project sees the family take residence and their comfort and enjoyment of the spaces and features is the greatest test of the design solution.

Andrew Maynard Architects (AMA)

▲ Hill House, completed. View from centre garden area shows corridor on right that connects the old and new dwellings.

Andrew Maynard Architects (AMA)

▲ The Hill House 'hill' adds a touch of fun to a sustainable and innovative design.

LANDSCAPE ARCHITECT ~ KATHARINA NIEBERLER-WALKER

Conrad Gargett Riddel Ancher Mortlock Woolley

Katharina Nieberler-Walker leads Conrad Gargett Riddel Ancher Mortlock Woolley's Landscape Department. She has considerable experience in management, design and documentation of a diverse range of landscape and urban design projects including public and educational buildings, health and defence projects. Katharina is a registered landscape architect and planner and has practiced both locally and internationally. She is highly skilled in stakeholder consultation and managing the complex requirements of large-scale landforms while coordinating detailed design issues. Her design approach embodies sustainability, practicality and longevity grounded in research and analysis. Katharina works collaboratively with clients and user groups, contractors, other design consultants and approval agencies to fully understand the objectives and parameters of a project.

When a client approaches you with a brief, what is the nature of the information provided? Is it detailed and specific, or is the brief developed in conversation with the client?

The landscape brief provided by a client can vary significantly. It can be detailed and list functional or operational requirements or it can be as broad as 'provide appropriate treatment for hard and soft landscape elements'. As a general rule the brief is developed during the design process and refined during the review stages of the project. The brief can include the original written client design brief, subsequent agreed design development outcomes and approved drawings from the schematic, concept and developed design stages. Written statements, hand-drawn sketches, image boards, three-dimensional images and technical drawings can form part of the brief. Character images, for example, photos of a similar setting, in combination with design sketches are an effective way to quickly illustrate design ideas. Often drawings and sketches are preferred, as they better communicate the design ideas; words are open to interpretation and can be easily misunderstood. It is important to clarify design opportunities and design constraints at the beginning of the design process to minimise misunderstandings. The brief is very much about understanding what the client wants. Clear communication is key. It is useful to capture the client's requirements in a quick sketch during the briefing meeting and include it in the formal return brief. The brief clarifies the design but also establishes processes, project budget and a project program.

How do you research ideas for a design brief? What resources do you usually call upon when doing research for a brief?

Professional journals, such as *Landscape Australia*, *Topos* and *Architecture Australia*, are good sources for design research as these journals feature the best and newest design ideas for the built environment. Articles in one issue are often related to a particular topic, such as public open space or health, and provide a good insight into a particular area of design. We were lucky to have a dedicated research person engaged during the early design stages of the design for the Queensland Children's Hospital. This helped the design team to be well informed on the latest research and to make evidence-based design the cornerstone of the hospital design. Extensive research is often limited to complex projects but research can be as simple as finding the right plant for a specific site condition. We recently did a literature search on 'senior living' to identify key factors and trends in the planning and design of retirement villages in Australia and overseas. Sometimes design ideas come from places I have visited, from the historical or social context or are inspired by the built or natural environment.

Thinking of a recent project, what were some of the constraints that you faced in this design brief?

The design brief for the new Queensland Children's Hospital – now renamed

ISBN 9780170349994

Lady Cilento Children's Hospital (LCCH) – called for a world-class public hospital that would make a significant contribution to the public realm. This resulted in the idea of a public plaza associated with the hospital within a constrained urban site in South Brisbane on the edge of Brisbane's CBD. Traffic was the main design constraint in designing the LCCH Plaza, which is the main public space in front of the hospital. A major intersection had to be redesigned to provide better vehicular access to the new hospital, adjacent school, car parks and the existing Mater Hospital. The challenge was how to design a significant public plaza divided by a road. A primary component of the design solution is the planting of six 30-year-old fig trees on a grid over the space. These 10-metre high trees would create a unifying urban forest across the plaza and road from day one. The urban forest on the LCCH Plaza was inspired by the landmark figs in Memorial Park across the road from the new hospital and the fig trees in Aubigny Place in front of the old Mater Private Hospital.

Conrad Gargett Ancher Mortlock Woolley

▲ Computer-generated 3D image, Lady Cilento Children's Hospital, Brisbane

What part does drawing play in your design process?

Drawing is everything to design. Until you draw what you are thinking you cannot understand what is going on, and you cannot convey the idea effectively to others. The process of drawing a sketch helps refine the design and identifies opportunities and constraints. A drawing illustrates and clarifies complex issues such as scale and interrelationship with other design components. If you work within a team of designers, drawing is the best way to convey an idea and coordinate the design across the design professions – architects, landscape architects, engineers and others. Plans and sections are a quick way to convey the design ideas. The use of 3D graphics – to visualise an object, spin it around to look at it from all sides, zoom in and out, and look at it from eye level – is an invaluable tool for illustrating design concepts and results. We frequently use drawings in meetings to clarify what has been said. Drawing is a skill that needs practice – so keep drawing!

Conrad Gargett Ancher Mortlock Woolley

▲ Freehand sketch, North Gheran, Libya

How involved is the client throughout the process? What participation do they have?

The involvement of the client throughout the design process varies greatly from project to project. Personal designs, such as a garden design, are the most challenging so it helps if you can put yourself in the client's shoes to understand what is important to them and where they are

ISBN 9780170349994

coming from. A good client relationship is vital to a good design outcome. It is important to remember that you are designing for the client and not for yourself. So make sure you ask plenty of questions and listen carefully to what they are trying to get across. To 'read' the client is a real skill and takes some time to master. After completing many contracts with the Australian Defence Force, it is now much quicker and easier to come to an agreed design outcome because we both have a clear understanding of what is required.

How are decisions made about design concepts?

In our office we have regular design reviews for each project. This means the designer presents the design to a group for robust discussion. The group critiquing the design is made up of senior and junior staff. There is no such thing as a stupid question and the critique should never be personal but focus on design issues. Trying to explain why you have designed something in a particular way helps to identify gaps, and a query from the reviewers often identifies areas in need of further research. A design review can also turn the design 'on its head' and lead to a totally different approach. I remember a design review for a competition that led to a redesign that ultimately won us the job. For design reviews to be most beneficial, participants need to be open-minded and ready to accept a good argument when they hear it.

Conrad Gargett Ancher Mortlock Woolley

▲ Landscape Master Plan, Australian Catholic University, Banjo

What factors inform the selection of materials, planting and structural elements in your designs?

The intent of the design, the context, longevity, use and, ultimately, the budget all inform the selection of materials. What is the nature of the design – uplifting or merely practical? Can the materials be exposed to the weather? How well does the material wear? And how does it relate to the existing context? Or is the selection of material intended to change the character of a place? Landscape architecture distinguishes between hard (paving, seating, built structures) and soft (trees, shrubs, grass) landscape elements. For example, natural and easily maintained materials are preferable for a bushland setting, whereas the design of a 'hip' inner city environment calls for a different approach, where finishes relate to the built environment, can make a statement and can convey a sophisticated urban life style. Considerations for plant material selection include, amongst others: growing conditions, climate, maintenance and quality of the urban environment. Because a lot of knowledge and experience is required to select suitable plants for locations in urban environments, we often seek horticultural advice to ensure the desired outcome. There is so much to know about soil types, nutrients, water requirements, structural integrity of trees and so forth, which justifies the involvement of plant experts such as horticulturalists and arborists. Safety is also an important consideration. Is the paving slippery when wet? Are structures safe and fit for use? Are trees marked for retention healthy and structurally sound? As required, we work with engineers and tree experts to ensure structural and services requirements are carefully considered. The palette for soft and hard landscape elements is developed in collaboration with the project team and the client as the design progresses and details are resolved.

What other professionals do you often work with on a design?

We work with a great variety of professions, depending on the project needs. Typically on a project we would work with architects, engineers (civil, structural, electrical and mechanical),

ISBN 9780170349994

quantity surveyors (costing), project managers and, of course, the client. On health projects, such as hospitals, this may include clinicians and other health experts. We use horticulturalists (advice on plant selection and soils), ecologists (endangered flora and fauna) and arborists where large trees are involved. Other collaborators can include planners, traffic engineers and economic consultants. Most enjoyable is working with artists when the project requires – artwork in the landscape, or furniture design, or graphic designers to create the layout of reports or the design of signage and so forth.

Conrad Gargett Riddel Ancher Mortlock Woolley is a multidisciplinary studio. What is the relationship between the landscape design, architectural design and interior design aspects of the studio?

Our practice offers a highly integrated approach to urban design, architecture, landscape architecture and interior design. This is fun and makes designing so much more interesting. Working as a team ensures the best possible outcome and also reduces the risk of mistakes, which can be very costly in the construction industry. There is always something you can learn from the other disciplines or contribute to improve the design and create a better place for people and the environment. The different disciplines at Conrad Gargett *et al* work collaboratively, not only on large and complex projects but also on small and intricate projects, to deliver the right results. Urban design is an area of work that requires expertise from a variety of disciplines. There is an overlap between architecture, landscape architecture and planning in urban design and this is a good thing. After all, urban design is a 'team sport' and no discipline should claim it outright.

What design technologies are used in the execution of a design?

Our office has recently converted from AutoCAD to Revit. AutoCAD, a 2D drafting program, was our drawing tool for the last 15 years or so. Revit is a 3D drawing program where you draw in three dimensions right from the outset. The Revit model can be shared by many users including other consultants such as structural and hydraulic engineers. This makes the detection of 'clashes' easier to see; for example when a stormwater pipe 'clashes' with tree roots or a structural footing. In the early stages of the design we use Sketch Up because it is easier to set up and use. We only commence modelling in Revit when a commission has been confirmed to avoid a time-consuming set up in Revit. A lack of three-dimensional graphics for Australian plants requires us to Photoshop over the top of Sketch Up or Revit images to achieve a more realistic representation of the vegetation for presentation. Rhino is another architectural program that we have used in the past, which is good for finishes. New programs are becoming available all the time and finding the right design tool can significantly improve the quality of presentation. Freehand sketches are still used for early design presentations but with the complexity of our designs, computer-generated three-dimensional models are more accurate, can easily be modified and can generate a schedule for costing purposes.

How is a final decision about the design direction arrived at?

After the briefing, the design idea is developed by the landscape architect in the form of a sketch design and then formally (design review) or informally discussed with the architect or other designers. The sketch design is then costed and presented to the client. The sketch design can be a simple hand drawing or a series of drawings plus a site assessment, design rationale, character images, scaled sketch design, hard and soft landscape palette and of course a preliminary cost estimate. The client has the ultimate say in what is adopted or modified. Once the sketch design has been signed off by the client, the designer can move onto the next stage of design: refining and developing the intent. This process is repeated until the developed design is signed off and then documented for construction.

What legal and ethical issues are considered in the design work you do?

Of course, while designing we have to comply with the general law and particularly those regulations and standards applicable to the area of design we are working in. This includes, for example, compliance with the *Disability Discrimination Act* (DDA) and the Australian Federal *Environment Protection and Biodiversity Conservation Act 1999* (EPBC Act). The Australian standards for design and construction set out a minimum standard that is to be followed in the design work – Conrad Gargett Riddel AMW pride themselves on far exceeding these minimum standards and there are ethical standards for designers that require professional (honourable and honest) conduct at all times in dealing with the client, community and colleges. The Australian Institute of Landscape Architects (AILA) has its own code of conduct, which has been expanded by design and environmental policies. This provides guidance to members but can also inform clients on best practice landscape design and construction.

On completion of a project, how is the success of that project evaluated?

As part of our Quality Assurance (QA) system we are required to do a post-construction evaluation, which is a series of questions answered by the client with regard to rating our performance and the project outcome. The intent is to gauge the satisfaction level of the client but more importantly it is a time to learn from our mistakes. It is important to articulate mistakes or shortcomings and to note what and how it could have been done better. This process is most beneficial if done in a group. It has to be noted that this is not the time for blame, but for constructive suggestions on how to do it better the next time.

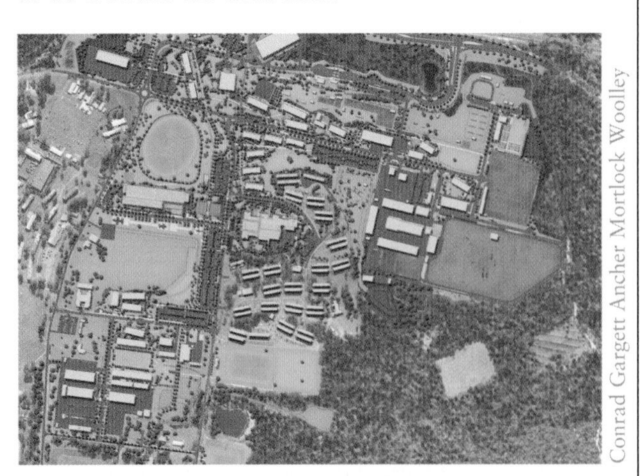

Conrad Gargett Ancher Mortlock Woolley

▲ Landscape Master Plan, Moorebank Units relocation, Sydney

What are some of your design inspirations? Where do you seek ideas and inspiration?

My design inspirations come from nature and landscapes and gardens I have experienced all over the world. Nature is a wonderful teacher where natural systems support a sustainable outcome. Nothing is contrived; there is a natural flow, logic and beauty that is awe inspiring. Climate, geography, plants and animals function together as one – interdependent and self-supporting. I take a lot of inspiration from these ecosystems and aspire to adapt nature's way to reverse some of the negative impacts humans have had on the natural world – baby steps nevertheless, but a way leading into the future. I feel there is so much more we can learn and adapt from nature, which will not only reverse some of the damage done to the natural world but also create wonderful and sustaining environments for people.

What advice would you give students seeking studies and a career in landscape/urban design?

The area of landscape architecture/urban design has a great future. Landscape architecture is a relatively young profession but has gained significant momentum in the last 10 to 20 years in Australia. It offers the opportunity to contribute to the environment and integrate development with the natural environment that sustains us all. Landscape architects understand natural systems, are capable of adapting them to the urban environment and are accustomed to working collaboratively with others to achieve the best possible outcome for people and the environment. What could be better?

ISBN 9780170349994

CHAPTER 3
GRAPHIC DESIGN

'You can have an art experience in front of a Rembrandt ... or in front of a piece of graphic design.'
Stefan Sagmeister

The origins of graphic design lie in the developments of typography in Europe from the 15th century onwards. The creation of the printing press and 'movable type' enabled mass production of printed materials and radicalised the way the written word was distributed. However, the design of graphic materials was not established as a separate and recognised practice until the early 20th century when attention began to be paid to the composition and layout of books, posters and other printed matter.

▲ Kleine Dada Soiree (Small Dada Evening), poster by Theo van Doesburg and Kurt Schwitters, 1923

The position of graphic design as a respected profession occurred later still, in the 1950s when the design principles of the International Style were reflected in the corporate logos, film posters and publication designs of the era. Highly influential designers, such as Saul Bass, Paul Rand and Milton Glaser, were seen as pioneers of modern graphic design and established many of the fundamental elements and principles of design that are in use today, such as the use of Swiss typefaces including Helvetica, the use of white space and grid systems.

The rise of graphic design as a profession saw its significance explode in the late 20th and early 21st centuries. It was during this time that the significance of logos and corporate branding grew. Graphic designers were sought after to create corporate identities for businesses, government and the not-for-profit sector. A recognisable visual identity and brand become essential for organisations in an increasingly competitive and developing marketplace. The role of the graphic designer in brand identity, brand development and brand management remains a key aspect of the design area today.

The rise of the Internet from the mid-1990s onwards has seen online visuals evolve from clumsy HTML to sophisticated interactive sites where graphics are designed for user interaction as a priority. The most successful online brands that apply effective user interface design illustrate the skills of talented graphic designers. The expanding area of user experience design is established in the realm of graphic design. However, access to design software, stock images and social media has meant that the creation of a logo, illustration or graphic is accessible to many, including non-designers with mixed results.

Graphic design is an area where innovation and experimentation have reflected social change, from album covers and rock posters to political slogans and online campaigns, graphic design has expressed dissent, new ideas, political affiliations and protests. It has been used to create things of minimal importance, such as memes of disgruntled cats that proliferate on the Internet, to tools of mass change, such as Shepard Fairy's 'HOPE' poster for the 2008 Obama presidential campaign.

The design of visual messages and meanings in our highly visual culture has become integrated into our lives; consider street signage, warning labels and advertising, which serve to inform, instruct and entice viewers respectively. In its short life, the profession of graphic design has become a respected and influential one.

▲ *Emigre* magazine was published from 1984–2005 and was a leading graphic design magazine. It was one of the first publications to use computers for layout and used many different typefaces and layout styles in each issue.

Emigre Magazine

3.1 GRAPHIC DESIGN PROFESSIONALS

WHAT DO THEY DO?

Graphic designers work with type and image to create a wide range of graphical products in print and digital media. Common projects for graphic designers include, logos and corporate branding, packaging, posters, signage/way finding systems, publication design, web design and interactive multimedia.

Graphic designers work for a range of clients across all business, government and not-for-profit sectors. Their work is diverse and often offers opportunity for creative design solutions. Advertising and corporate branding are two key areas for graphic designers. Some graphic design studios specialise in branding, offering businesses and organisations a comprehensive design focus to suit their needs.

WHAT SPECIALIST SKILLS DO THEY HAVE?

Universities and TAFEs offer qualifications in graphic design (also called communication design

ISBN 9780170349994

or visual communication) and each course offers a variety of subjects in areas such as traditional print and digital media, typography, branding and identity design, two- and three-dimensional design, motion graphics, illustration and photography. Designers have opportunities to build knowledge of design theory and analysis as well as practical skills.

In their professional life, graphic designers are often multiskilled with abilities in drawing, illustration, and design software, such as Adobe Illustrator, InDesign and Photoshop, and motion-graphics programs. As in other design areas, the ability to communicate well with clients and work with other design professionals is advantageous.

WHO DO THEY WORK WITH?

Although many graphic designers specialise in design areas such as print or digital media, they can be required to collaborate or seek assistance from specialist practitioners. They may work together as part of a cross-disciplinary team or require the service of a specialist to complete a final design product. Specialists include printers, exhibition and display designers, multimedia specialists and web designers, game and animation designers, illustrators, photographers, sign writers, industrial designers and advertising art directors.

WHAT RESOURCES DO THEY USE?

Contemporary graphic designers generally use computers for the bulk of their design work. Drawing may still form an important part of the design process, particularly in the early stages when visualisation and ideation are required. However, the main tools used in graphic design are design software packages that includes Adobe Illustrator, Photoshop and InDesign. Motion graphics may be created in Final Cut Pro or Premier Pro.

Designers often use digital tablets to input and edit design ideas digitally and 2D printing technologies are often used for proofing of final artwork. Additional resources that might be found in a graphic designer's toolkit are Pantone colour swatches for colour selection, a camera for collecting research and a library of books and magazines that document contemporary trends in global design.

3.2 SIGNIFICANT GRAPHIC DESIGNERS

+ Milton Glaser
+ Paul Rand
+ Saul Bass
+ Paula Scher
+ Stefan Sagmeister
+ Vince Frost
+ April Greiman
+ Michael C Place (Build)

3.3 GRAPHIC DESIGN CASE STUDY: WILLOAKS BED AND BREAKFAST

GRAPHIC DESIGNER: EMMA RICKARDS

Graphic designer Emma Rickards was asked to redesign the corporate identity for the WillOaks Bed and Breakfast business. The successful business wished to reposition itself at the luxury end of the market and required an identity and selection of design applications that communicated the nature of the business while expressing a warm sense of rural hospitality combined with sophisticated style. The task involved the redesign of a logo, signage, printed promotional materials and a website.

THE DESIGN BRIEF

The design brief outlined the requirements of the client. The brief was initially a verbal one so Emma created a return brief to ensure that she and the client had the same understanding of the task, the timeline, costs and deliverables. The return brief also offered detailed information about costs to the client and a list of the deliverables that Emma would present at the end of the process.

In the return brief (below) the steps of the design process were defined for the benefit of the client.

DESIGN BRIEF ~ RETURN BRIEF

STAGE ONE: LOGO CONCEPT DESIGN

After exploration of various approaches to the identity's development, two logo concepts will be presented for your review.

STAGE TWO: LOGO DEVELOPMENT AND APPLICATION

After receiving your feedback, the chosen logo will be further developed, and applied to a business card, 'with compliments' slip, voucher, brochure, website and signage. One concept design per item will be presented for your review.

STAGE THREE: DESIGN REFINEMENT AND CLIENT APPROVAL

The chosen items are further developed and refined in accordance with your feedback. Author's corrections will occur during this stage, and the developed designs will be presented for your signed approval.

STAGE FOUR: FINISHED ARTWORK

Digital artwork and files will be prepared in accordance with the printer's specifications and web requirements, and printer's proofs will be reviewed to check accuracy and ensure the quality of the reproduction.

RESEARCH

To familiarise herself with both the product and the area, Emma undertook a site visit and travelled to the bed and breakfast where she photographed the building and surrounds. Some unique features noted by Emma were the delightful gardens and classic Art Deco building details. These were recorded for inspiration. Emma also investigated similar properties in the area and researched the design offerings of accommodation at the luxury level.

As part of the research phase, it was also important to look at the existing identity used by WillOaks.

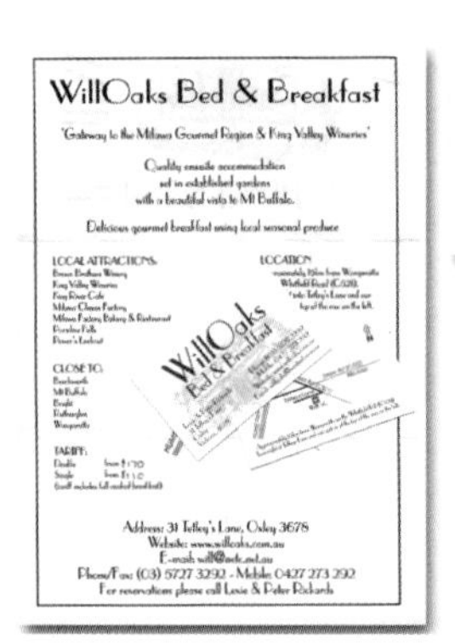

▲ Existing identity materials

GENERATION AND DEVELOPMENT OF DESIGN CONCEPTS

Drawing was an initial method used by Emma for generating concepts. A combination of quick sketches and observational drawings were used to stimulate ideas for further development.

▲ A range of media was used to generate initial concepts including pencil and ballpoint pen.

ISBN 9780170349994

HELPFUL LINKS FOR GRAPHIC DESIGN.

AGDA Australian Graphic Design Association: Information about the organisation and, most importantly, about their outstanding design conferences

Behance 'Served' Galleries: These online galleries showcase the best contemporary practices in all areas of design. For graphic design start with Typography Served, Branding Served, Packaging Served and Illustration Served.

It's Nice That: Diverse and up-to-the-minute graphic design site that addresses all aspects of the field including motion graphics and illustration

The Dieline: A comprehensive site featuring the latest packaging trends

Process Journal: Quarterly journal focused on discovering the practices of the world's top graphic designers

Access all weblinks directly at http://nsg.nelsonnet.com.au.

Using Adobe Illustrator to develop some of her preferred ideas, Emma was able to experiment with colours, type and patterns.

Emma Rickards

▲ Using design software, such as Adobe Illustrator, enabled Emma to repeat pattern elements and create a wide range of options for her design.

Choices regarding design elements were inspired by the Art Deco features of the bed and breakfast. The selection of type (Scala Sans) was made for two aesthetic reasons: firstly to reflect the art deco influence and secondly because of the existence of non-lining numerals.

Throughout the process, client input was used to clarify the design direction.

Once a final design for the identity was established, it was necessary to select final colours and stock. Opting for a blue palette inspired by the external colour of the bed and breakfast, Emma continued the colour onto the brochure and website design.

Emma Rickards

▲ Comparisons of colour and type. A digital presentation such as this can also be sent to a client for feedback.

ISBN 9780170349994

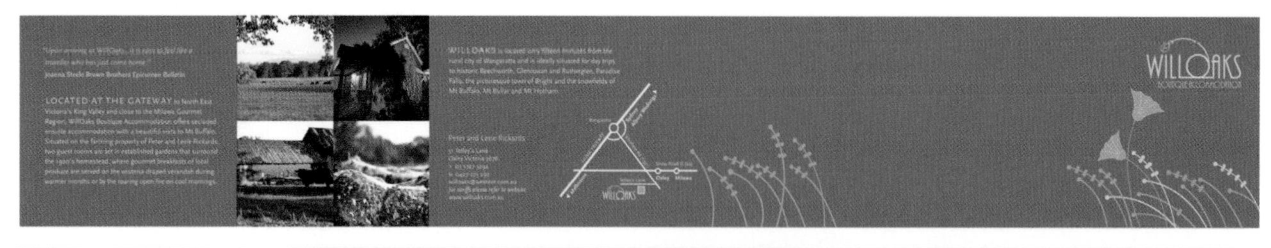

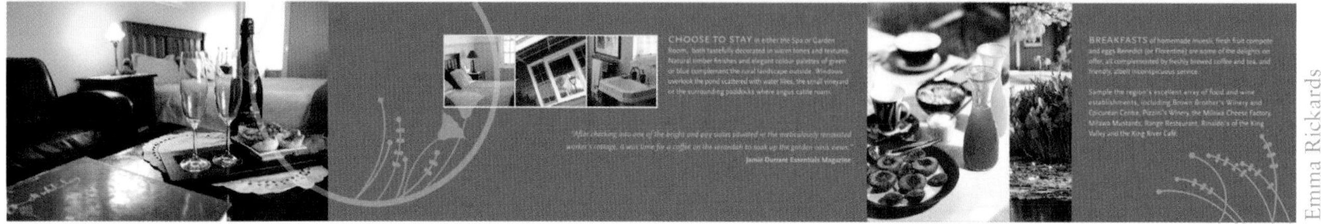

Emma Rickards

WORKING WITH SPECIALISTS

In addition to her own skills in design and layout, Emma enlisted the help of a photographer to provide images for use on promotional materials and the website. To follow copyright laws, the photographer's images were fully credited. A professional printer was involved in the printing of business cards, letterhead and other stationery items as well as the four-colour, three-fold brochure. Consulting with the printer ensured that digital files were appropriately prepared and the most suitable printing stock was used for each application. For the website, Emma employed a web designer to implement her design vision for the bed and breakfast; she provided artwork and images with a clear brief as to the look of the website and its functionality.

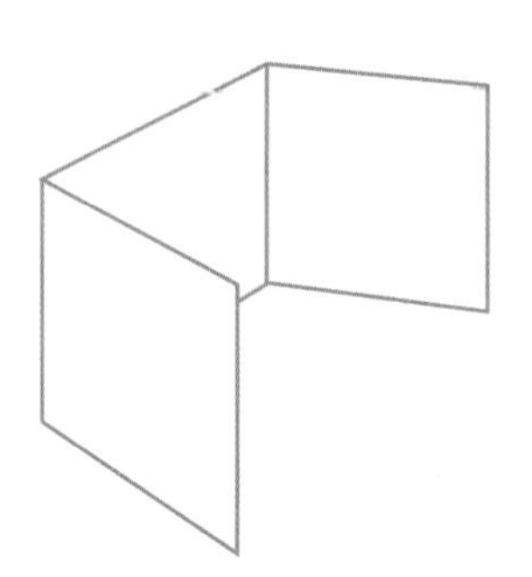

▲ Mock-up of the brochure design

EVALUATION

Positive feedback from her clients was the immediate response to Emma's design and an indication of its success. However, data from sales and the success of the WillOaks website provide clear indications that the redesign has had a positive effect on business.

Emma Rickards

▲ Final implementation of the WillOaks logo on a range of carriers

ISBN 9780170349994

GRAPHIC DESIGNER ~ ALEX FREGON

Alex Fregon is a graphic designer at one of Australia's biggest daily newspapers. He is also an illustrator and designer of music graphics for independent Australian musicians and music promoters.

When designing music graphics are you usually provided with a formal design brief?

Most of the time I'll get a short email from whoever is briefing the project stating what it is they're looking for, what (if any) 'look' they're after, and when they need it. The briefs are usually very informal – more of a casual 'do you have time to knock this over' than a formal job briefing. I will have a look at the brief, see what's involved and if I've got time to do it, and whether or not it's worth the amount of money being offered. I'll level with you: there's not a great deal of money in designing posters for record labels! Most people involved in the independent music industry are doing it for the love of good music, not the coin. So as a designer you have to love the music too, and just appreciate the opportunity to do some fun, creative work.

Alex Fregon

▲ Poster for Mistletone show

If you do get a design brief, what information does it usually contain?

If it's a poster, the initial brief will say who the band is, when and where the gig is, and what kind of feel the record label wants for the poster. I've done quite a few posters now, and they all have a different look and feel, so sometimes a label might say 'in the same style as that Bats poster you did last year' if they like something I've done in the past. Or they might reference another designer's style, or a style from a specific era. Sometimes they might just reference a colour palette, or say 'we want something fun'. It can be pretty vague, but thankfully the dialogue is very casual so you can figure out what they're after with a few friendly emails and a bit of Google image searching.

If it's something like an album cover, the details provided in the brief might include the format (jewel case, digipak, cardboard

ISBN 9780170349994

wallet, etc.), title, track list, liner notes, and any details the artist's label might want to include (logos, copyright information). You always have to leave space for a barcode too. In my experience, most musicians have studied art or design in the past, are mates with artists and designers, or are in a band with artists and designers, so they always know what they're looking for. Often they'll already have artwork that they want to use, so the designer's job is simply to bring everything together.

Alex Fregon

▲ Artwork for Anontendre by Brisbane-based musician Andrew Tuttle, aka Anonymeye. The client wanted something coastal, so Alex used a series of illustrations based on photos he had taken of various coastlines around Victoria.

Can you describe the relationship you usually experience between client and designer?

In the independent music industry, everything is very casual. By 'independent' I mean small inner-city record labels like Mistletone. They're usually groups of mates (or, in Mistletone's case, a couple) doing it for the love of it, so they really appreciate any support you'll give them. As a designer you have to weigh it up: you can do work for bigger companies with bigger budgets, but the work is not going to be anywhere near as much fun and you won't have anywhere near as much scope to experiment. Working for small labels means you get to try new things and push the envelope. Of course, the payoff is that you don't get paid as much. But the relationship is very good, and easy. More like a friendship than a business relationship.

That is not to say that you can't approach the work in a professional manner, however. Every client has a different approach to the briefing process, so part of the designer's job is to gauge the client's expectations, both in process and end product, and respond accordingly.

How do you identify the target audience for the music graphics you design?

Often I'll already know about the band that I'm designing for, but if it's someone I haven't heard of I'll Google them and have a listen, and see what kind of thing they're doing. It is very easy to identify an audience these days, with all the music blogs on the net. With my work for Mistletone, most of the bands I design posters for are listened to by the same 20 something inner-north music loving audience, and these people are open to pretty much anything graphic-wise. As long as it is interesting and doesn't use Brush Script!

Do you undertake any research when designing for a client? If so, what is the nature of that research?

Research-wise, I'll often have a look at what work the label or band has done previously and what style they've built up. Some bands are very picky about the fonts and colours used, and they get upset if you make them look too 'playful' and not as serious as they take themselves!

What are some of the constraints that you may have to deal with when designing music graphics such as posters or CD artwork?

Bands and labels don't always have a lot of money to spend on printing, so sometimes you have to take the quality of the stock and finish into

ISBN 9780170349994

▲ Poster for a Tones show, for the Mistletone label

With album artwork, you'll often find yourself working with different manufacturers, and they always have different template files. It's important to have a good look at the templates before you start laying out the design.

What are some of your inspirations in design?

I am inspired by a lot of things: artists, other designers, photography … I could go on and on. Some of my biggest inspirations are David Shrigley (a very funny Scottish cartoonist), Chris Johanson (an American painter with a fantastic drawing style and colour palette), Kim Hiorthoy (a Norwegian designer with a great minimalist approach to … everything) and Josef Muller-Brockmann (a Swiss graphic designer who wrote the bible on grid-based layouts, the romantically titled Grid Systems in Graphic Design). There are countless others but these ones stand out for me!

People aside, I am also inspired by colours and moods. And, of course, some of the best inspiration for music graphics is the music itself!

consideration. Often, the nicer stock options aren't the most expensive ones though – if this is the case I'll push the client to use a natural, uncoated stock if possible, as it soaks up the ink and can 'disguise' a cheap print job!

When designing posters, size is also something to consider. You're designing something huge, but doing it on a small screen, so it's important to proof it out at full size before you send it to print. Sometimes the font size will look perfect on the screen, but when you see it in print it will look too big.

You use a lot of hand-drawn type and illustration in your work – what is it about these techniques that appeals to you? What is its relationship to the design work you do?

I don't know why I love drawing so much, I just always have. When I was a kid I would sit at the table with a ream of A4 paper and just draw for hours. Footballers taking marks, Ninja Turtles, Voltron … I loved drawing all that stuff. And I guess that love of drawing has stayed with me right through school, uni and into my adulthood.

I still draw a lot these days, on the train to work, at work, at home ... Yeah, it's a bit of a problem.

Thankfully, the Mistletone guys always push for the handmade look with their posters, so I do a lot of hand-rendered stuff for them (pen and ink, pencil, texta, etc.). It is good fun. Doing the posters does give me the chance to try out new approaches to hand-drawn type and illustration too, which is great. For the Castle Tones poster I designed the whole thing in InDesign (laid out in Helvetica Neue Condensed), then printed it and hand-traced the type in pencil to get a neat, hand-drawn look. And I also do a fair bit of 'digital hand-drawing' with the Wacom tablet, as seen in my work for Anonymeye and my Harvest magazine cover illustration. The Wacom is a fantastic tool, once you get used to it.

How are decisions made about the use of elements such as imagery, type, colour and layout?

When I'm starting a job, I'll usually start by putting the text on the page. If it is a poster, I'll work out the hierarchy and see how it all fits on the page, allowing some space for an illustration. If I've already got an illustration in mind, I might put that on the page first and see how the text fits around it.

At first I'll work in black and white – this is a general rule for me, whether I'm doing something illustrative or designing a logo. Colour is easy, it can be added later. The way I see it, if a logo or poster works in black and white, it'll definitely work in colour, but it doesn't necessarily work the other way round.

When doing a poster you can afford to make decisions about imagery, type, colour and layout on the fly, as it only really relates to the piece you're working on. When working on something more substantial (a booklet, brochure, or corporate identity), however, you need to get these things right from the start, as any changes you make down the track will need to be reflected across everything you produce.

How do you use drawing in your designs? What materials, media do you use?

I usually draw with a pencil or a felt-tip pen, or with a brush and a pot of ink, and then scan it in and play around with the contrast in Photoshop.

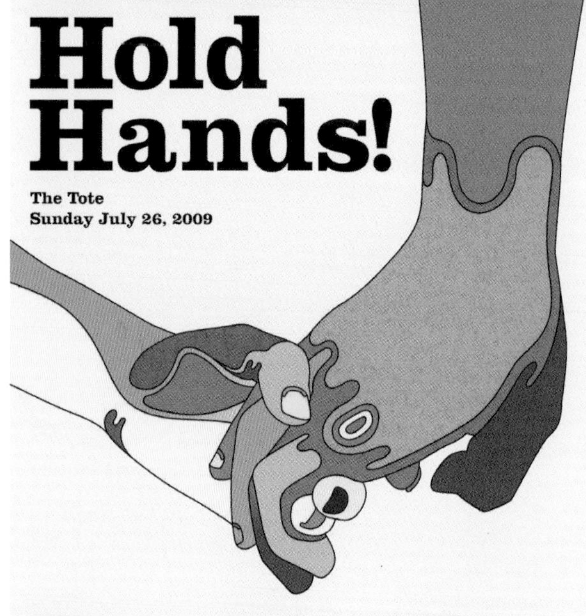

▲ Poster and CD artwork for fundraising event

I have a pretty crappy scanner, so I usually need to clean up the scan a bit. But it is good fun.

What skills do you, as a designer, apply when creating each design?

Having worked at a newspaper for the past seven or eight years, I have a very good knowledge of the entire Adobe suite (InDesign, Photoshop, Illustrator), and I use all three programs when doing my freelance work. Technical skill and

ISBN 9780170349994

knowledge of software is important, but I believe the most important skill a designer can have is an understanding of layout. Being able to take a number of elements, figure out the hierarchy and put them on the page in a balanced, aesthetically appealing way is a real skill, and something that can only be learned by looking at a lot of quality design. I am also a big advocate for the grid: setting up a rigid, versatile grid before you start designing is very important, and though it may take a little extra time at the start of a job, it will save you hours in the long run. Grids are fantastic. Don't start me on grids, I can go on for hours!

How do you stay up to date with trends and changes in design?

Read, read, read. I just go into bookshops and newsagents and pick up anything that looks half interesting. There is so much design out there these days, it is scary and more than a little daunting. But it is a really exciting time to be involved in design. People are willing to pay money for good design and the general public's appreciation of design is growing more sophisticated by the year. Magazines are becoming more like websites and websites more like magazines, and both are placing a heavy emphasis on clean, sophisticated design.

How do you evaluate the success of a design?

With music graphics, it is pretty simple: social media. Once a poster is finished, the label will post it around town and all over the Internet (Facebook, Twitter, and on music blogs like Mess And Noise), and if people like a poster they will usually leave a comment. Sometimes a member of the band will comment themselves, and the label will pass it on to me. That's always nice. But for me, I'm usually my biggest critic. I'll finish a design, send it off, and then forget about it for a couple of days. Then, I'll be in town and I'll walk past a copy of my poster somewhere, and I'll know instantly if it works or not.

I guess there are a few ways to define 'success'. It could be that the client is happy with your work. Or it could be that the public seems to like it, and you've got some good feedback from your peers. Or it could be that the gig sells out in record time. I guess you never know for sure. But if I'm happy with the result, and I look back on it a few weeks down the track and think, 'yeah, that came up alright', I'm happy.

ISBN 9780170349994

PART B

DESIGN PROCESSES

CHAPTER 4
THE DESIGN PROBLEM

'Design is an art of situations. Designers respond to a need, a problem, a circumstance that arises in the world. The best work is produced in relation to interesting situations – an open-minded client, a good cause, or great content.'

Ellen Lupton

In the production of design products, a design process is usually applied. The design process involves stages of concept development and production, initiated and guided by a design brief.

Designers in different design areas use design processes, which may vary in structure and terminology, but generally speaking, the basic components of each design process remain consistent. The design process is a cycle. The cycle begins with the identification of a design problem, which is subsequently outlined formally in a design brief. Information, such as target audience, user characteristics, function and constraints, is usually identified in the brief. Moving through the cyclical process, research identifies further information and data, which focuses the direction of the design development.

Throughout the design process there is scope for imagination and creative risk taking. Experimentation is a key component of the development phase. Consideration of design factors becomes increasingly important during the process; the process allows for freedom and flexibility within the framework of the brief, and encourages experimentation with ideas, materials, elements and principles of design. Constant evaluation is an essential part of the process. It is through this process of inspiration, experimentation, evaluation and elimination that effective designs evolve.

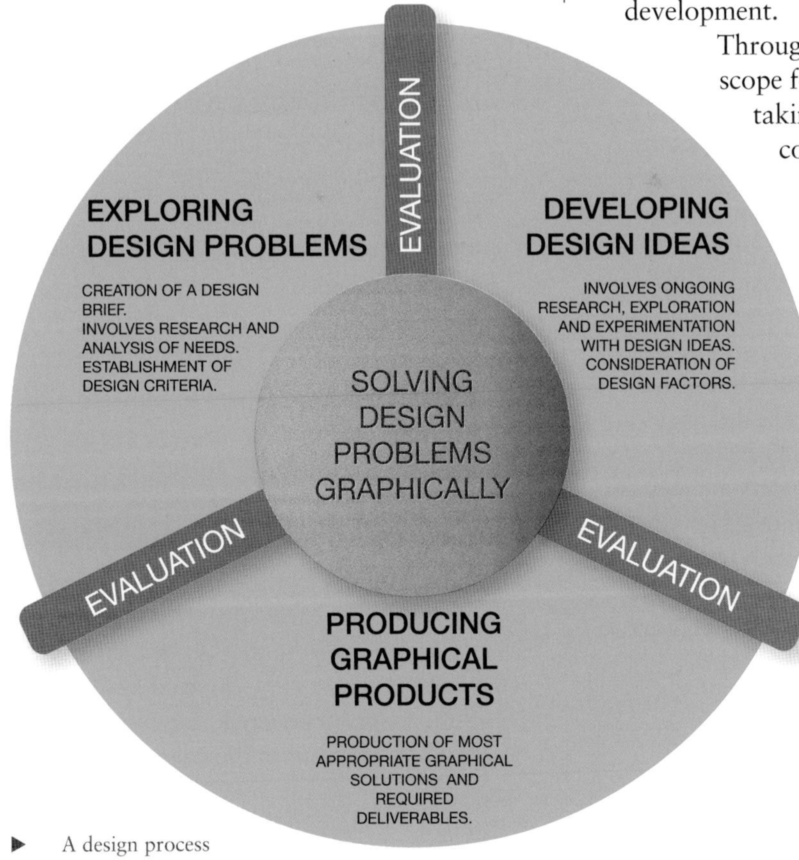

▶ A design process

4.1 THE DESIGN PROBLEM

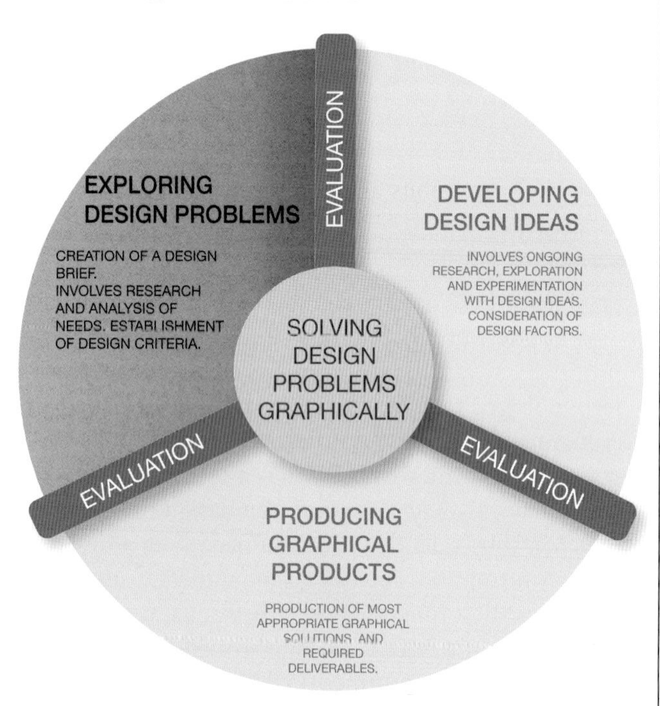

▲ Exploring design problems

A design problem is an identified need that requires a designed solution. Design problems can range in scale from the small, such as a brochure design for a small, gourmet grocery store, to the very, very large such as the design of an Olympic stadium.

In a professional design environment, a client may approach a designer or design firm with a design problem because they were recommended to the client, or because the client was interested in work previously completed by that designer or firm. Initial meetings take place to establish what the client sees as the design problem. At this stage, both client and designer can judge whether a suitable working relationship is likely. Some designers specialise in particular fields, so it is important to determine at an early stage whether the client's requirements and the designer's expertise are a match.

Clients may have a limited knowledge of the design process so initial meetings can facilitate discussions about cost, timing and possible design outcomes. A detailed written brief is then developed. The brief identifies important details such as the audience or market, the design criteria, the function of the final design, and any constraints – such as cost, timing and the 'deliverables' or final outcomes required of the final design.

Some designers use a 'return brief' in the early stages of a project. Taking the information gathered from the client, the designer writes the brief as they interpret it and then asks the client for comment. This technique allows the designer to craft a brief from a design perspective; returning it to the client ensures that both parties have a clear understanding of the final design deliverables.

WHAT IS A DESIGN BRIEF?

A design brief is the starting point of the design process, and is usually a written document that includes key information. The brief outlines the design problem that requires an effective visual solution. A brief might be a detailed written document but can also be a verbal or even visual explanation of the design problem.

The design brief:

+ describes the design problem – the need of the client, design criteria and deliverables
+ provides information about the initiator – the client
+ describes who the product is to be directed at – the user/audience
+ explains how the product will be used – the context and function of the design
+ identifies constraints and limitations – including cost, time etc.

THE USER/AUDIENCE

An understanding of the end user of the design is essential in any successful design process. The user or audience is the market or target group to whom the design will be directed and information about the targeted group is usually included in the design brief. Sometimes the client will have a clear idea of the market but may be seeking to expand it or attract a new audience to a location, product

ISBN 9780170349994

or service. The designer's task is to identify the specific characteristics of the target group; the characteristics will subsequently affect how the design is developed.

User characteristics are often divided into specific types of data such as age, gender, socioeconomic status and interests. Other factors such as location, cultural background and religious affiliation can also affect the content, form and appearance of a design.

Age

Age can be identified in very specific terms or more broadly, and is often classified by arrangement into groupings. For example, 18–25 year olds might be more loosely defined as young adults, 40–55 year olds might be classified as middle-aged adults, and so on. Terms such as Baby Boomers, Generation X or Y and Millennials, used by marketers to establish generational groupings, are somewhat helpful, but can be overly broad categorisations.

Gender

Graphical products can be targeted at a male or female audience or may be gender non-specific. The influence of gender is very strong in defining personality traits and consumer preferences, and will dictate the use of many design elements and principles in a visual communication.

Socioeconomic status

This usually refers to the financial and social position of an audience. In Australia we perceive ourselves to be an egalitarian society and have little interest in a 'class' structure. Rather than defining social groups as working class, middle class, and so on, we tend to identify ourselves by our level of financial income. Employment status, salary level and educational background can be factors in identifying the socioeconomic status of an individual or group. The amount of money people earn determines their 'disposable' or discretionary income (that is, the income remaining after essentials are covered). Groups with a high disposable income are attractive to marketers.

Interests

This is a vast category of great importance to designers and market researchers. The interests of a specific audience may include music and fashion, for example, but there exist subcategories of those interests that can define an audience in even more detail. The specific style of music and the fashion labels that are preferred by the end user will influence their habits as consumers.

Interests may also refer to specific professional or personal interests. A design may be targeted at small associations of professionals, such as surgeons, carpenters or chemical engineers, or at an organisation based on shared interests, such as the Veteran Car Club or the Surfrider Foundation Australia.

Cultural background

The content or appearance of a design may be influenced by the belief system of the audience. The appropriateness of imagery and content will be defined by cultural and religious traditions. If a brief addresses the needs of a culturally specific audience, it is essential that the designer builds an understanding of what visual material is and is not appropriate to use.

Location

Where an audience lives can have an impact on the effectiveness of a design. A target audience in a remote area will have different opportunities to access or view designs compared with an audience in an urban location. Their needs may vary substantially so a design may be quite specific to a region or area. Location can also affect the socioeconomic status of an audience, as some areas offer different opportunities for employment or professional advancement. Location can also affect the language used in a graphic design, for example as the appropriateness of colour and images, and the scale or proportions of the design may be impacted. An environmental design is invariably affected by its location; the use of materials and the appearance of the structure may be impacted by geographical and planning boundaries.

THE CONTEXT OF THE DESIGN

Where a design will be used has a major impact on the content, appearance, materials and format. The physical location will determine scale, materials and the design elements and principles to be used. A billboard displayed at the edge of a freeway, for example, will be viewed by drivers and passengers passing by at 100 kilometres per hour. The content will be read from a distance, so a design that is heavily dependent on text and small detail would clearly not be suitable.

Design products found in magazines and journals are tailored to suit the purchasers of the publications. A fashion magazine features quite different advertising content to a magazine about cars or aviation. The content of the magazine itself may dictate how the visual communication should appear. Architecture and design magazines often use large areas of white space and apply fashionable design methods such as line illustrations and sans serif typefaces. Magazines directed at young people interested in surfing or skateboarding may contain page layouts that are action-packed, with large photographs and contrasting typefaces.

THE FUNCTION OF THE DESIGN

All designs have one or more functions (or purposes), which have a major impact on their content and appearance. The function/s will define the content of the design and establish where and how it will be seen or used, who will see it/use it, and how often.

Often, a design will have more than one function, but it is usually possible to identify these as the primary function and additional, secondary functions. For instance, a poster advertising a music festival may include the date and time, ticket prices and booking information, along with a map of the location to guide the attendees. The primary function of the poster is to advertise the music festival and sell tickets, but the other functions such as informing about the cost and location and guiding with the inclusion of a map are also important.

The function of the design is usually outlined in the design brief and will be affected by the design area. Of course, there are many functions that can be identified in each design area – some of the common functions are listed here:

Design area	Possible function	Example
Industrial design	To facilitate	Domestic appliance
	To transport	Motor vehicle
	To display	Shelving
	To protect	Packaging
	To interact	Digital media product
	To accommodate	Furniture design
Built environment design	To shelter	Public transport shelter
	To engage	Playground design
	To accommodate	Residential building
	To make sustainable	Commercial building
	To beautify	Landscape design
	To decorate	Interior design
Graphic design	To advertise	Advertisement
	To inform	Road sign
	To promote	Poster for event
	To explain	Diagram
	To identify	Logo
	To illustrate	Illustration
	To highlight	Signage
	To decorate	Textile design or pattern
	To guide	Map
	To teach	Visual instructions
	To symbolise	Icon or symbol
	To communicate	Webpage
	To attract	Packaging
	To persuade	Advertisement

ISBN 9780170349994

CONSTRAINTS

Constraints can have a considerable impact on the final outcome of a design brief and it is important to identify them as early as possible in the design process. During the initial meetings between client and designer, the possible restrictions on the project are identified. These include:

+ time
+ location
+ cost
+ materials and technologies.

Common constraints such as time and cost have a major impact on the design outcomes. A large project requires a longer time frame and usually needs a bigger budget, so a clear understanding of the cost and time frame must be established in the early stages of the client–designer relationship. The location of a design task can also provide challenges; in a global marketplace, many designers find themselves working in overseas locations, where language and cultural differences can affect the flow of a design process. The scale of a design task and the materials required might also be factors in determining the success of a final design. If a designer is unfamiliar with new materials, technologies and processes, then training and education must be addressed.

In developing a design brief, the following information is important to establish before undertaking the practical aspects of the design process:

The design problem

+ Describe what are you designing.
+ What are the design criteria? For example, design features, functions and characteristics required by the client.

Client *(if relevant)*

+ Describe the client's background, business and location, and any other factors relevant to the design brief/task. List previous or existing designs.
+ List any non-negotiable inclusions such as logo, corporate colours, branding and so on.
+ List any general client preferences for elements, such as colour, appearance, type and so on.

Function

+ What is the primary function of the design?
+ Are there additional functions that the design must address?

End user/audience

+ Who is the end user/audience for the design? Describe their age, interests, gender, location, socioeconomic status and any other relevant factors.

Constraints

+ What restrictions or limitations are there on the design?
+ Consider: cost, time, location and accessibility

Context

+ Where will the design be seen/found/located?
+ What effect might this have on the design?

Design factors

Consider the implications and impact of design factors (see Part B of this student book) including:

+ user-centred design
+ elements and principles of design
+ design technologies
+ legal responsibilities
+ design strategies
+ project management
+ sustainability
+ materials.

Deliverables

+ Describe the final design deliverables, mentioning the materials used, size, format and possible content.

SAMPLE DESIGN BRIEF

If you are required to create your own design brief, the following template (overleaf) may assist you in outlining the key information required.

ISBN 9780170349994

<table>
<tr><td>Describe the design problem</td><td colspan="6"></td></tr>
<tr><td>Client name (if relevant)</td><td colspan="6"></td></tr>
<tr><td>Client information (if relevant)</td><td colspan="6"></td></tr>
<tr><td>What is the main function of the design?</td><td colspan="6"></td></tr>
<tr><td rowspan="2">End user/audience information</td><td>Age</td><td>Gender</td><td>Interests</td><td>Socioeconomic status</td><td>Cultural or religious factors</td><td>Location</td></tr>
<tr><td></td><td></td><td></td><td></td><td></td><td></td></tr>
<tr><td rowspan="2">Constraints</td><td>Time factors</td><td>Cost factors</td><td>Location</td><td>Materials</td><td colspan="2">Technologies</td></tr>
<tr><td></td><td></td><td></td><td></td><td colspan="2"></td></tr>
<tr><td>Context</td><td colspan="6"></td></tr>
<tr><td rowspan="8">Design factors</td><td colspan="3">User-centred design
How should the design address the direct needs of user? (See Chapter 7.)</td><td colspan="3"></td></tr>
<tr><td colspan="3">Elements and principles of design
What are the required design elements and design principles. (See Chapter 8.)</td><td colspan="3"></td></tr>
<tr><td colspan="3">Design technologies
What design technologies will be required to meet the needs of the brief? (See Chapter 9.)</td><td colspan="3"></td></tr>
<tr><td colspan="3">Legal responsibilities
What are the relevant legal responsibilities? Copyright? Standards? etc. (See Chapter 10.)</td><td colspan="3"></td></tr>
<tr><td colspan="3">Design strategies
What design strategies could be applied for the best outcome? (See Chapter 11.)</td><td colspan="3"></td></tr>
<tr><td colspan="3">Project management
What techniques can be used to ensure effective project management? (See Chapter 11.)</td><td colspan="3"></td></tr>
<tr><td colspan="3">Sustainability
What are the impacts of the design and what sustainable strategies can be applied? (See Chapter 12.)</td><td colspan="3"></td></tr>
<tr><td colspan="3">Materials
What materials characteristics are required in the design?
How can they be represented? (See Chapter 13.)</td><td colspan="3"></td></tr>
<tr><td>Deliverables</td><td colspan="6"></td></tr>
</table>

ISBN 9780170349994

Once the design problem has been identified it is important to build an understanding of the needs, scope and potential of the design problem as outlined in the design brief. This may involve the collection of visual information, data and other research materials to assist in the understanding and development of the brief. In researching, a designer might use a wide range of sources for inspiration and ideas, including books, the Internet, direct observation, market research and focus groups.

Investigation and research is undertaken to assist in identifying trends, tastes, preferences, historical information and possible design ideas. These resources then assist in defining the needs and likely parameters of the task ahead.

4.2 RESEARCH

Research is essential and provides important information about current trends, the work of other designers, and essential information about the preferences of the target audience.

Observation and investigation are key methods of research. An architect will visit the site of a building to study the location of the structure and the surrounding environment. A graphic designer may meet with members of the target audience to gather first-hand information about trends, fashion and attitudes. An industrial designer may research how a consumer uses a product. Many designers will investigate existing products and designs in the marketplace for inspiration and an understanding of where a new product might be positioned.

To gain very specific information about the target audience, designers may use market research companies. Small groups of people who fit within the target audience range are observed during discussions or asked specific questions about their tastes and responses to existing products. Trend-forecasting firms also offer vast amounts of information about future developments in colour, styling, product forms and fashion.

The collected research data is analysed carefully to establish the accuracy of the information and its value to the brief. The verification of research is very important as styles and trends can change quickly, relegating what was once thought to be innovative to the out-of-date bin. Fashion design, advertising and graphic design are particularly susceptible to changes in public attitudes and fashion trends.

Research might continue throughout the design process. A design brief can demand more than one deliverable. For example, the main design brief may be for the design of a corporate logo but the brief may also require that the logo be used in a wide range of applications. These applications may include vehicles, clothing, stationery and advertising. The designer, in this scenario, would need to undertake research into the best means of applying the logo to these diverse carriers.

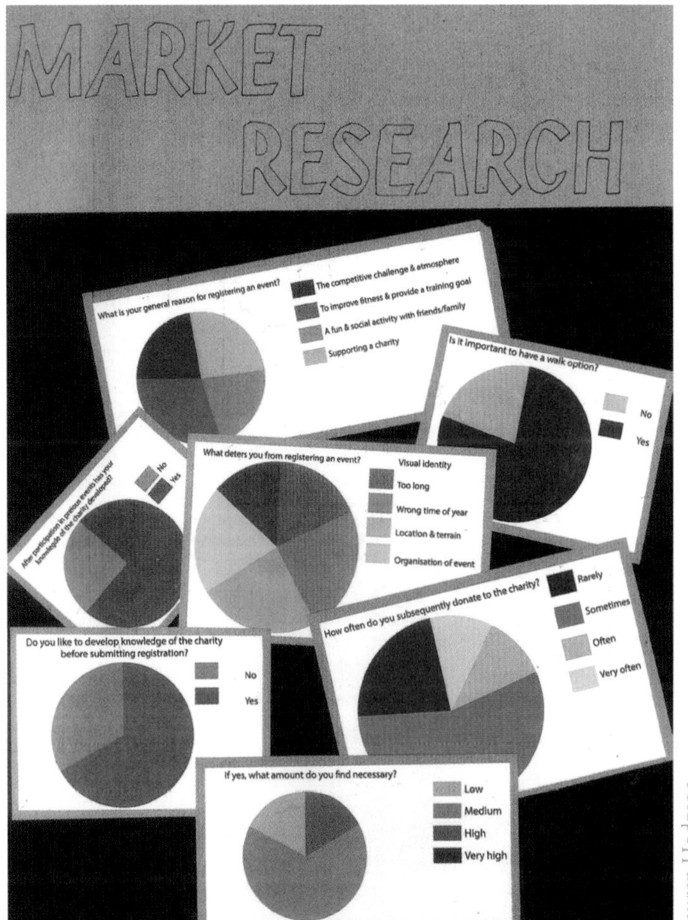

Lauren Hodgson

▲ This student undertook her own market research in designing products and an identity for a charity fun run. She conducted surveys and then organised her data visually to help direct her initial design ideas.

ISBN 9780170349994

Research can include:

+ investigation of the client history and existing products or services
+ investigation of direct competition to a new product or service
+ observation of the target audience in order to understand trends and preferences
+ observation of the location or context of the final design
+ analysis of data about future trends, new materials, community attitudes
+ collection of visual information to inspire new concepts.

UNDERTAKING RESEARCH

Research is an integral part of the design process. Research provides a window to essential information about important aspects of the brief, such as the audience and the relevant historical background to the task. Researching all aspects of the design brief is an important undertaking. Knowing about the client and their design history can help to ensure a new design is original and not repetitive. Research of the target audience helps to identify interests, preferences and trends. Research of similar design products is important to establish a point of difference and to view competitors' products.

Gathering inspirational material is also an important function of research. Inspiration may come from anywhere and anything. Many designers find inspiration in random 'found objects', colours, words, textures, landscapes and intangible experiences such as events, interactions and moods. Be open-minded and aware that inspirational material is all around you and may capture your interest and creativity at any time! Be prepared to collect items that might inspire the form, function or appearance of a design product.

Research is valuable for concept inspiration and it is essential to a response to a brief. How thoroughly you understand the design problem will direct the success of your final design concepts.

RESEARCH TYPES

Quantitative and qualitative research

In the discussion of research there are two common categories into which most research will fall: quantitative and qualitative research. The key to the difference between each category is in the name.

Quantitative research is concerned with statistical data and measureable, objective information. For example, quantitative information may be gathered by a web designer who is interested in establishing the success of interactive elements on a website. Gathering data related to the number of times users click on web elements may assist the designer in creating appealing and successful online content.

Qualitative research is subjective and deals, in many cases, with images, words and behaviours. Qualitative research methods include interviews and observations. For example, in the design of a children's playground, the designer may apply qualitative research by observing and documenting the behaviour of young children at play. Using the observations noted, the designer may be able to establish patterns and commonalities to assist in designing the most appropriate and engaging playground.

It is likely that you will apply predominantly qualitative research methods to your design work but it is also very help to use quantitative methods when gathering factual data to inform your design concepts.

Examples of quantitative and qualitative data and research methods are given below:

Quantitative	Qualitative
The company has eight employees.	The company is small.
The house has four bedrooms.	The house is spacious.
She is 178.5 cm tall.	She is tall.
The document contains 150 words.	The document is brief.

 ISBN 9780170349994

Quantitative	Qualitative
Surveys	Observations
Data	Interviews
Measurements	Market research

Primary and secondary research

Primary research and secondary research refer to the source of research materials. Primary research involves collecting information specifically for a design brief. It is gathered from original sources through methods such as interviews and market research. Your observational drawing of related imagery is primary research.

Secondary research involves the use of data or other information that has been collected by another source. Census data, books and articles are secondary sources. Secondary research may not have been created for the design brief but may still offer relevant insights and information, and usually contains analysis of primary sources. Use of design magazines is considered secondary research.

Examples of primary and secondary research are given below:

Primary	Secondary
Interviews with members of the target audience.	Review of articles describing the characteristics of the target audience.
Photographs and collage images of objects related to the design brief.	Use of stock photography images related to the design brief.
A focus group of users to discuss how effectively a design functions.	Use of census or similar data.

COLLECTING RESEARCH

It is important to research intelligently. Research should be collated and sorted carefully, and used throughout the design process. Although the research stage is identified at the very beginning of the design process, its influence extends throughout the design task.

Where should you start?

How do you know where to start? It is important to have a purpose when researching. Using the design brief, you may choose to focus on the following areas to initiate your research.

Research the client

Who are they? What is their background? Their location? What previous designs have they used/ created?

Research the user

Who are they? What do they look like? What do they do? How old are they? Where do they live? What designs appeal to them? How will they use the final design?

Research existing and past design products that are similar

Which are most effective and why? Where are possible points of difference?

Research other designs that have a similar function and context

How do they achieve the function required? How are they placed/used/displayed in their given context?

RESEARCH RESOURCES

Where to look

The Internet is usually the first port of call for research and it offers a wealth of information – sometimes too much. It is advisable to look

ISBN 9780170349994

at a range of different sources for research and not rely on one source only. You also have the opportunity to create your own research by taking photographs, making notes and creating drawings.

Electronic resources

The Internet is one of the most powerful information resources available. The sheer volume of available information from a wide range of sources makes this an invaluable resource.

Online databases provide access to articles and papers that might otherwise be unavailable to individuals outside specific industries and professional fields. Most of these services are subscription based, but many libraries have access to them.

There is a wealth of free information and images online provided by individuals and organisations. However, when seeking factual material, it is important to verify the source of the information given, as there are essentially no overall rules to ensure that all material published online is accurate. Email and social networks are important communication tools that can assist in the collection of information. Many manufacturers will provide company and product information in response to a request via email or offer information via their website or social media pages.

The availability of digital image technology makes the sharing of still pictures and video online quick and straightforward. Keep in mind that many web images have a resolution of 72 dpi and are small in size to ensure that download times are minimal. Such images print poorly. Web images that have been enlarged can appear pixelated and blurred, making them unsuitable for presentation purposes, although they may be quite suitable for research. There are some valuable online libraries of stock photos that are free to use; amateur photographers and illustrators post their images and allow their work to be used for non-commercial purposes. The quality of these images can be quite high. Most content on these sites requires appropriate attribution of the source so ensure that you follow the guidelines set out by the copyright owner of the image.

STOCK IMAGES

freeimages: This free stock image site offers high-quality photographs for use by the general public. Attribution requirements are specified by the owners of each image.

Access all weblinks directly at http://nsg.nelsonnet.com.au.

When accessing online images and information, a search engine is often the first port of call for users, and search engines such as Google provide lists of many sites based upon defined search criteria. Helpfully, Google Images allows users to search by image size, which make sourcing high-quality images easier.

Social media have made understanding the preferences and tastes of an audience even more accessible. It is possible to evaluate interest in ideas, products and fashions by gathering information about 'trending' topics online. Trending is a term used to explain the popularity of key words and terms used on blogs and on social media sites such as Facebook and Twitter. The proliferation of an idea online is known as a 'meme' and a popular meme can spread rapidly via social networking. An example of a successful meme might be a popular culture reference, a quote from a TV show or a lyric of a song.

There are many online design blogs that are a rich resource for design ideas and provide an insight into current trends in design. Too numerous to mention here but easy to find online, design blogs offer opportunities to see evidence of contemporary international design at both amateur and professional levels.

When using the Internet for research, begin with an idea of what it is you would like to find. The information available online can become absorbing and time consuming, so beginning any search for images and information with a clear plan will make the most efficient use of your time online. As you search for information, record useful sites and email addresses in your notepad or sketchbook, or copy and paste the relevant links into a document or notes file. It is often easier to

ISBN 9780170349994

refer to your own notes later, rather than a 'favourites' menu, particularly if you use more than one computer. It is also important to retain URLs of images that you have gathered to ensure that your sources are clearly identified in your work.

Found materials

Look around you. Don't underestimate the usefulness of junk mail, direct marketing and free promotional material. Such publications can provide an insight into different markets and interest groups and may even inspire compositional ideas.

Free postcards, street newspapers and brochures that use illustration, photography and other methods to promote and advertise events and products may be readily available. Many corporate and non-profit groups use direct marketing as a relatively cheap means of gaining access to a broad section of the community. Often found in cafes, music and fashion retail stores, and entertainment venues, free postcards provide a wealth of visual material and potential research.

Ben Jennings

▲ This student took photographs of patterns, forms, shapes and structures that inspired him. He later used the images to initiate design ideas for product forms and surface patterns.

Verbal resources

Valuable information about the design brief can be gathered from the source of the original need itself. Guidance from your teacher will be great help throughout the design process; suggestions from another source can offer different directions and alternative interpretations of aspects of the design brief.

Discussion is a worthwhile research technique. Market researchers use questionnaires and surveys to gather information about the preferences and attitudes of a specific audience. Face-to-face discussion provides not only written data but also allows for the interpretation of vocal inflections and the physical body language of participants. Sometimes written material can be informative but it lacks the extra detail of a verbal or physical response.

Discussions with experts in professional fields are an important resource when responding to a design brief. An expert can provide first-hand knowledge about techniques, the viability of production methods and information about product history. Initially, it may seem difficult to find an expert in the particular area of the task, but asking parents, friends and teachers may lead you to people with expertise in similar or related fields.

ISBN 9780170349994

IMPORTANT COMMENTS: NUMEROUS DESIGNERS, SIMILAR BELIEFS.

Due to certain constraints, I was not able to process all of the interviews. However, I thought it was important to show these responses, not only because they were from valid designers, but to show these similarities of ideas that stretched right across the industry spectrum.

THEREFORE MY PRODUCT SHOULD AIM TO CONNECT ON A MORE PERSONAL LEVEL, OR ATLEAST TRY AND DROP THE SEVERE PROFIT-MAKING FACADE THAT SURROUNDS SOME PRODUCTS OUT THERE...

BOOKSHELVES/CABINETS ARE A GOOD IDEA, AS GRAPHIC DESIGNERS/DESIGNERS TEND TO COLLECT ALOT OF MATERIAL FOR RESEARCH + INSPIRATION.

TALKING TO ANDREW MADE ME REALIZE THAT MY BRIEF WAS NOT RELEVANT TO LARGER DESIGN COMPANIES (SUCH AS HOYNE). RATHER, FREELANCE SEEMS TO BE THE MAIN NICHE OF MY AUDIENCE. TO ANOTHER EXTENT, SMALLER GRAPHIC DESIGN FIRMS (SUCH AS BUILD).

COMFORTABILITY HAS BEEN MENTIONED ALOT. NOT SO MUCH ERGONOMICS, BUT THE PHYSICAL APPEARANCE IN THE WORKPLACE SHOULD BE SOMETHING TO KEEP IN MIND.

UPCOMING MAGAZINE ART DIRECTOR, DREW TAYLOR.

PERSONALLY I THINK THESE THINGS ARE USELESS.

Our setup is made up of really bad, modern, modular, corner workstations. They're too small (probably 1.5 metres of usable space in both directions) and they have that lovely fake wood veneer. An adjustable keyboard shelf is in the middle (though is completely impractical) and a movable set of drawers (on coasters) sits under the desktop. A shelf (about forehead height), lined with folders, is mounted along the edge of the partition that's hard up against a wall.

My tower PC sits just off centre and a 19" flat screen sits on a couple of stock art books to raise it to the right height. Plastic in, out and shake-it-all-about trays (five of them all up) sit to my right taking up most of the usable desk space there and the deskspace on my left is littered with paper, burnt CDs, plates and past publications. (Not to mention a box of half-eaten Pizza shapes).

Two filing cabinets are further down the wall. And our photographic cabinet (full of camera gear) is over on the left.

Oh, please, design me something to inspire me. Not these awful, thought-less bastions of self sacrifice! Give me some space, some mood, some surfaces that aren't coated in veneer or cheap fuzzy material.

A desk that is acutally designed to take a computer and get it out of the way would be nice. More space for folders, magazine holders, CDs and unsorted paper. A desk that isn't designed so that you lose half the deskspace behind the monitor. Better lighting. Not these awful fluoros. Lighting built into the desk would be interesting. A desk alignment that doesn't have my back facing the rest of the office would be good too. An intuitive desk would be one that considers technology, workflow, lighting, 'creative space' (ie. enough space for laying out paper and doodling, or something similar), colours, alignment to the room and others, storage and more.

As a designer, if you were in the market to purchase to purchase a product like the one described. What factors would you consider? Such as ergonomics, features, styling and how would you prioritize these?

Strangely, how it made me 'feel' would be the first consideration. Does it make me want to be there. Does it inspire design. Does it create that 'hush' in your mind and your soul. Price would be second. Features and usability third. Storage, ergonomics, portability, and size fourth.

How would you define innovation?

The ability to take the way something is done and to do it differently. Design is innovation. It's that simple. Take innovation out of design and you have manufacturing.

HOYNE DESIGN, ST.KILDA, MELBOURNE. ANDREW HOYNE.

Your brief of a manufactured integrated desk set does not really apply to the way we use space. To be completely honest, there is little about our desk arrangement that is any different to any professional person who uses a computer. We have desktop space to use for drawing visuals and developing ideas. However, there is nothing unique about it. My workplace is somewhere I want to feel comfortable. It's more emotional and less structural.

Who or what inspires you?

The people I surround myself with, especially my good friends. My staff and team. They are all talented individuals. And they have vibrant, energetic personalities.

How subjective is design? Is there such a thing as good design?

It's VERY subjective. It's all about opinions; popularity of a style at the given moment, fashion etc. That's what many of these issues are about. THere is often not a right or wrong answer. Just what you believe. And what you can substantiate or support can only help you.

What is innovation? Innovation is being inspired by a respect for existing standards then surpassing them in your work. Your work should be a product of your passion; the engine-room of your enthusiasm and the personal pride in what you do. And it's fun when you believe in what you are doing. Work is an opportunity for us to be our best.

ANOTHER EXAMPLE OF THIS FACTOR OF 'COMFORTABILITY'. DREW RAISED A GOOD POINT OF HOW IT SHOULD MAKE HIM 'FEEL' AND WHETHER IT INSPIRES HIM TO DESIGN, OR JUST WORK. THIS TIES IN WELL WITH THE IDEA OF A 'MUSE': A GODDESS THAT ENCOURAGES INSPIRATION.

A GOOD POINT HERE, A DESK THAT TAKES A COMPUTER AND GETS IT OUT OF THE WAY. MOST COMPUTERS THESE DAYS (DESIGNERS WITH MACS) HAVE THE TOWER IN BUILT TO THE SCREEN OR ARE MARKEDLY SMALLER THAN THEY USED TO BE. HOWEVER, I'VE GOT TO KEEP OPTIONS OPEN, THINGS LIKE G5 (TOWER MACS) HAVE QUITE A LARGE TOWER THAT COULD EASILY BE PLACED ELSEWHERE.

KATALYST WEB DESIGN ADELAIDE, SA. KIPP BRADY.

My workspace is one big square desk (like an IKEA desk) that has my computer, scanner, cutting matt and associated junk on it. A bookcase full of design mags and books etc. A small filing cabinet which sits next to the desk full of good paper, files, and cd's with a printer on top of it. Any drawing I do is on whatever space I can find.

I believe the bigger the workspace the better, at the momtment I have no room for drawing, what with my computer taking up all the desk. More organization of the stuff on my desk would be nice also as pens and stuff are every where as well as cds and notes all over the place. If I were to buy a desk though it would be primarily for computer use so it would of course have to cater for one. Crappy Office Works computer desks dont cut it though. Finally a large area under the desk would be preferable so I can work on all sections of it without my legs getting in the way.

A customizable deskset would be of interest to me, although the basis of it would have to meet my basic demands without requiring any other parts or extensive customization. Thats not to say that I wouldn't appreciate any addons to it, its just that it would have to be of good use in its basic form.

A good solution would have a simplistic, minimal design, comfortable too use (for those long hours) good computer support (such as easy access to powerpoints) a large amount of work area and specific areas for paper, pens, rulers etc, whether these be in draws or on the desktop itself. The ability to combine more elements too it easily, which add to the style/ergonomics of the unit. Multi computer support should be acheived by the addition of further units to desk rather than supported in the one unit alone. Also an area for desktop speakers would be nice.

In the long run, if I was to buy a product like this, it will come down too price. I will of course pay more for better things but outlandish design or over the top features, which may provide a slightly better unit though add considerably to the price will be a turn off. Apart from that my list of preferences are styling number one, then features and finally ergonomics. The unit should also be quite sturdy as I plan to put alot of stuff on it.

SIMPLISTIC, MINIMAL DESIGN COMMENTED ON HERE ONCE AGAIN. KIPP WOULD RATHER PURCHASE A PRODUCT FOR ITS LOOKS RATHER THAN ERGONOMICS.

ALTHOUGH KIPP DOES MAKE MENTION OF LEG SPACE UNDER THE DESK, WHICH IS RELATED. I MYSELF AGREE WITH HIM 100% AND WILL FOLLOW IT THROUGH IN THE DESIGN OF MY PRODUCT.

Ben Jennings

▲ In researching his design brief, this student used email interviews. The design brief specified a desk design for graphic designers, so the student made contact with designers in Australia and overseas. He asked pertinent questions to gain information about the most appropriate design directions.

Literary resources

Shutterstock.com/DavidPinoPhotography

The library is a key resource in the collection of information. Libraries offer access to books, magazines, journals and the Internet, and library staff are extremely knowledgeable about the resources they provide.

Shutterstock.com/racorn

Many libraries subscribe to a range of journals and online archives. It is possible to investigate the background of a company, product or service using resources such as these.

ISBN 9780170349994

Journals and magazines, such as *Choice*, offer objective testing and analysis of products, which could be a valuable resource in the development of product design concepts. Magazines that focus on popular culture provide insight into trends, tastes and fashion related to their specific market.

The Australian and international magazine markets are vast and there are magazines suited to all ages and interest groups, offering information about diverse audiences.

Books are, of course, an essential resource to any research, whether for information specific to the design brief, such as a book about the history of a product, or information that is related to the task, such as a reference book about ergonomics.

Many design and industry magazines produce an annual edition, giving an overview of design developments over the previous 12 months. These publications often select key movements, fashions and innovations, and showcase emerging and established designers and associated professionals. Similarly, compendium volumes of design examples – such as logos and typography – provide topic-specific overviews, as do some country-specific volumes, which focus only on designers from a defined region. The concentrated nature of these publications makes them valuable resources.

Historical resources

Historical resources are not necessarily dusty old books in the archives of a library, but may relate to designs from last year or last century. The history of previous designs is a good place to start researching the success and failures of the past. From existing constructions, products or compositions it is possible to analyse the application and effectiveness over time of construction materials, content and the elements and principles of design.

'Retro' references are often made in design so it is good to familiarise yourself with past representations of your design task.

The National Library of Australia has archives of old photographs, newspapers and journals for reference, and permit viewing of many primary sources. Photographic and physical records of social, environmental and geographical changes can be found at local museums and historical societies. Such material can offer insight into not only the appearance of past designs but also their application and context.

Your local area may also provide clues to past designs. Using a camera and a notebook, you can collect images and ideas from local buildings, antique or junk stores, and landmarks.

Demographics

Demographics involve the analysis of statistical information about a population. Demographic data provides information about trends, ethnic diversity, average age, education levels and current interests. Companies use demographic information to make decisions about product development. This data provides the company with clues to common sets of values and attitudes of consumers. When marketing professionals use buzzwords and phrases such as 'the target demographic', they are simply talking about an audience that shares common characteristics.

Government agencies also collect vast quantities of data about Australian citizens. This information is then used in the development and distribution of services and the deployment of government funding. A national census is held every five years to gather information about employment, household characteristics, education and lifestyles. Census data can be viewed at the Australian Bureau of Statistics website.

Remember that research is often ongoing throughout the design process. It is important to be aware of trends and developments. So, as you design, keep an eye on your target user and the factors that influence them.

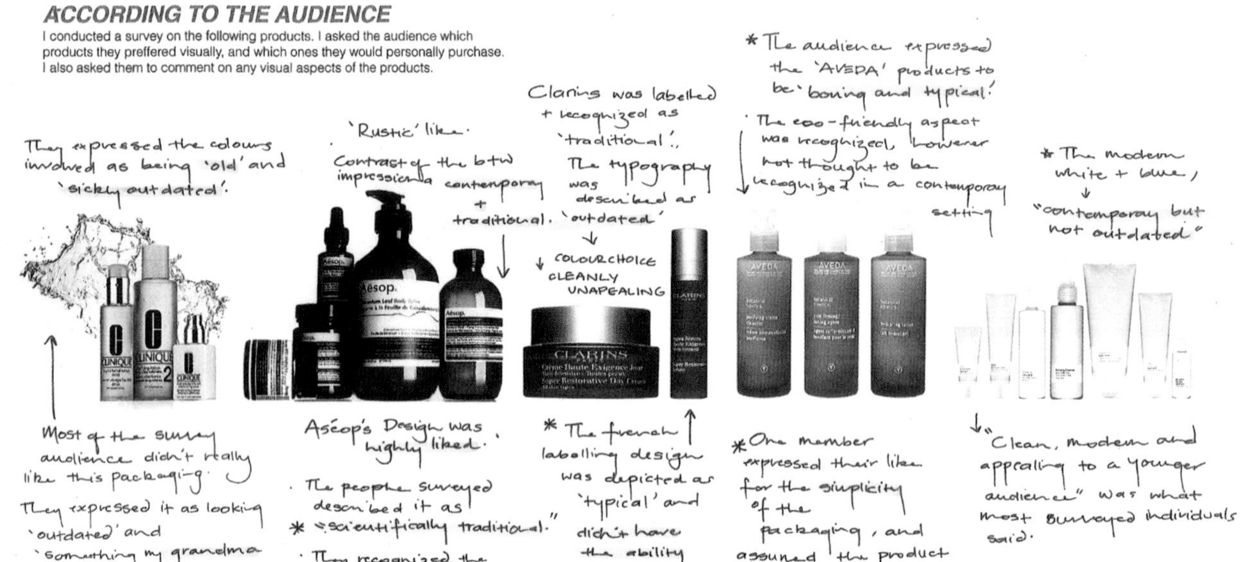

Lucy Boehme

▲ The design brief required this student to design packaging for organic skincare products. In gathering research, the student showed a selection of existing packaging designs to her target users and noted their responses. This informed some of her decision making about packaging forms, textures and materials.

ORGANISING AND INTERPRETING RESEARCH

We are surrounded by a wealth of visual, verbal and written information – so, how do we determine what research will be of value to the design process?

Organisation is the key to using researched material effectively. Organise your research into clearly defined categories, such as:

+ research specific to the end user
+ research of similar products with a similar function
+ research of materials and media
+ inspirational material: layout, design elements and design principles, media, materials.

You will discover that your research has provided you with information directly related to your user, the context of the planned design and some items that have a similar function to yours. This material will assist you in understanding the design problem and will allow you to make educated decisions about the suitability of proposed graphical representations.

ANNOTATING RESEARCH

Annotation is the best means of indicating – to you and to others – the value and meaning of researched materials.

Insightful annotation indicates that you have the ability to analyse researched material.

+ Write about why you have chosen to use the research you have.
+ Explain how you might use your research in the design process.
+ Indicate the aspects of the research that may give you starting points to generate ideas.

 ISBN 9780170349994

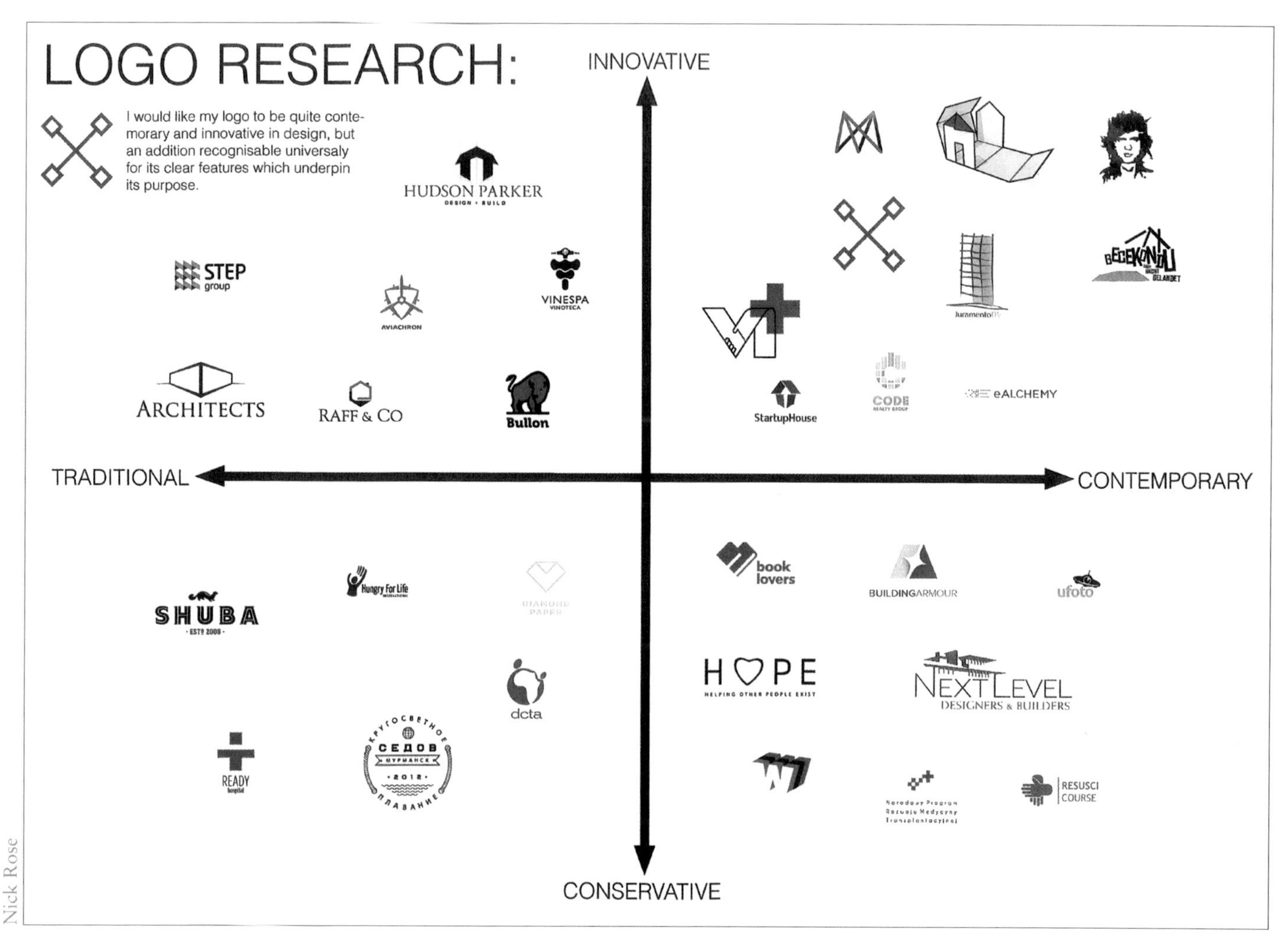

▲ As a tool to assist in organising a variety of researched images, this student used a scale to organise researched images of corporate identities. By organising the collected logos into four quadrants, the student was able to identify which were the more innovative, conservative, traditional and contemporary of the images. He was then able to identify where he would like his own design to fit. This organisational tool can be very helpful in defining the aesthetic qualities and direction of a design idea.

ISBN 9780170349994

CHAPTER 5
DEVELOPING IDEAS

'It doesn't matter how long something takes. All that matters is how the end user perceives the design.'

Paula Scher

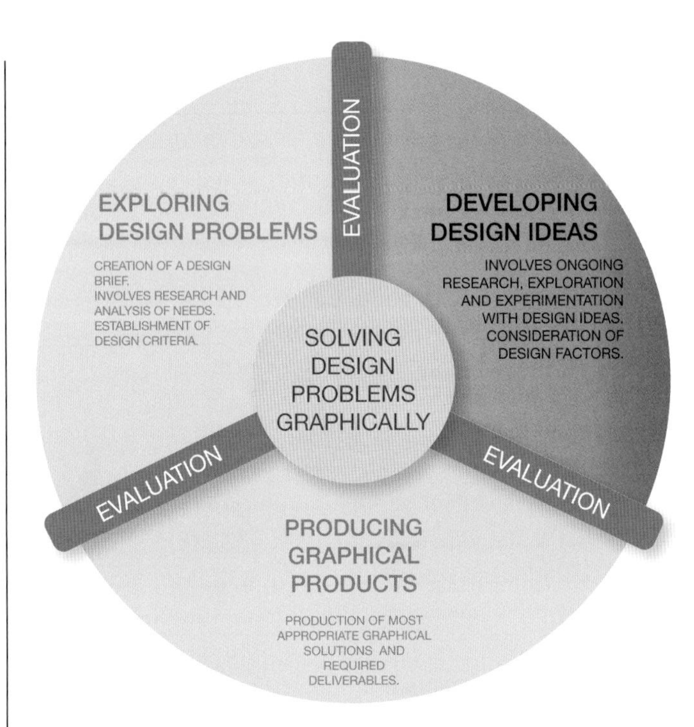

▲ Developing design ideas

The development of ideas involves brainstorming, idea generation and decision making. It is here that a broad base of creative concepts is generated leading to the selection of a concept that best solves the original design problem as outlined in the design brief.

During the developmental stages, designers continue to look to the brief and the initial research for inspiration. The design brief is the core that travels through the entire design process, ensuring that even the most experimental ideas refer back to the brief for guidance. Decisions regarding the application of design elements and principles are based on not only what is the most creative solution, but what is also the most effective solution in line with the brief.

Throughout the development of a design concept, the designer will continue to ask questions: Does it fulfil the needs of the client?

ISBN 9780170349994

Will it appeal to the target audience? What materials and media will be best suited to the final design?

Testing of design elements and design principles, materials and graphical representations is a constant process. Numerous methods are used to validate or reject ideas. Discussions and focus groups may continue to be used to assess and evaluate ideas.

If required, professional specialists such as photographers, illustrators, animators, Photoshop artists and web developers may be briefed on aspects of the design during this phase. Creative professionals who specialise in highly specific areas play an important part in the development and production of visual communications.

Development of ideas may involve:

+ brainstorming
+ sketching of ideas
+ selecting the most suitable concepts for development
+ testing media, materials and representations
+ experimenting with elements and principles of design
+ making reference to the brief to ensure that the designs are on the right track
+ evaluating and assessing the most effective design solutions
+ forming relationships with external design professionals.

5.1 GENERATING DESIGN IDEAS

The design process is extremely flexible – it allows creative ideas to evolve from diverse sources. The design process has a definite structure but it is also 'elastic', allowing designers to experiment with countless media, materials, methods, elements and principles to achieve a suitable design solution.

The way a designer responds to a brief can depend on personal and professional preferences. Some designers work in teams where different tasks are taken on by co-workers. Some designers work alone, or may call on the assistance of external specialists.

In the initial stages of the design process, designers use design strategies including brainstorming, sketches and design concepts based on research and inspirational material.

In seeking inspiration, a designer might use concept maps and word lists as well as thumbnail sketches. To give the designer a large pool of ideas to work from, design companies often use group discussions during the developmental stages of the design process. The use of a concept board (also known as a 'mood board' or a 'look and feel' board) is a common method of gaining ideas and inspiration. The concept board is a collage of images, words and ideas based on the brief or a related theme. Images, swatches and words are combined to suggest possible design directions.

In the initial stages of a brief, designers respond in a variety of ways. Personal preferences dictate

Johanna Schreiner

Nick Rose

▲ These 'look and feel' boards (also known as 'mood boards') provide a visual guide to the aesthetic appearance of a likely future design. The boards can be used to guide idea generation and inspire the 'look' of the design product.

ISBN 9780170349994

▲ In the design of a desk, this student used sketches to brainstorm as many initial ideas of possible.

the methods used to explore and present initial concepts, but most designers begin the design process with simple thumbnail sketches in pen or pencil on paper. Drawing is a quick and hassle-free way of getting early ideas onto paper; it also provides a means of communicating concepts with other professionals involved in the process. Freehand sketches may contain minimal visual detail at this stage, and are often accompanied by notes (annotations) explaining the concept in more detail.

This stage is also referred to as the 'ideation' stage, a point where the ideas – rather than any concrete concepts – are the primary focus. This phase of the design process invariably provides a lot of freedom and allows designers to explore many possibilities. It is important to ensure that the initial focus is broad rather than narrow at this point in the design process.

Generation of ideas involves:

+ concept and mood boards
+ brainstorming
+ initiating ideas through sketches
+ communicating ideas with other design professionals
+ beginning with a broad base of ideas.

Throughout this stage of the design process you will be applying different thinking skills to the practical tasks you undertake. These skills assist in helping you generate the most creative options for your design. There are many techniques and strategies available to brainstorm ideas. See Chapter 11 for more detailed information about brainstorming, including word lists, graphic organisers and other tools, and helpful templates to get you started.

Brainstorming is the application of small, stimulating tasks that tap into the imaginative resources of your mind. These are designed to exercise your brain or to help expand on an idea that needs a bit of a push. There are software programs that can assist you in brainstorming ideas and help to build concept maps and idea diagrams; there are also many helpful books available on the subject.

SKETCHING

Freehand sketching of initial ideas created during brainstorming is, perhaps, the most valuable method of idea generation. 'Visual thinking' is a term used to describe visual brainstorming. Thoughts, ideas and concepts may be extremely

ISBN 9780170349994

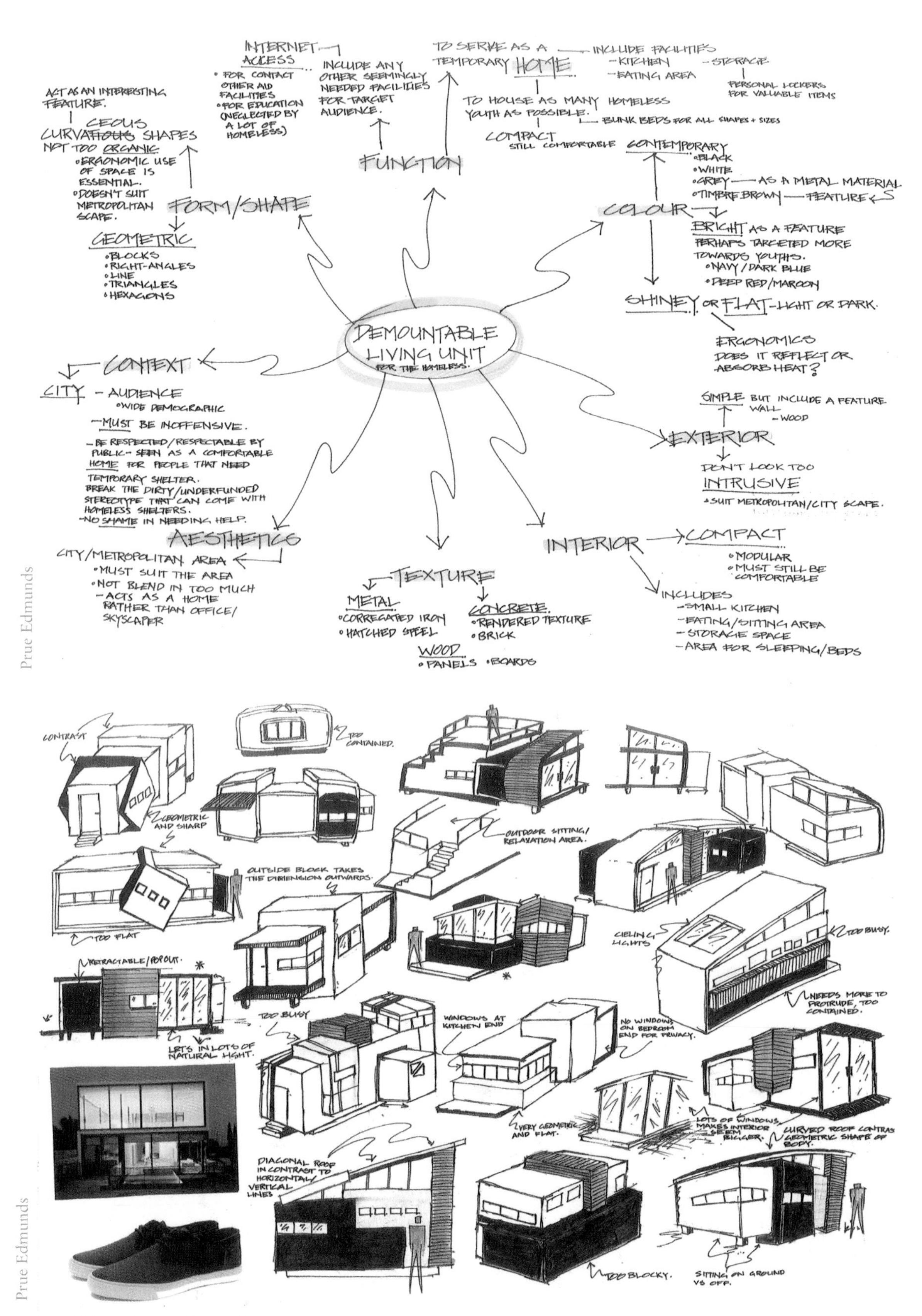

Prue Edmunds

Prue Edmunds

▲ This student used several techniques to stimulate design ideas for a small residential structure. The concept map (top) provided starting points for the generation of ideas including the form and function of the building. In sketching design ideas (bottom), the student used inspirational images of related and unrelated objects. The use of the sneakers can be linked to an exploration of colours and textures.

ISBN 9780170349994

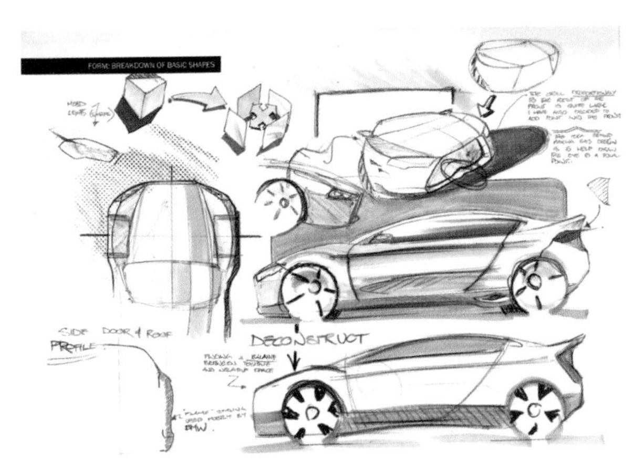

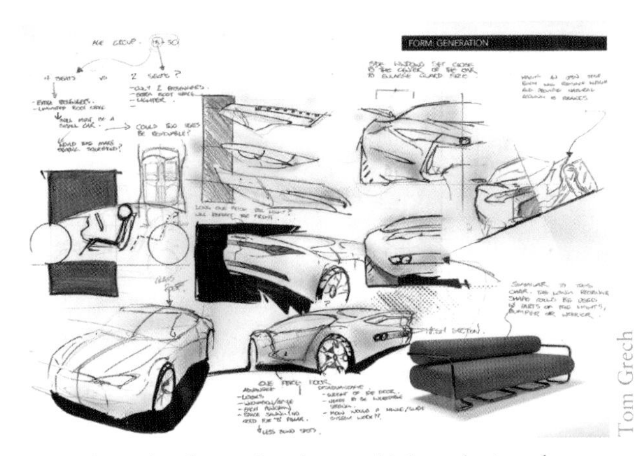

Tom Grech

▲ Generation of ideas for a vehicle design. Note the way that this student has drawn his design ideas from multiple angles in order to visualise different design possibilities. Drawings in this stage do not need to be finished, polished presentations. The focus is on getting ideas onto paper.

vague in the initial stages of the design process but through drawing they can be more clearly explored. (See Chapter 15 for more information about sketching.)

There is no formal method of learning to think visually, but drawing expands our ability to represent not only what we see around us but also what we see in our imaginations. Practice is key to building skills in sketching. The more you draw the better your drawings will become; it really is that simple.

This stage of the design process is about the generation of ideas and involves quick drawings drawn directly from observation or from material found during the research stage. Thumbnail sketches allow for the exploration of initial ideas that can later feed the development of more formal design concepts.

Some students find generating ideas the most intimidating stage of all. Faced with a blank sheet of paper, coming up with dozens of designs can be challenging. However, the generation of ideas can be structured to help you produce a large number of possible concepts and design directions.

ANNOTATING YOUR DESIGN IDEAS

It is helpful to explain your concepts as you research, generate, develop and refine them. Annotations are notes placed beside images

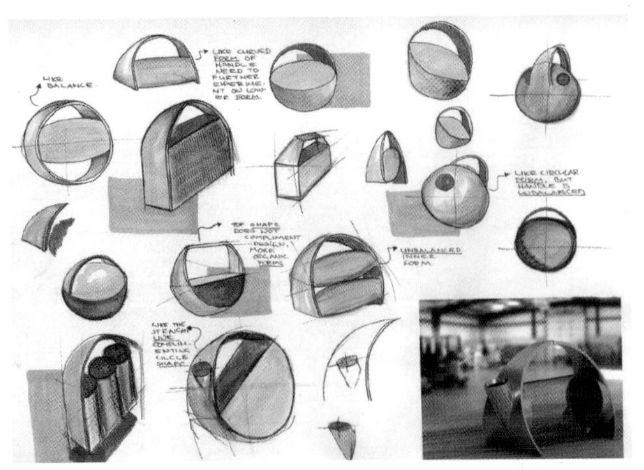

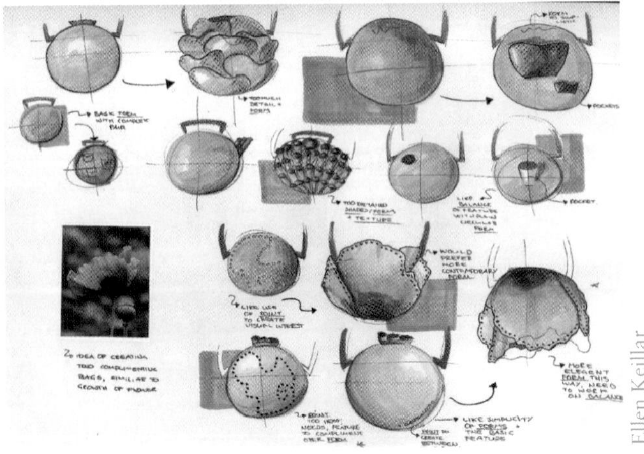

Ellen Keillar

▲ This student used inspiration images of a flower and kettle to inspire her first design ideas. Many contemporary designers look to unconventional imagery for inspiration.

ISBN 9780170349994

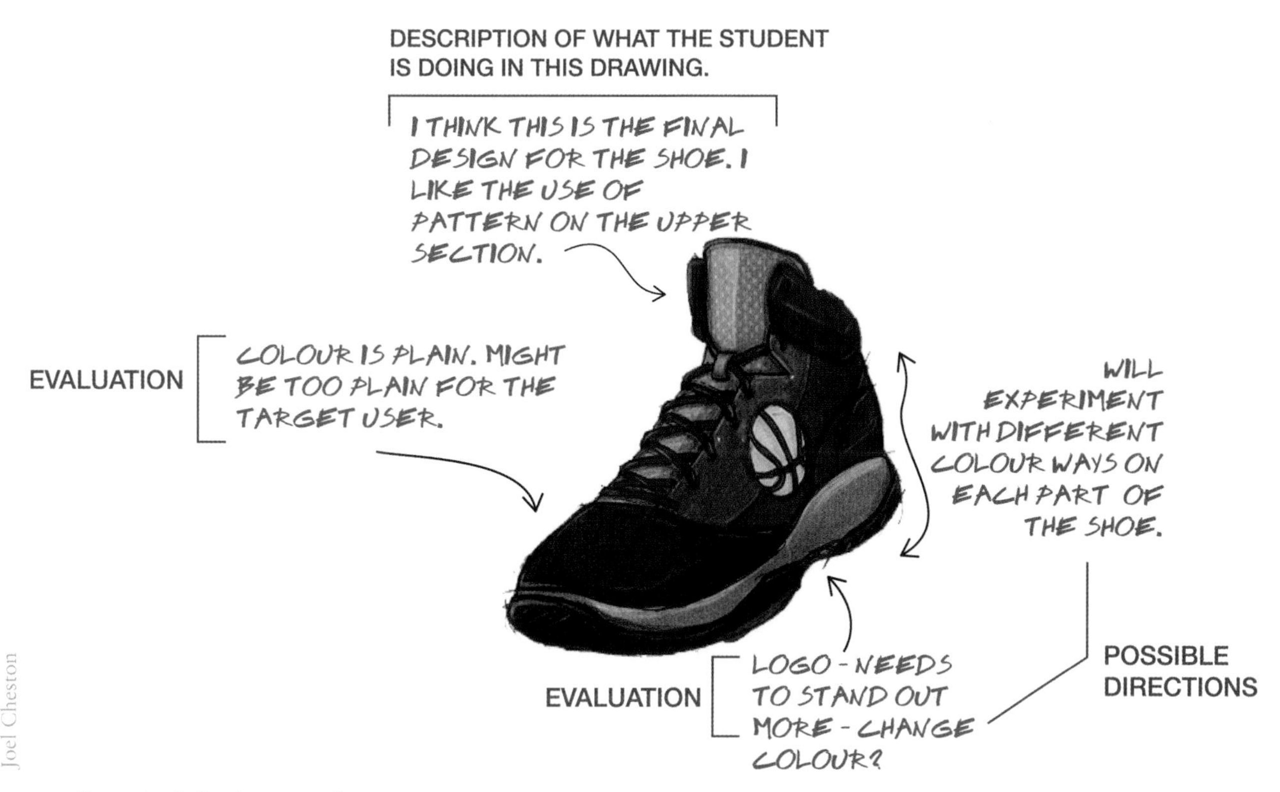

▲ Example of effective annotation

that explain the concept that is being shown. Annotations are written reflections on your design ideas; actively making notes involves thinking about your design thinking. Good annotations are reflective, succinct and relevant. They convey evaluation and suggest possible directions for further development.

These key concept questions will help you to make effective annotations:

+ Descriptive: What were you doing? What design factors are being applied/considered?
+ Analytical: Why did you do it?
+ Predictive: Where might the idea lead? What design strategies could be applied?
+ Reflective: Is it a good idea? Does it fulfil the design brief? Does it appeal to the end user?

Good and bad annotation

Annotation is a demonstration of your reflective and critical thinking skills. Effective annotation is not an essay; it is succinct and insightful, demonstrating an understanding of why a design is effective. Good annotation is a balance between description and evaluation.

Effective annotation should be DESCRIPTIVE (of what you are doing), REFLECTIVE (how successfully you are doing it) and PREDICTIVE (of where the next direction might be).

Bad annotation is too brief and tells us nothing. Bad annotation is judgemental and subjective, referring to personal taste rather than objective analysis.

Keys to effective annotation

1 Write about WHAT you are doing:
Include a summary of what the drawing represents, for example, 'This design shows the details of the product features'.
2 Include an EVALUATION of the concept/image:
'The design works well with the addition of pattern on the surface but the colour needs to be adjusted to suit the target audience.'
3 Note POSSIBLE DIRECTIONS from this point:
'Next step is to try to use warmer colours on some of the details to fit the brief better.'

ISBN 9780170349994

5.2 DEVELOPING YOUR IDEAS

Throughout the design process, it is essential to refer back to the design brief. By always considering the brief, you will keep both the end user and the requirements of the client clearly in mind. This is particularly important at this stage of the design process as you begin to make decisions about which conceptual path to follow.

So far, you will have thoroughly researched the end user and, more than likely, examples of similar designs. In combination, these two factors will help you grasp what the end user may find appealing. This knowledge allows you to make educated design decisions about concepts that fulfil the purpose of the brief. This is the stage for selecting and developing the most fitting concepts from the sketches and thumbnails you produced earlier. In answering questions that relate back to the design brief, your choices will become clearer. You may select more than one design concept to follow, but the choice will be based on your sound knowledge of the design problem.

After you have selected the most suitable design concept or concepts for further development, you will begin a process of experimentation and elimination. Experimentation – with elements, principles, materials and graphical representations – will help you to eliminate the least suitable concept.

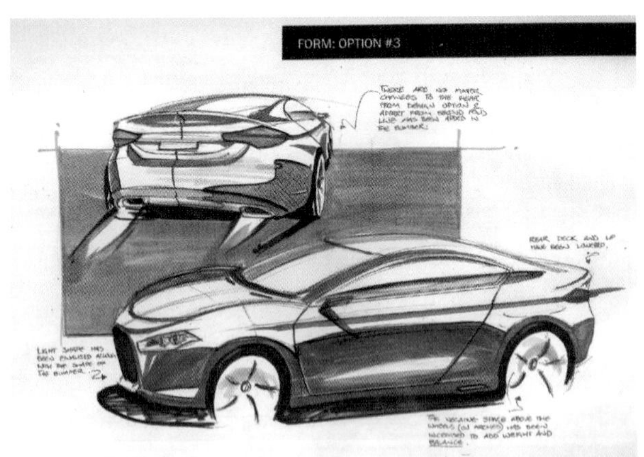

▲ In generating different design ideas, this student devised three options in his vehicle design. The design process enabled him to experiment with different details, colours and materials as well as design principles, such as balance, scale and proportion, until he decided on his preferred final concept.

Tom Grech

EXPERIMENTATION

The developmental stage of the design process is about testing the feasibility or effectiveness of ideas. Through a series of tests and experiments, this stage leads to the discovery of the most effective and suitable means of solving the original design problem. You may choose to trial the effectiveness of different media, materials or elements and principles of design. Experimentation with different graphical representations would also be viable at this stage of the design process.

Experimentation with design alternatives is a form of visual decision making. Tests can be highly experimental and there are no set rules for the trialling of design concepts, but it can help to be quite systematic in your approach. For example, you could allocate one page or screen to each trial. Each series of tests should provide you with further information about the possible solution(s) to the design problem.

ISBN 9780170349994

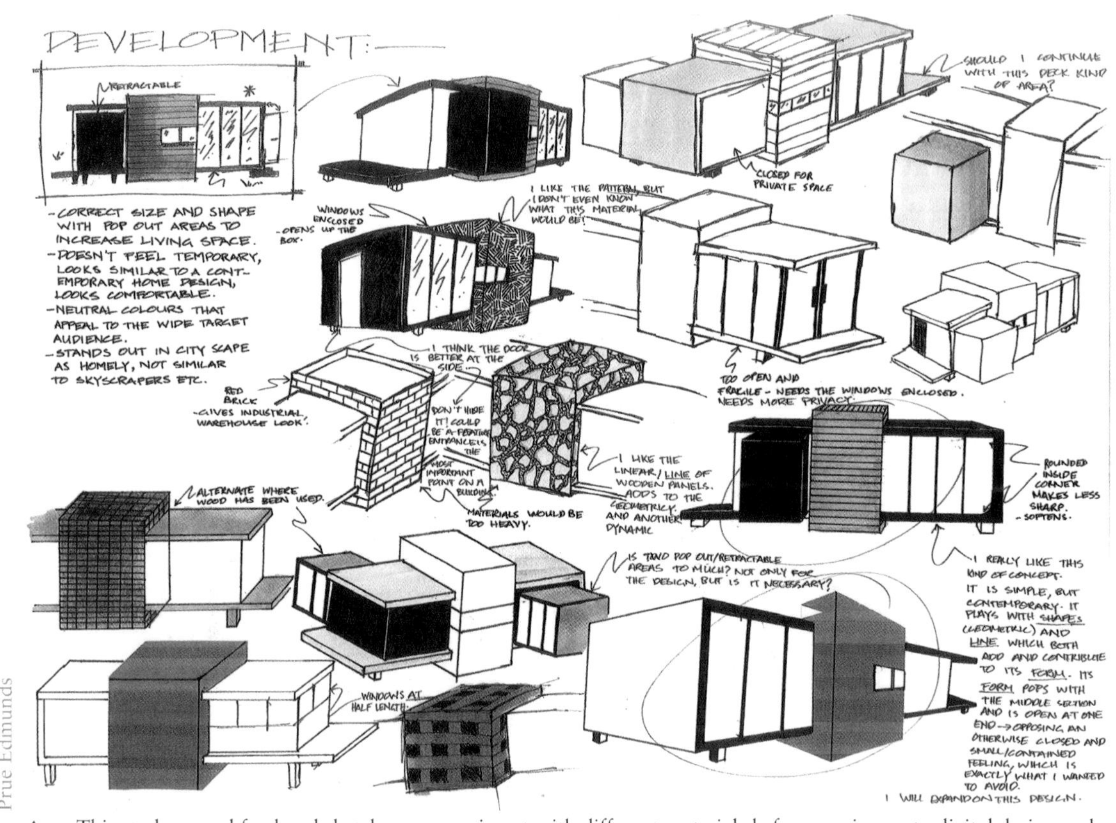

▲ This student used freehand sketches to experiment with different materials before moving on to digital design and rendering of her structure.

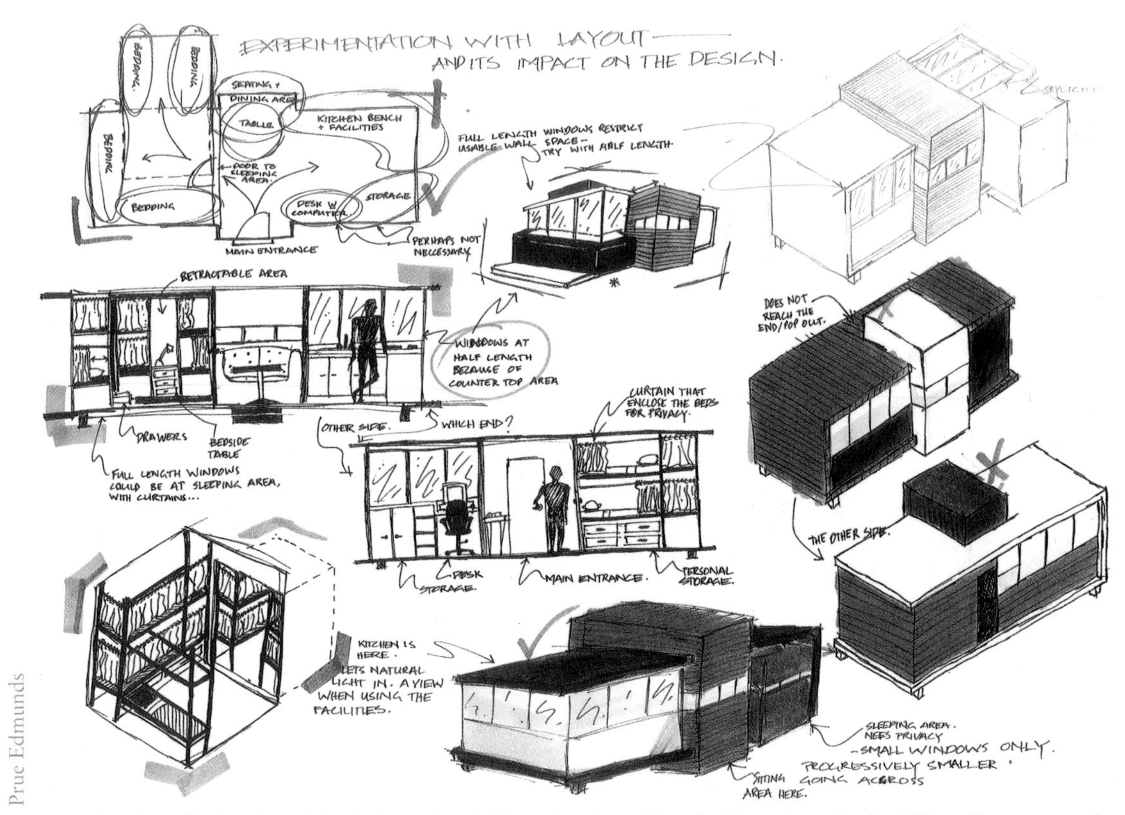

▲ Sketching the interior of the design assisted this student in making decisions about the feasibility of her structural design.

It is helpful to continue annotations throughout this stage of the design process. Notes can be useful to observe the efficacy of your tests, trialling and developmental work. To be helpful, your annotation should be concise and analytical; evaluate how effective your development and experimentation is.

Through experimentation, you can analyse the best results and make a firm decision on which direction to proceed. Never underestimate the value of testing ideas. Your knowledge of the design problem and analysis of your research might lead you to favour a certain visual direction, but concepts always benefit from development. It is essential to keep an open mind at this stage of the design process.

▶ In developing the details of his designer's desk, this student focused on the functionality of the product using knowledge gained from his extensive research and creative generation of concepts.

Ben Jennings

ISBN 9780170349994

In diagrammatical form, the testing and decision-making process might look something like this:

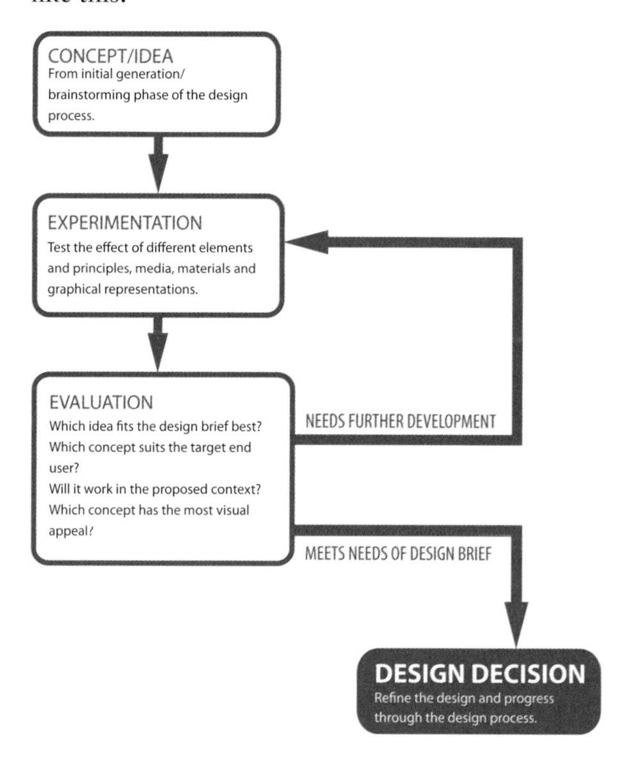

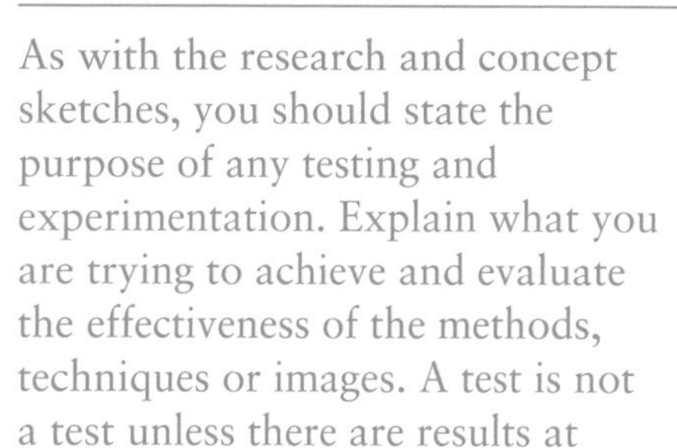

TESTING, TESTING!

As with the research and concept sketches, you should state the purpose of any testing and experimentation. Explain what you are trying to achieve and evaluate the effectiveness of the methods, techniques or images. A test is not a test unless there are results at the end!

REFINEMENT

After the development phase, the design concept that is most effective in fulfilling the requirements of the brief is selected. The question now raised is: What needs to be done to ensure that the final design fully reflects the needs outlined in the brief?

Refinement is the icing on the cake, the polish on the car. It refers to the finishing of the most successful concept, and leads to the production of a mock-up. During this stage, the appearance of

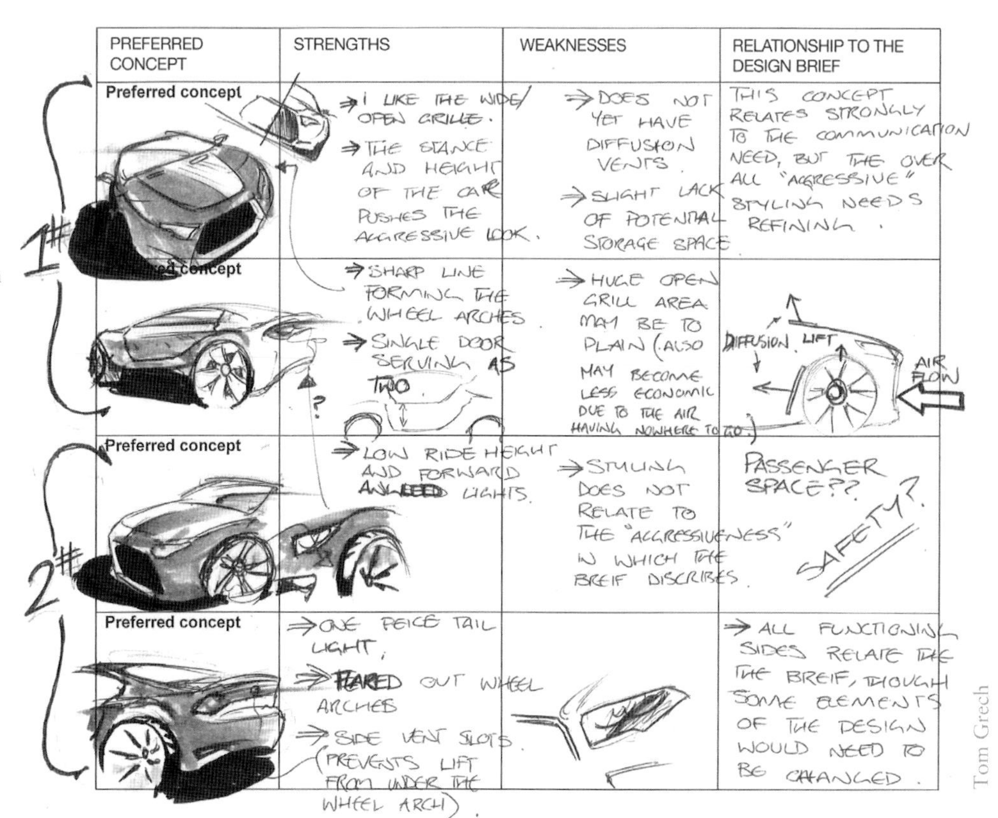

PREFERRED CONCEPT	STRENGTHS	WEAKNESSES	RELATIONSHIP TO THE DESIGN BRIEF
Preferred concept	→ I LIKE THE WIDE/OPEN GRILLE. → THE STANCE AND HEIGHT OF THE CAR PUSHES THE AGGRESSIVE LOOK.	→ DOES NOT YET HAVE DIFFUSION VENTS. → SLIGHT LACK OF POTENTIAL STORAGE SPACE	THIS CONCEPT RELATES STRONGLY TO THE COMMUNICATION NEED, BUT THE OVER ALL "AGGRESSIVE" STYLING NEEDS REFINING.
Preferred concept	→ SHARP LINE FORMING THE WHEEL ARCHES. → SINGLE DOOR SERVING AS TWO	→ HUGE OPEN GRILL AREA MAY BE TO PLAIN (ALSO MAY BECOME LESS ECONOMIC DUE TO THE AIR HAVING NOWHERE TO GO.)	DIFFUSION. LIFT. AIR FLOW
Preferred concept	→ LOW RIDE HEIGHT AND FORWARD ANGLED LIGHTS.	→ STYLING DOES NOT RELATE TO THE "AGGRESSIVENESS" IN WHICH THE BREIF DISCRIBES.	PASSENGER SPACE?? SAFETY?
Preferred concept	→ ONE PEICE TAIL LIGHT, → FLARED OUT WHEEL ARCHES → SIDE VENT SLOTS. (PREVENTS LIFT FROM UNDER THE WHEEL ARCH).		→ ALL FUNCTIONING SIDES RELATE THE THE BREIF, THOUGH SOME ELEMENTS OF THE DESIGN WOULD NEED TO BE CHANGED.

▶ This student used a simple grid to identify the best ideas that he developed during this phase of the design process. He was able to identify the strengths and weaknesses of each concept and select the most appropriate design to move forward.

ISBN 9780170349994

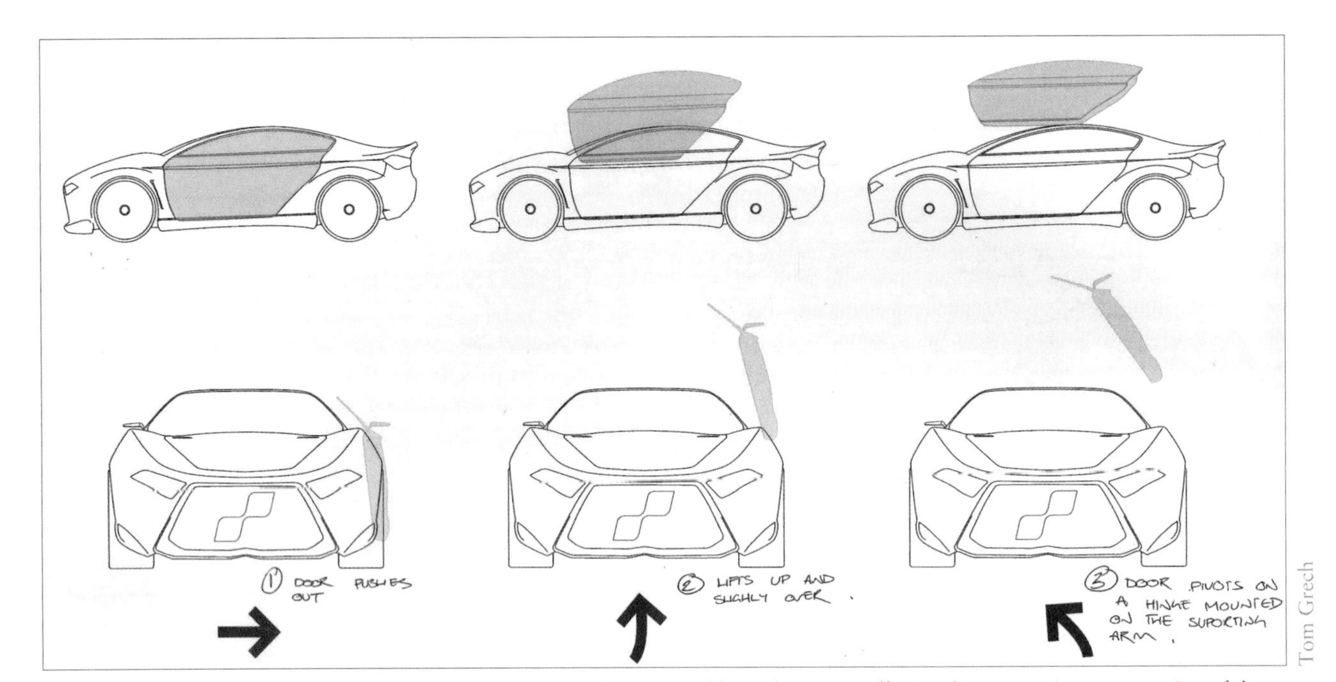

▲ Final details of the vehicle design, such as door functionality, were addressed using an effective diagrammatic representation of the student's design idea.

the final design is clarified and refined, and final decisions are made about elements such as colour or the most appropriate grades of materials and media that will be used. Methods of production are finalised at this stage; decisions are made about the most appropriate form of printing, presentation, construction or manufacture.

If external professionals have been involved, this stage may require that all aspects of the project be pulled together as a cohesive whole. Evaluation continues as the final product takes shape.

Refinement of ideas can involve:

+ selecting the final concept
+ finalising the choice of materials, design factors and representations
+ refining the application of the chosen elements and principles of design
+ ongoing reference to the brief to ensure the design is accurate.

As you become more certain of the suitability of a particular design concept, you should begin to refine it. In refining the design concept you will focus on the best possible design solution, ensuring that it fits with the requirements of the brief.

Ask yourself whether the expectations of the client and the needs of the end user have been addressed. If not, return to the developmental stages and make the appropriate adjustments to some of your concepts. If your preferred design concept covers the design problem, then proceed with refining it.

Refinement may involve the reproduction of your concept through the use of design technologies to create precise line work, to render materials and textures or to integrate other imagery. Alternatively, you may need to clarify features and functions of your concept to make them more apparent.

ISBN 9780170349994

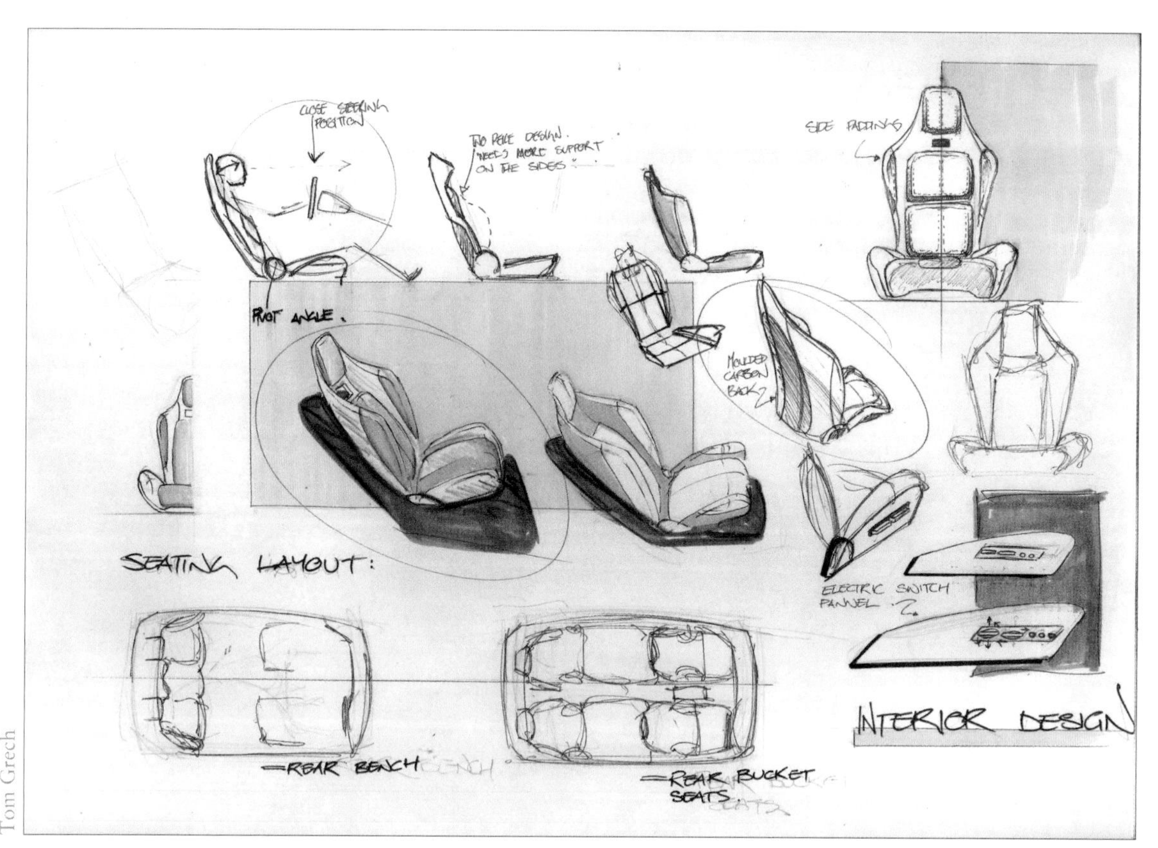

Tom Grech

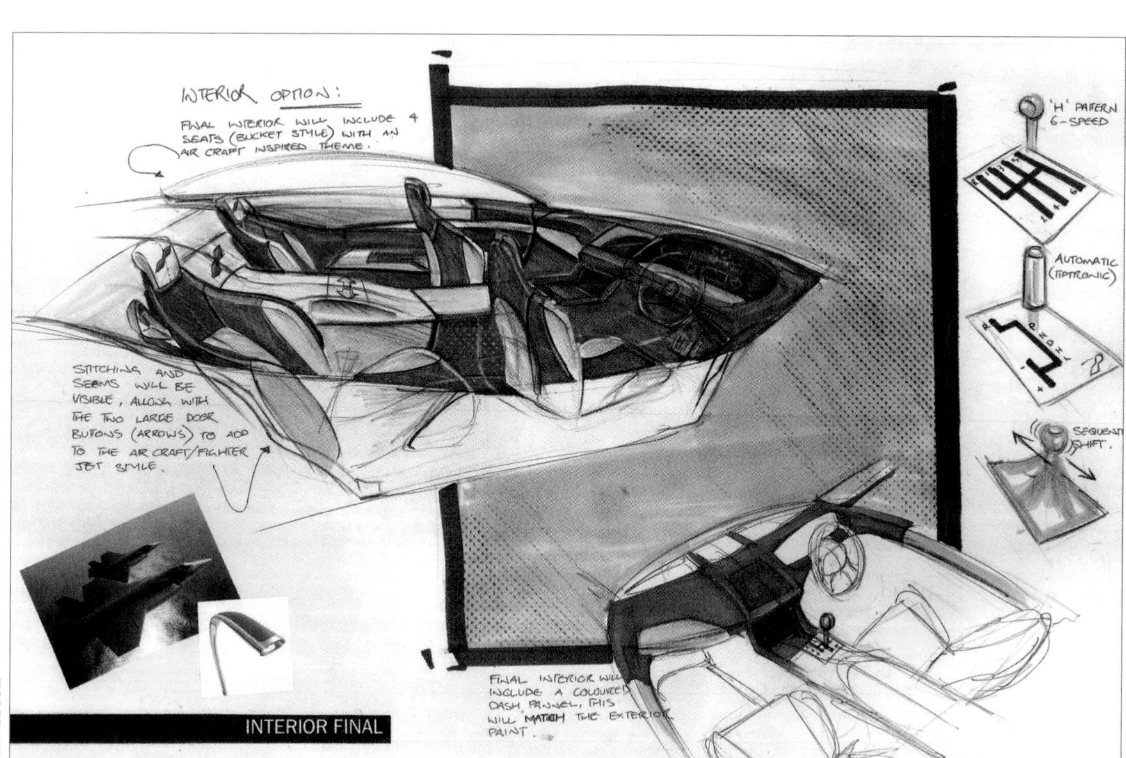

Tom Grech

▲ The refinement of this student's vehicle design meant resolving details such as the car interior. Note the use of inspirational images to assist in creating the final interior design.

ISBN 9780170349994

CHAPTER 6
PRODUCING GRAPHICAL PRODUCTS

> 'When I am working on a problem, I never think about beauty but when I have finished, if the solution is not beautiful, I know it is wrong.'
>
> *R. Buckminster Fuller*

6.1 EVALUATION

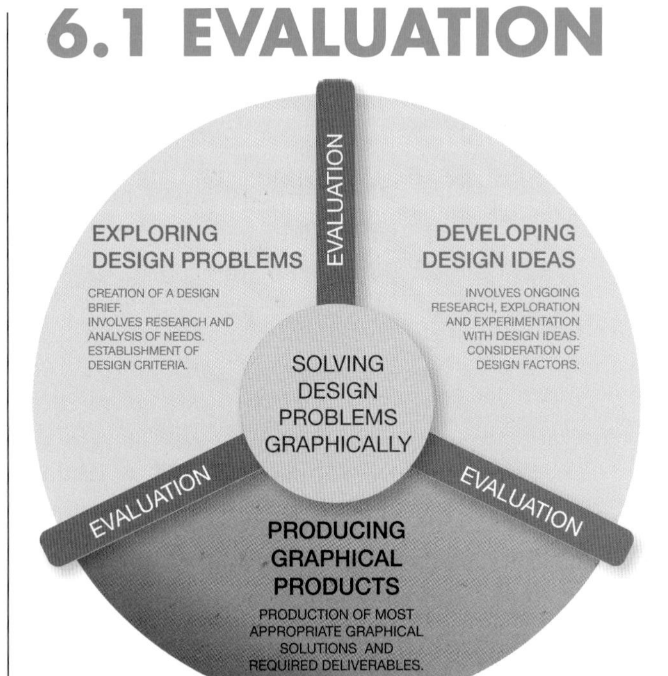

▲ Producing graphical products

In deciding on the most appropriate graphic product, it is essential to evaluate design ideas and assess how effectively they meet the needs of the design brief and solve the original design problem. Evaluation occurs throughout the design process to ensure that the requirements of the design brief are being met. The costs and processes involved in the production of most design products are often expensive and complex, so it is essential that the designer identify strengths and weaknesses in a design before final production commences. The cost of destroying final products due to design flaws is prohibitive and must be avoided at all costs.

Throughout the design process, the designer continues to evaluate the direction of the design solution within the parameters of the design brief. Regular reflection on the progress of the design

ISBN 9780170349994

is essential to ensure that it meets the needs of the design brief and addresses/solves the original design problem.

In the later stages of the process, a mock-up, model or draft of the design concept is used to ensure compatibility with the brief.

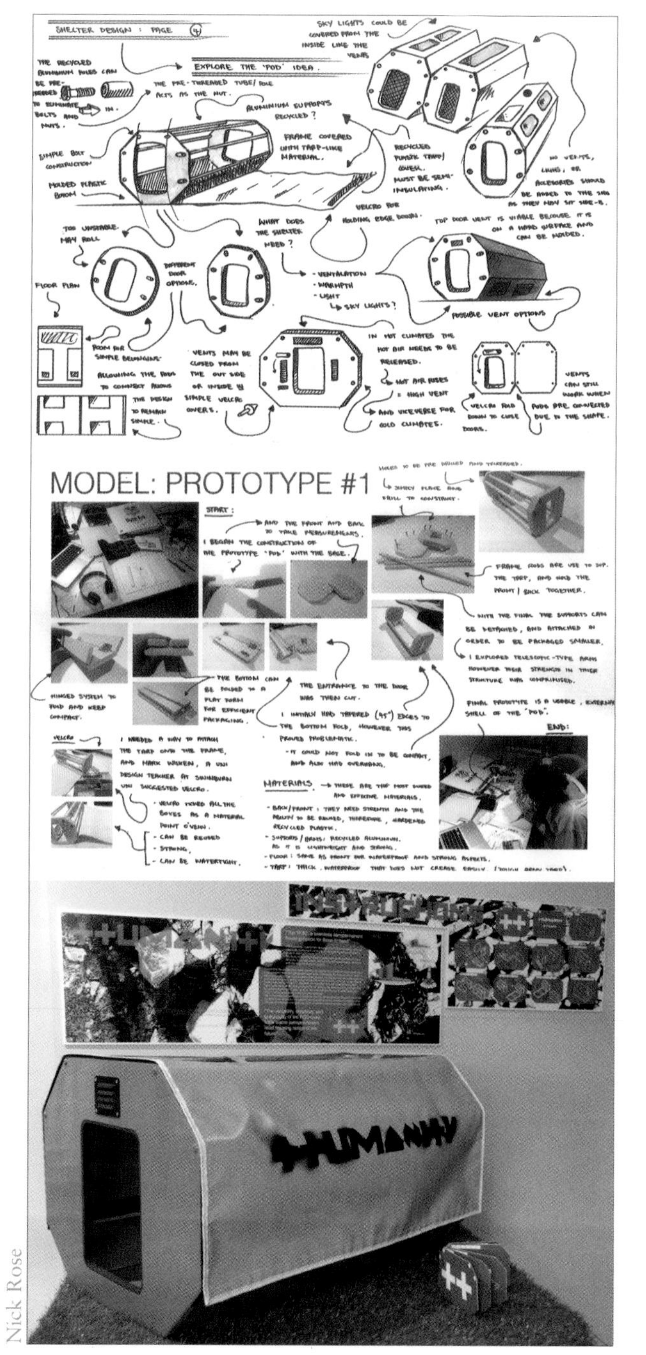

Nick Rose

▲ This student used annotation to explain and evaluate his design thinking. At every stage of the process, he posed questions about the direction of his ideas and evaluated the most appropriate materials, functions and graphical representations for his design. The final production was a 1:15 scale model and display.

EVALUATION TECHNIQUES

A mock-up or draft provides the opportunity to test the final product as it will appear after production. If this stage proves successful, the designer will proceed to the final production. If the mock-up is tested and found not to fulfil the needs of the brief, the designer will return to the design process and either refine the concept further or scrap it and start again.

Mock-ups come in all shapes and sizes. A mock-up can be a scale model, which is common in architectural and product design. Set and theatre designers regularly create models of sets, which enable others involved in the production to visualise the use of props, lighting and the position of actors before the actual set is constructed.

Design technologies including CADD and 2D or 3D printing methods enable small-scale prototypes or computer-generated mock-ups to be used in evaluating graphic designs, products, interiors and buildings. Software can simulate a three-dimensional 'walk-through' of a building or other structure and may include all relevant details such as lighting, decor and materials.

Prototypes and mock-ups are often presented to the client or to a select group of end users, as part of an evaluation process. This provides the opportunity for modifications to be made at a stage where changes are affordable and achievable. This stage is often the final opportunity for evaluation before the production process begins.

This phase is important for designers in communicating ideas. It allows non-designers to visualise design concepts in a familiar three-dimensional way. In property development, this has become a marketing tool. By viewing models through 3D modelling software in addition to more traditional two-dimensional architectural plans, potential buyers can view a house or apartment before it is built.

The purpose of evaluation is to:

+ ensure that the design is in line with the original design brief
+ inform decisions about directions taken during the design process

ISBN 9780170349994

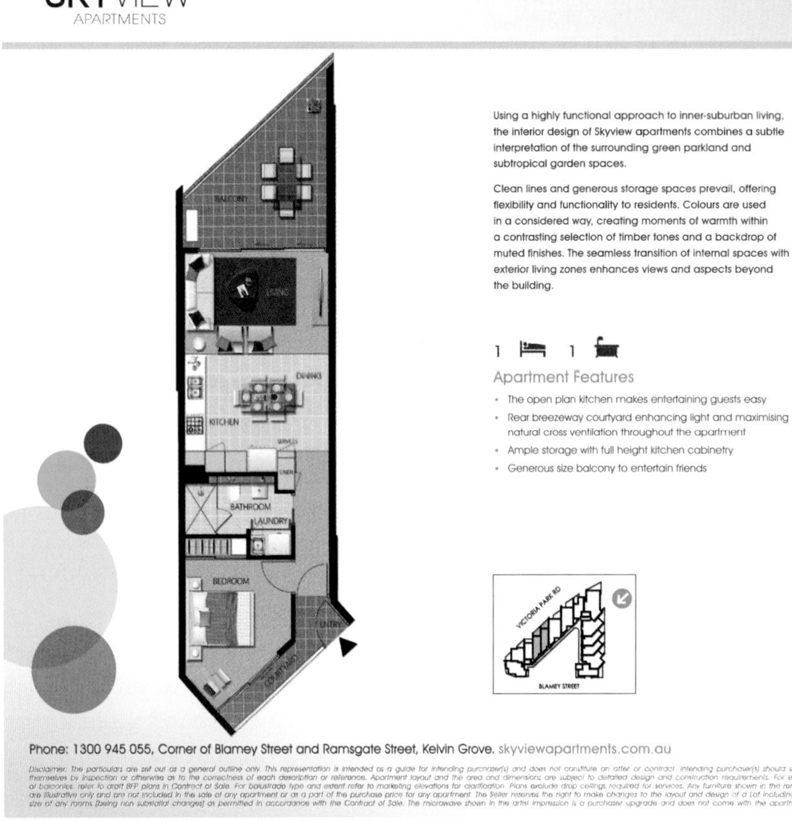

Artist Impression - Typical Bathroom - Sky Scheme

PRADELLA

Pradella Developments – Skyview Apartments

▲ This brochure for Skyview apartments provides both a two-dimensional and three-dimensional preview of the interior design of each residence. The use of digitally generated imagery is used to depict with great detail, a space that is yet to be constructed.

+ develop a mock-up or draft to test the suitability and effectiveness of the final concept or to provide the client with a preview of the final design
+ provide an opportunity for change and alteration throughout the design process.

In evaluating your own design work you will apply critical and analytical thinking to determine decision making and design directions.

As in a professional design environment, the most common design evaluation tools that you will find helpful in your own work are mock-ups – models/prototypes that will allow you to evaluate and test your final concept. These tools allow you (and others) to examine and judge the effectiveness of the concept in line with the design brief. These may be three-dimensional or digital representations.

These drafts should be as close in appearance as possible to the final deliverable. However, the materials that you choose to use may be different. Plain paper, for instance, may be used in the mock-up as an alternative to more expensive printing paper with a special coating. You may choose to create an entirely digital model to assist in making decisions about final colours, textures, forms and materials. The scale of a model will most likely differ from the plans for production and this is often a practical consideration in a school setting.

The effectiveness and suitability of the draft will determine if and when production goes ahead. It is essential to have a critical eye when judging the success of a design concept at mock-up stage. Be objective and, if necessary, use the opinions of others to gain feedback.

If your evaluation is a success, then proceed to the production stage. If not, it means quite literally going back to the drawing board. An unsuccessful design idea can usually be traced back to poor understanding of the original design brief.

ISBN 9780170349994

...CONTINUED:

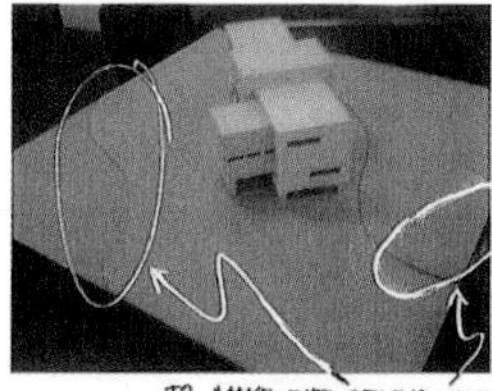

▲ This student documented the development of a scale model with photography. Each stage of the construction process was annotated and illustrated the evaluation process that occurred at every step.

ISBN 9780170349994

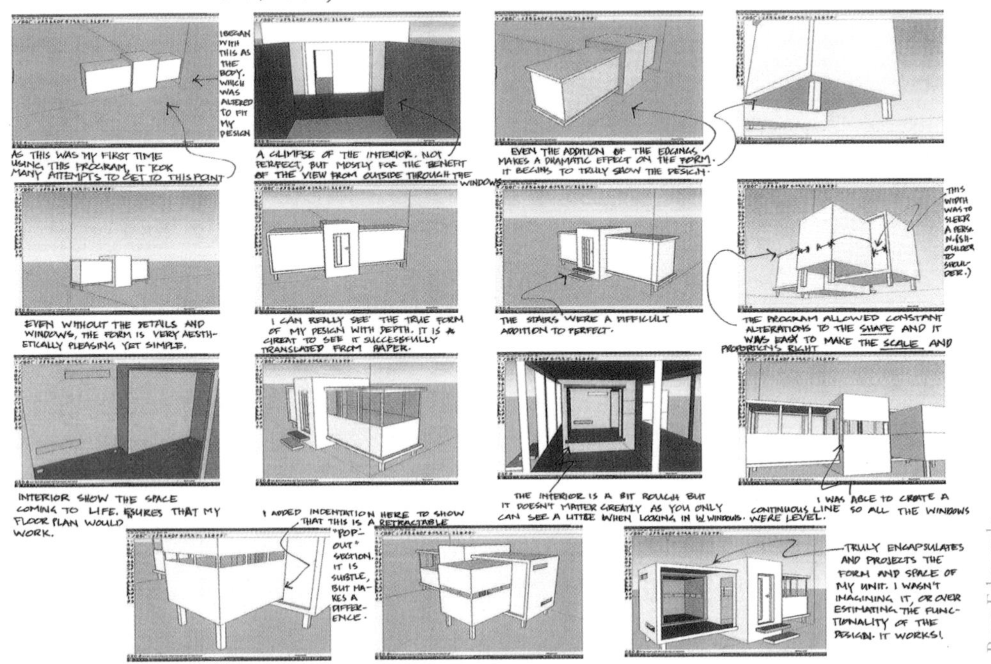

Prue Edmunds

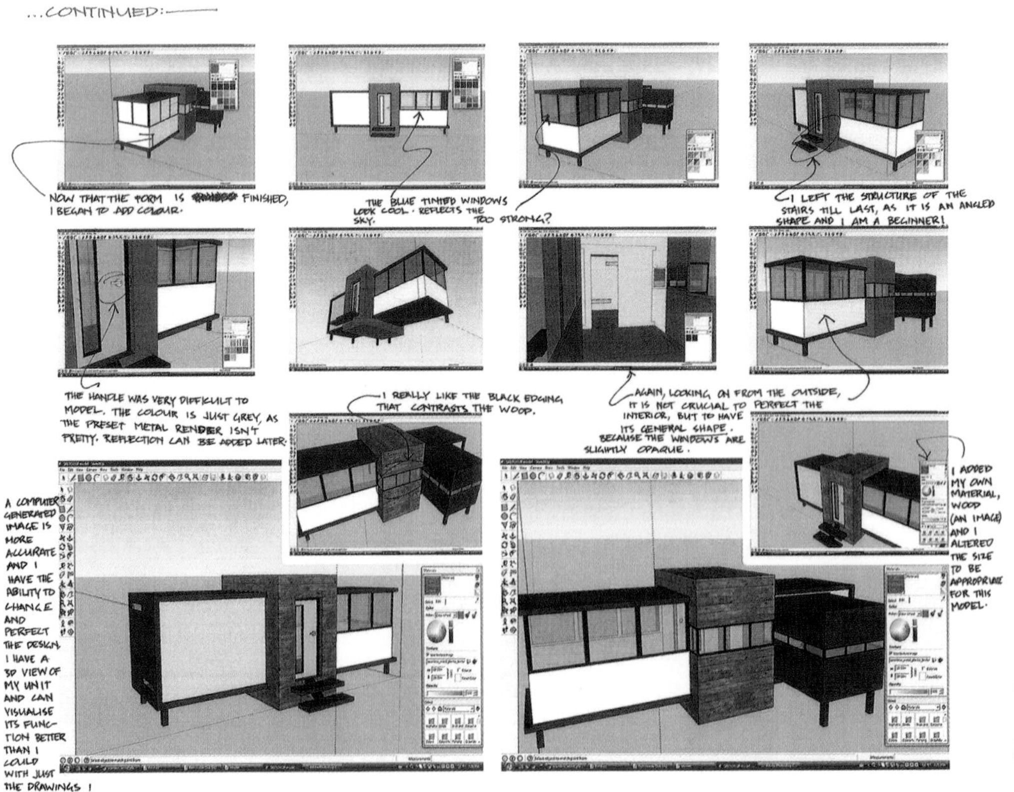

Prue Edmunds

▲ The same student depicted the steps involved in devising a computer-generated three-dimensional model of her design. Each step of the building process was analysed and annotated, enabling her to consider her design decision making and evaluate the most effective details in producing a final rendering.

ISBN 9780170349994

EVALUATION TEMPLATE

Evaluation should occur throughout the design process. This template may help you reflect on your work and progress at any stage of the design process.

Describe the design concept/ idea that you are evaluating:			
Is the concept suitable for the intended user?	❑ Yes Describe how it is suitable:	❑ No Describe why the current concept does not suit the intended audience:	Suggest how it could be changed to better suit the user.

Application of design elements (as relevant)			
	Where applied	How effective is the application?	How could it be changed/ improved/ altered?
❑ Colour			
❑ Form			
❑ Line			
❑ Shape			
❑ Space			
❑ Texture			
❑ Tone			
❑ Type			
❑ Point			

ISBN 9780170349994

Application of design principles (as relevant)			
	Where applied	What impact does it have on the design?	How could it be changed/ improved/ altered?
❑ Alignment			
❑ Balance			
❑ Contrast			
❑ Harmony			
❑ Unity			
❑ Hierarchy			
❑ Proximity			
❑ Repetition			
❑ Consistency			
❑ Proportion			
❑ Scale			

Is source material identified e.g. copyright?	❑ Yes (Ensure correct attribution using Harvard Generator)	❑ No	❑ Not relevant	❑ In progress

Does the design show consideration of sustainable practice?	❑ Yes Describe where and explain the impact it will have on the design.	❑ No Explain how sustainable factors could be applied.

List the graphical representations that have been used in the design.	Explain why these graphical representations are the most appropriate selection.

Explain how the concept/idea addresses or solves the design problem.	

ISBN 9780170349994

6.2 PRODUCTION

The final graphical product is produced only after extensive evaluation of the preferred design concept. Production can vary from the printing of 500 single-colour business cards to the construction of a city building, and is often expensive.

The general nature of the deliverable is usually defined in the design brief, but the design process often allows room for creative interpretation and decision making about the most appropriate graphical product to meet the original need. Some of the most remarkable or exciting designs are those that have been developed within a very flexible brief for a client willing to take an innovative and even risky design approach.

Designers can be involved in the production of diverse designs, making it difficult to define the boundaries of the final product. Production methods can be as varied as the design products they produce. They might range from the printing of a durable waterproof logo decal for an ocean-going yacht, to the creation of a three-dimensional type designed to sit on top of a hotel, to the production of injection-moulded plastic forms for a child's toy.

It is possible to categorise, by design areas, just some of the final presentations that might be produced.

In creating your own final design deliverables, practical considerations must be taken into account, as these will influence your decisions on the form of the final. Practical issues to consider may include choice of materials, scale and form of the final presentation, as well as the presentation space and location.

When it is not feasible for you to actually manufacture the design you have designed, a concept presentation may be the most appropriate means of communicating the final concept.

Tom Grech

▲ This student created a concept board as a final production to explain his design ideas to his client. The board uses a combination of hand-drawn and computer-generated imagery illustrating details about the appearance and features of a new car design.

Prue Edmunds

▶ Computer rendering and scale model presented as final design deliverables. The design brief required the design of a demountable housing option for homeless people in urban areas.

ISBN 9780170349994

Concept presentations present visual information about what the final product would look like were it to be produced. A concept presentation may include a range of graphical representations, which explain the features, and appearance of the final design.

A combination of the requirements of the original design brief and the available resources determine the production of your design concept. It is an opportunity to respond creatively and to present your thoroughly researched and developed work. Your work may be assessed on a range of criteria including the imaginative and creative response you bring to the task.

Production	Examples
Graphic design These presentations provide information for the viewer and can convey complex information and messages clearly. They can be used for the advertising, promotion and depiction of products and services. The application of design factors including the elements and principles of design are central to the effectiveness of these designs, and the application of media and materials is diverse. Presentations may take a two-dimensional or three-dimensional form. Designs may vary from a corporate logo to the signage for the Olympic Games, from a map of bicycle paths to a multipage interactive website.	Maps Packaging Symbols Advertising Charts Logos Illustrations Brochures Freehand drawing Posters Diagrams Publications Graphs Infographics Clothing Signage Exhibition displays Multimedia Motion graphics
Industrial design These presentations conform to rules and conventions that define the arrangement of images and the presentation of visual data. They may convey two-dimensional and/or three-dimensional information. Final designs may be manufactured from a diverse range of materials and involve combinations of design factors including sustainability, materials and elements and principles of design. Products can vary from small-scale domestic items to automotive design, from fashion to aircraft.	Two-dimensional and three-dimensional instrumental drawings Engineering drawings Concept presentations Three-dimensional scale models
Built environment design These presentations present information about the construction of designs within a built environment. The presentation of information may be two dimensional or three dimensional. The final production of the finished product is usually three dimensional. Some environmental designs are purely conceptual and are designed to inspire ideas rather than conclude with a finished product. Designs can vary from small residential projects to large apartment complexes, from courtyard landscaping to the design of a space station.	Architectural drawings Instrumental drawings Three-dimensional scale models Multimedia Maps Diagrams Concept presentations Plans Digital 'walk-through'

ISBN 9780170349994

PART C

DESIGN FACTORS

Design factors are the considerations that designers take into account when exploring, developing, evaluating and producing design products. They also form some of the core content in the Senior Graphics syllabus.

Design factors range from the essential visual components of a design (elements and principles of design) to the strategies undertaken to ensure that a project meets its deadline (project management). Design factors play a part in almost every design process, they may play a small role or a significant one but they account for some of the most essential facts and information that designers consider and subsequently apply to their work.

There are many factors that play a part in the creation of a design but the eight that are explored in this section are some of the most significant:

+ User-centred design
+ Elements and principles of design
+ Design technologies
+ Legal responsibilities
+ Design strategies
+ Project management
+ Sustainability
+ Representing materials

ISBN 9780170349994

CHAPTER 7
USER-CENTRED DESIGN

'I never design a building before I've seen the site and met the people who will be using it.'
Frank Lloyd Wright

Next time you use your smartphone or access an app on a tablet or computer, consider that the interface that you see was designed with you, the end user, in mind. 'User-centred design' is a term used to describe the consideration of the needs of the end user as a central focus in the design process.

If you type 'user-centred design' into a search engine, you will find that it is a term most commonly used for describing the design of software, operating systems, apps and websites. Terms such as 'graphical user interface' (GUI) or 'user experience' (UX) pop up and indicate the importance of efficient interaction with digital products. However, user-centred design is an approach that can (and should) be applied to any design product where the priority of the final design product is ease of use for the end user.

Designing with the end user in mind appears to be common sense but it takes considerable research and thought to ensure that a design meets the needs and accounts for the abilities of its audience. Good user-centred design is about asking questions in the early stages of the design process to build a solid understanding of who the end user is. Understanding the characteristics of the audience is essential to designing products that are effective and meaningful.

Questions might include:

+ Who are the end users of the design?
+ What are their key needs in using, interacting with or viewing the design product?
+ What are the limitations of the user?
+ How can the design be made accessible to a range of abilities, ages and body types?
+ What are the key functions that the user requires the design to provide?

7.1 UNDERSTANDING THE USER

As discussed in Chapter 4, design research is an integral aspect of the design process. Research of existing designs, research of the marketplace, the client history, materials and processes are all important. Equally so, is the research about the user of the final design product.

The end user is the target group to whom the design will be directed. The characteristics of users are often divided into specific types of data such as age, gender, socioeconomic status and interests. Other factors such as cultural background, educational level and religious affiliation can also affect the content, appearance and functionality of a design.

USER CATEGORISATION

It has become popular over the past 20 years for marketers and social commentators to categorise consumers into birth-related groupings such as:

+ Mature consumers: born between 1900 and 1945
+ Baby Boomers: born between 1946 and 1964, in the era after the Second World War
+ Generation X: born between 1965 and 1980
+ Generation Y: born between 1981 and 2001
+ Generation Z: Born 2000 onwards.

USER CHARACTERISTICS

Characteristic	Profile
Age	Age groups may be identified in very specific terms (e.g. 18–25 years) or more broadly (young adult, Baby Boomer, etc.).
Gender	The user may be male, female or gender non-specific. Gender can be a very strong influence on consumer preferences.
Socioeconomic status	This refers to the financial and social position of the user, usually identified by employment status, salary level or educational background.
Interests	This covers a vast range of categories and subcategories, including music, sport and fashion. Target users might be an association of professionals or an organisation for people who share a common interest.
Cultural and religious background	Content may be influenced by the belief system of the user. The appropriateness of imagery and content will be defined by cultural and religious traditions.
Location	Where a user lives can have an impact on their opportunities to view visual communications and on their employment or socioeconomic status. Location can determine visual and oral language and be linked to cultural or religious factors.

ISBN 9780170349994

7.2 METHODS OF USER RESEARCH

Research is important because it can not only provide important information about WHO you are designing for but also help to identify fashion and trends in design. Contemporary designers understand that it is important to stay up to date with changes in tastes, preferences, technologies and materials. Very often, these changes can be identified through the behaviour and feedback of the target market.

There are many different research techniques that can be utilised when investigating the user of a design. Sometimes the client provides designers with detailed analysis of the end user while other briefs see the designer generating their own research. Some use a range of techniques while some may use only one or two to familiarise themselves with the characteristics of the end user. In your own work it is important to first establish what it is that you need to know about your target users and then select research methods that are best suited to gathering the relevant information that will help to propel your design to a successful resolution.

OBSERVATION

One of the simplest methods of research available to a designer is direct observation of the target user. Watching the end user enables the designer to see behaviour and interactions in environments that may be familiar to the user. Casual observation – of how an environment or space is used, how a product is handled or responded to and the way a user reacts to a graphic design – provides data that may be helpful in identifying the needs of the user. Observing how people react to visual stimuli can also occur in a more controlled environment; for example, focus groups provide a sample of users brought together to discuss and respond to design concepts. Observations of the dynamics and reactions of the focus group may be made discreetly via video or a two-way mirror.

Built environment designers make regular use of observation. Known as site analysis, architects, interior architects and landscape designers will visit a proposed site, not only to observe important physical details about a structure but also to gather information about the space it might be utilised by the end user. Factors such as accessibility, pedestrian flow and traffic movement within the site are all important considerations in environmental design.

MARKET RESEARCH TECHNIQUES

Market research is widely used in business to gain information about consumer preferences. In the form of interviews, email surveys, questionnaires or telephone canvassing, companies collect data about the likes and dislikes of people from a wide range of backgrounds. As audiences evolve and become more aware and sophisticated as consumers, data collection can be helpful in gaining insights into cultural and societal changes.

User collage

A user collage is a visual representation of the characteristics of the end user/market. Using images, colour swatches, patterns, typography and symbols it is possible to create a visual overview of the people that make up the target users of the final design. This can be beneficial when presenting research findings to a client or as a concept to share between co-designers working on the same design project. At a glance, the user collage can represent visually what might take pages to describe in words.

ISBN 9780170349994

Prue Edmunds

▲ This student designed a multipurpose shelter for homeless people living in the inner city. The user collage presents a clear image of the likely shelter users.

Johanna Schreiner

▲ In designing packaging of food products for food lovers and the gourmet home chef, this student created a collage that clearly captures the interests and aspirations of her target users.

Nick Byrne

▲ In researching his target user for innovative wetsuit designs, this student identified an active and youthful market with clear interests in surfing and surfwear.

HOW MUCH IS YOUR SELFIE WORTH?

FYI

Your personal data has a dollar value that you might not be aware of. Economists often attempt to place a numerical value on personal information about users of popular sites such as Google and Facebook. Influenced by a range of economic factors, 'the implied market capitalisation per Facebook user has fluctuated between USD 40 and USD 300 at different times between 2006 and 2012, with a recent valuation in May 2012 approaching USD 112 per user.'

Source: OECD (2013), 'Exploring the Economics of Personal Data: A Survey of Methodologies for Measuring Monetary Value', OECD Digital Economy Papers, No. 220, OECD Publishing. http://dx.doi.org/10.1787/5k486qtxldmq-en

Personas

A persona or persona profile is a representation, both written and visual, of a 'typical' user. Ordinarily, they are fictional and personify the characteristics of the most recognisable audience members. A persona can capture the 'essence' of the end user and assist in guiding the design process by targeting the appearance, functionality and design ideas towards the preferences of this fictional character. When working in a team, designers may use multiple personas to ensure that all team members are designing for the same set of audience interests and needs.

A persona will often include data about the character, a visual reference and detailed summative information about their key characteristics such as age, gender, employment, location and so on.

It can be helpful to build one or more personas in your design work to identify who your target user is throughout the design process. Keeping the end user in mind throughout a long design process can be challenging so creating a character to reference at different stages can be very helpful in staying focused.

ISBN 9780170349994

PERSONA (typical audience profile)
NAME Ingrid Pertunia
AGE 22
STATUS Single
INCOME Disposible
OCCUPATION Student
INTERESTS
Socialising with friends, at cafes, bars & clubs.
During the day Ingrid is at Uni or working.
She enjoys listening to alternative, electro, folk and rock music, depending on her mood.
She appreciates good design in architecture and advertising.
Shopping is a common pastime in her breaks at Uni and one day she would like to travel around Europe and South America.

Sarah Mason

▲ This student used a persona profile to assist in identifying the target audience for the interior design of a fashionable new bar to be located at music festivals.

AUDIENCE PERSONAS:

TYPICAL AUDIENCE PROFILES

Shutterstock.com/ Wilson Araujo

PERSONA ONE:
Name: Liz Myers
Age: 42
Status: In a relationship
Income: High Disposable
Occupation: Writer/Columnist
Description: Liz is a busy professional, yet prioritises her time in order to pursue her passion for cooking and appreciation of fine gourmet produce. Quality ingredients are essential to Liz's cooking and she is prepared to pay extra to ensure the quality of what she is purchasing from providores and delicatessens in and around inner city Melbourne. Liz is well known for her great cooking and love of experimenting with new flavours and culinary techniques.

When she isn't working or cooking a home for friends and family, Liz enjoys catching up with old friends at 'quirky' cafes and restaurants as well as bars and other live music venues. Liz also spends time reading novels and appreciating art and design- both traditional and contemporary.

Shutterstock.com/ wavebreakmedia

PERSONA TWO:
Name: Johnny Walshe
Age: 30
Status: Single
Income: Moderate Disposable
Occupation: Chef
Description: Johnny loves good, high quality food and only uses the best ingredients in his cooking - both at work and at home. He works for a boutique restaurant in Southbank Melbourne and is responsible sourcing quality ingredients and pantry staples for the restaurant. He frequents wholesale hospitality stockists who offer high quality spices, herbs and imported produce from all over the world in order to purchase the best; Johnny purchases these expensive ingredients on the restaurant's account. When buying ingredients for his own cooking at home Johnny budgets for buying high quality produce and tries to save money elsewhere, food is his passion and he does not compromise on quality ingredients.

When not at work, Johnny enjoys socialising at popular alternative bars and clubs in and around Melbourne, and eating at simple yet quality restaurants. Listening to live music is also something Johnny enjoys in his spare time.

Shutterstock.com/Goodluz

PERSONA THREE:
Name: Frank Schnoit
Age: 52
Status: Married
Income: High Disposable
Occupation: Gourmet Providore/Delicatessen owner.
Description: Frank's family were immigrants from Germany after WW11, and have owned their delicatessen/providore for decades. Frank inherited the business from his parents and now runs it with his wife. It is a very successful business as it has created a boutique, 'high quality' reputation for itself. Frank only stocks quality fine gourmet ingredients and produce, he has tailored his business to suit his customers who are also passionate about quality ingredients and gourmet food.

Johanna Schreiner

▲ This student used three different personas to describe the broad audience of her packaging designs and cited her online sources beside each image.

ISBN 9780170349994

Social media

▲ These social media icons are instantly recognisable. Logos reproduced with permission of Instagram, Inc.; Pinterest; Twitter, Inc.; Facebook; Tumbler, Inc.; and Shutterstock.com/Lucian Milasan.

Analysis of social media sites offers a wealth of information, both visual and written, for the design researcher. Many demographic groups, but especially people in younger markets who traditionally have a high disposable income, are heavy users of social media and appear willing to share a wealth of personal data publically. Using sites and apps such as Instagram, Pinterest, Facebook and Google+, it is possible to develop detailed profiles of users within a particular age range or interest group. Be aware of privacy considerations when accessing information about real individuals; seek permission if you plan to use photographs that belong to others (see Chapter 10 for more information on legal responsibilities including privacy and copyright).

Usability testing

Later in the design process, when you have progressed towards a design solution, it can be helpful to research the responses of your target users. Designers apply usability testing to assess the progress of a design concept. Most commonly used in industrial/product design, usability tests for suitability, functionality and aesthetic appeal. Participants may be given the opportunity to handle prototypes and scale models to develop an understanding of form and function. Data collected from trials and tests with members of the target user group can help resolve design issues, address ergonomics and apply changes to meet an appropriate design solution. Graphic designers may also apply usability testing to focus groups, asking participants to comment on a range of design ideas for advertising, packaging, websites or other forms of visual communication.

FASHION AND TRENDS

Recognising and responding to fashion and trends in design is big business. Very often, users demand the newest and most up to date designs and the marketplace moves quickly to meet that demand. Many areas of design, such as fashion design, textile design and interior design, undergo seasonal shifts in colour, styling and theme. These cycles can occur very quickly and what was fashionable last month may no longer be seen as desirable next month. Blogging and the visual feast that is social media can influence what is and what is not fashionable over relatively short periods of time. Designers stay up to date with trends and developments by reading widely, attending expos and conferences and observing cultural shifts. In many cases, innovation and trends are set in motion by talented and innovative designers themselves.

FASHIONABLE COLOUR! FYI

Colour forecasting is big business. Professionals working in fashion, interior design, product design and manufacturing often begin the design of a concept many months – and even years – in advance. Predicting colour trends is therefore very important in ensuring that a design is relevant and marketable in the future. Companies such as Pantone, Edelkoort and Fashion Forecast Services provide clients with reports that analyse trends in fashion, accessories, textiles, paint colours and furniture.

Access all weblinks directly at http://nsg.nelsonnet.com.au.

ISBN 9780170349994

PANTONE® 18-3224 Radiant Orchid	PANTONE® 17-5641 Emerald	PANTONE® 17-1463 Tangerine Tango	PANTONE® 18-2120 Honeysuckle	PANTONE® 15-5519 Turquoise	PANTONE® 14-0848 Mimosa	PANTONE® 18-3943 Blue Iris	PANTONE® 19-1557 Chili Pepper

PANTONE® 13-1106 Sand Dollar	PANTONE® 15-5217 Blue Turquoise	PANTONE® 17-1456 Tigerlily	PANTONE® 14-4811 Aqua Sky	PANTONE® 19-1664 True Red	PANTONE® 17-2031 Fuchsia Rose	PANTONE® 15-4020 Cerulean

▲ Pantone releases a 'colour of the year' that reflects current trends in design, fashion and interior architecture. This illustration shows all Pantone colours of the year from 2000 (bottom right) to 2014 (top left).

Although trends appear more slowly in built environment design, the design of spaces and structures is also affected by changing preferences in the application of materials, colours and textures. Product design may also be influenced by colour trends, changes in the desired appearance, form and shape of consumer products and a demand for innovative materials and technologies.

Aesthetic preference plays a major role in user decisions when faced with a choice of designs. Aesthetics relate to the physical appearance of a design; a well-designed product that makes effective use of principles of design, such as harmony and balance, is naturally more appealing.

7.3 HUMAN FACTORS IN DESIGN

Not only do the preferences and tastes of the target users need to be taken into account, the physical characteristics of the human body have an impact on industrial design and built environment design in particular. The way humans 'fit' with a design, how they interact both physically and with their senses are important considerations in design.

ERGONOMICS

Ergonomics is the study of human factors in design and how human beings interact with products and environments. This scientific discipline looks at the functions, limitations and needs of the human body in relation to product design. Ergonomists often work with designers to design products that take into account the physical, organisational and psychological effects on the user. You may be familiar with the term 'ergonomic furniture', which is often a selling point of chairs and desks for a home study or office environment. Standard ergonomic height requirements exist so that the user is most comfortable when seated at a desk for a prolonged period.

Ergonomics is concerned with the interactions between a user and a product. It relates not only to physical and biomechanical interactions with design but also to cognitive processes such as memory and decision making. Good design takes such factors into account and ensures that a product is not so difficult to use that a user cannot operate it or remember simple functions.

Ergonomic principles are embedded in the publications of organisations such as Australian Standards whose guidelines cover the design and manufacture of products and built environments throughout Australia. All products sold and

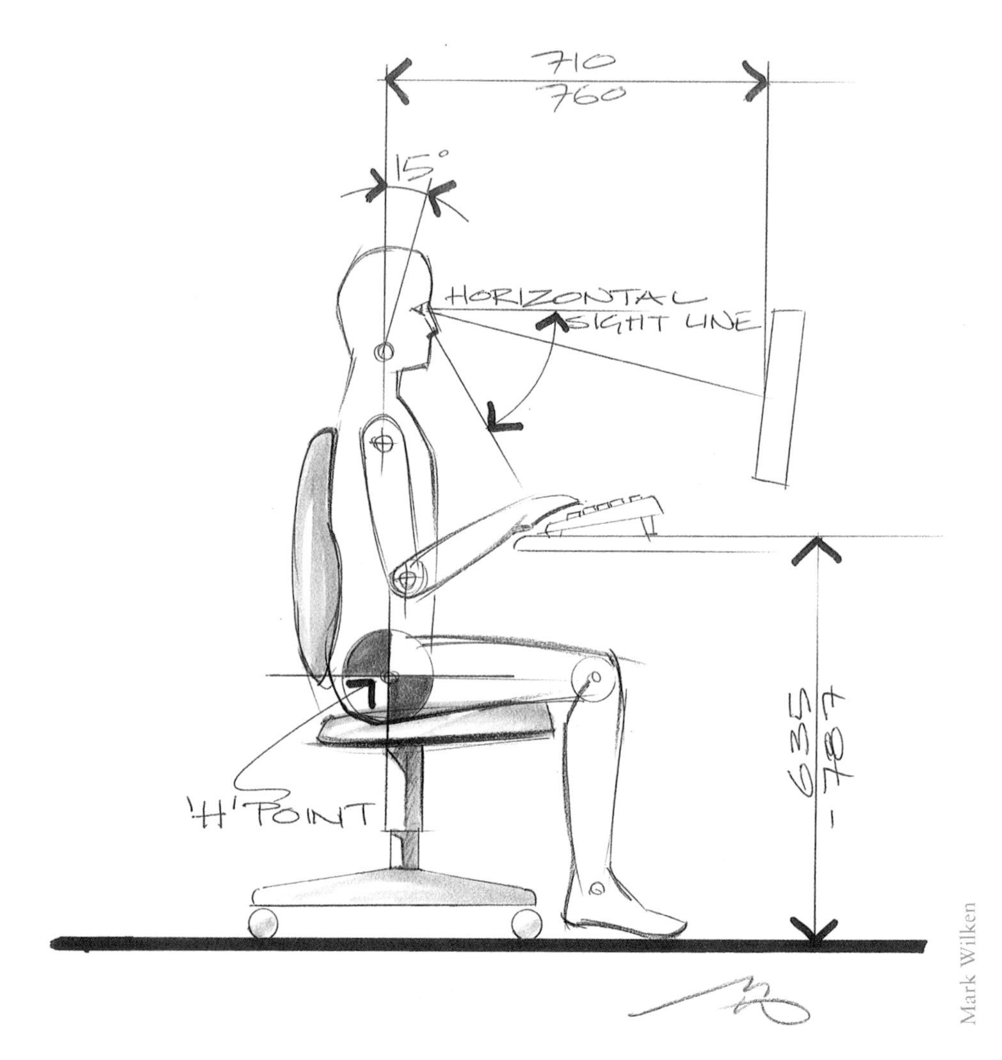

▲ This diagram indicates the ideal height of the desk and chair as well as the preferred position of the user. Anthropometric data about human body size is taken into account in the design of ergonomic products (anthropometry is the study of human body size, posture, movement, surface area, volume, and mass). Due to the vast variety of human shapes and sizes, many designs need to take into account the physical characteristics of the average user and apply proportions that suit a wide range of people.

used in Australia must meet the relevant industry standards, which range from the paper pulp used in packaging to water quality, domestic appliances and road vehicles, mining technologies, smartphones and food. The purpose of standards is to ensure safe design and manufacturing practices lead to the safety of the end user.

ERGONOMICS

For more detailed information about ergonomics, visit the International Ergonomics Association website. Access all weblinks directly at http://nsg.nelsonnet.com.au.

USER INTERFACE DESIGN

In digital design, users interact with a screen so the design of intuitive and functional interaction is very important to the success of computer operating systems, apps and software products. With the growth of smartphones and tablets, user experience design or UX design is a field that has developed to meet the needs of increasingly educated technology users. UX designers focus on developing digital products that are easy to use and make the most of users' existing familiarity with technological functions.

An important aspect of user-centred design in digital media is to enable users to access and use new applications without significant levels of

ISBN 9780170349994

learning. Users prefer to pick up a device being confident that fundamental functions operate in the same or similar way to previous devices. This is a challenge for digital designers and reinforces the need to have a good understanding of user needs before commencing the design process. Users will learn new functions but it is the designers' challenge to present innovations in an accessible manner and it is here that the application of design elements and principles is of paramount importance.

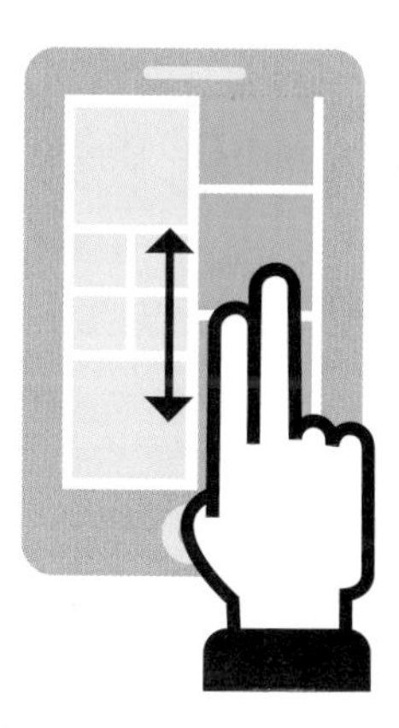

Shutterstock.com/tovovan

▲ This diagram illustrates some of the finger gestures required to navigate a tablet by touch. Interestingly, the first tablets established these gestures as standard and subsequent tablet designs have applied them in the same way. Ensuring consistency in interactivity means that users do not have to relearn complex processes when using a new product or software system.

Shazam

▲ The popular Shazam app that identifies music via smartphone or tablet uses design elements, such as shape, type and colour, to instruct the user. It is the use of clear and effective design elements and simple instructions that make Shazam an easy app for users to learn.

Smart Design

▲ OXO Good Grips are a suite of domestic products originally designed for people with limited functionality in their hands such as those suffering with arthritis. Interestingly, the success of the designs has been due to users of *all* abilities recognising the excellent ergonomic properties and high level of comfort.

ACCESSIBILITY IN DESIGN

From the humble potato peeler to low-floor buses, the design of products, the environment and graphics is constantly evolving to meet the needs of people with a range of abilities and disabilities. In particular, the designs of many public environments and some product designs are required to accommodate the needs of users with disabilities. Designers research the capabilities and limitations of users and strive to design products that are inclusive and accessible to users.

Designers use research and observation to understand the requirements of users with special needs and respond with solutions that use form, space, texture and colour among many other elements and principles of design. Consider that braille on lift controls, reflective textures on road signs and lighting controls that can be distinguished by texture were all designed in response to users with limited or poor vision. Assistance for users with sensory limitations, such as hearing loss or vision impairment, and physical disabilities are often incorporated into design fields such as transport, the built environment, way finding (signage), digital design and many others. Although Australian Standards usually require accessibility considerations to be factored into designs, it is important that designers consider all potential users of their final design product and explore innovative means of meeting their needs.

HUMAN-CENTRED DESIGNERS

Smart Design and IDEO are two large design and innovation firms that focus on the user as the priority in their design processes. The websites of both organisations document their successes in focusing on the user as part of their 'human-centred' approach to designing products, brands and experiences.

Access all weblinks directly at http://nsg.nelsonnet.com.au.

7.4 USER ANALYSIS

The following questions are designed to help you document and build your understanding of the end user. You might choose to use this as a template or develop your own questions that are directly relevant to your design problem.

ISBN 9780170349994

Age range of user: ______________
Socioeconomic status: ❑ High ❑ Medium ❑ Low
Disposable income (if relevant): ❑ High ❑ Medium ❑ Low
Interests and tastes:

Brands Which?	Music What?	Magazines and TV What?	Social life What and where?	Shopping What and where?	Hobbies and Sports What?

Where does your user spend most of their time? (Circle all that apply.)

+ Alone
+ Home
+ In an office
+ In company
+ Indoors
+ Outdoors
+ School/university
+ Work
+ Working as part of a team

Circle the key words that describe the user.

+ Adventurous
+ Alternative
+ Carefree
+ Cheap
+ Conservative
+ Contemporary
+ Cutting-edge
+ Elderly
+ Environmentally focused
+ Ethical
+ Expensive
+ Extravagant
+ Family focused
+ Fashion conscious
+ Natural
+ Politically minded
+ Safe or cautious
+ Social
+ Stylish
+ Technology savvy
+ Youthful

List any other words that describe the user.

....

....

....

....

What is important to the user?

+ Accessibility
+ Contemporary design
+ Ease of use
+ Following trends
+ High quality
+ Innovation
+ Learning/education
+ Low price
+ Luxury
+ Manufactured things
+ Natural things
+ Problem solving
+ Saving money
+ Security
+ Social and political issues
+ Social networking
+ Spending money
+ Technology
+ The environment
+ Tradition
+ Traditional design

Describe the aesthetic preferences of the user.

Colours	
Shapes and forms	
Patterns and decoration	
Materials and textures	
Specific abilities and/or disabilities	
Any other preferences	

ISBN 9780170349994

EXTEND YOUR UNDERSTANDING

ACTIVITY 1: USER COLLAGE

Using digital images or images collected from magazines and other print sources, create an user collage that represents the key characteristics of the target user.

ACTIVITY 2: MARKET RESEARCH

Using one of the following tools, collect information or data regarding the preferences of your target users. Data should be highlighted and annotated to indicate how it might be used to inform/direct your design ideas.

+ Questionnaire/survey (paper or survey software)
+ Interview
+ Analysis of article (news/journal/government statistics).

ACTIVITY 3: PERSONA

Create one or more personas using visual and written information that represents a typical user of your design.

ACTIVITY 4: SUMMARISE

Write a 2–3 paragraph summary of your user research. Focus on what you have learned about your target users and describe their characteristics, interests, preferences and tastes. Describe how this information might influence the direction of your design task.

ISBN 9780170349994

CHAPTER 8 ELEMENTS AND PRINCIPLES OF DESIGN

'Design is the method of putting form and content together. Design, just as art, has multiple definitions; there is no single definition. Design can be art. Design can be aesthetics. Design is so simple, that's why it is so complicated.'

Paul Rand

The elements and principles of design are integral to the design process and, when applied effectively lead to harmonious and successful design products. They are the fundamental tools applied in the creation and production of all design and it is important to build a solid understanding of their application, significance and influence.

This chapter is structured to help you understand how elements and principles of design are applied in a range of contexts and design areas.

There are many design elements and design principles but the Queensland Senior Graphics syllabus suggests the study of the following:

Elements of design

The elements of design are the visual components used in the creation of a design.

+ Colour
+ Form
+ Line
+ Shape
+ Space
+ Texture
+ Tone

Additional elements of design that will be covered in this chapter:

+ Type
+ Point

Principles of design

The principles of design are the tools used in the application, arrangement and manipulation of design elements.

+ Alignment
+ Balance
+ Consistency
+ Contrast
+ Harmony
+ Hierarchy
+ Proportion
+ Proximity
+ Repetition
+ Scale
+ Unity

Additional principles of design that will be covered in this chapter:

+ Cropping
+ Figure-ground

ISBN 9780170349994

8.1 ELEMENTS OF DESIGN

COLOUR

Colour is a very powerful design element. For 90 per cent of the population, colour is perhaps the most dominant and influential of all the design elements. Colour attracts us, warns us, calms and soothes us – it can influence our moods and our behaviour.

In traditional colour theory, there are three pigment colours that cannot be mixed or formed by any combination of other colours. All other colours are derived from these three colours: red, blue and yellow. Mixing the primary colours creates the secondary colours of green, purple and orange. Mixing the secondary and primary colours further creates multiple tertiary colours. Hues of all colours can be modified via the addition of black and white.

The colour wheel

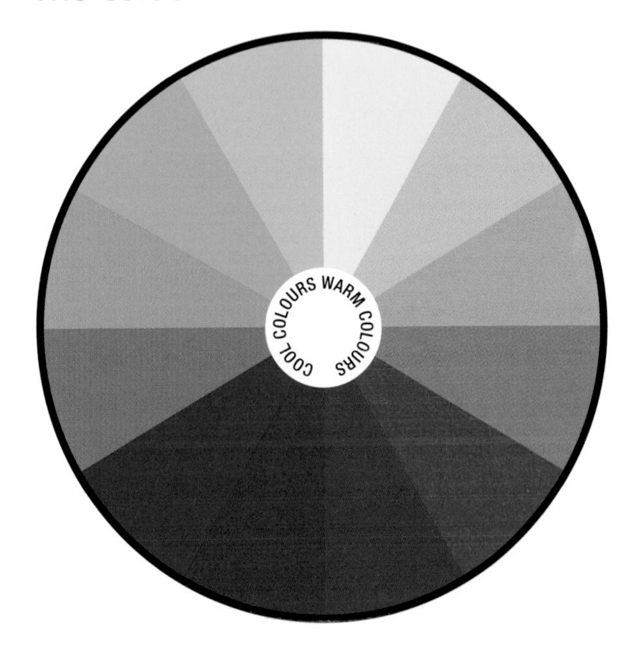

Primary colours

The primary colours are yellow, blue and red.

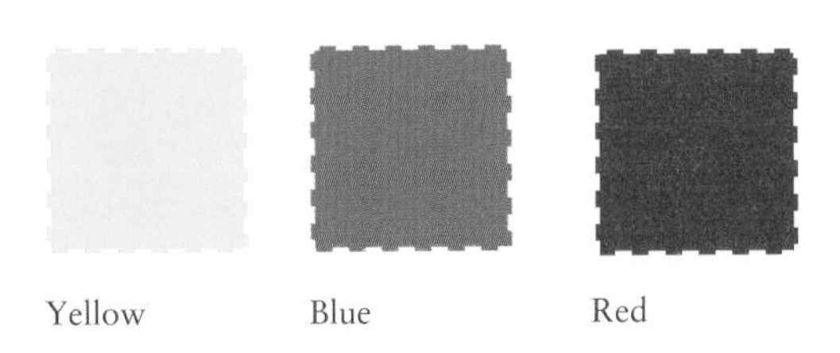

Secondary colours

The secondary colours are green, orange and purple. Secondary colours are created by mixing combinations of primary colours.

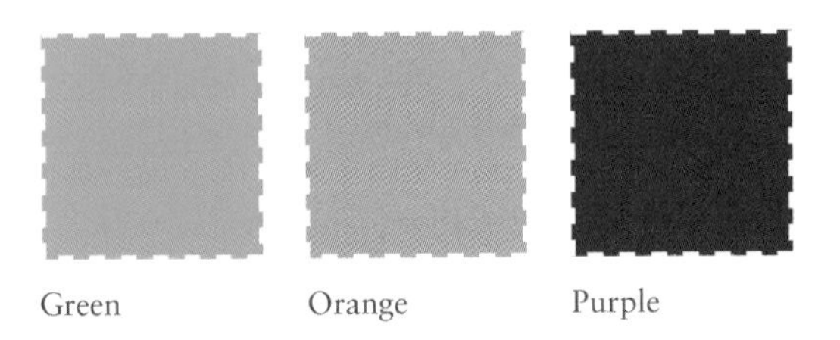

Tertiary colours

Tertiary colours are created by mixing a primary colour and a secondary colour.

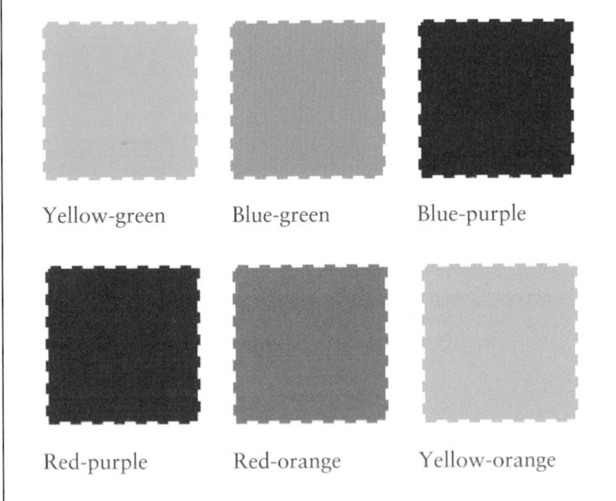

ISBN 9780170349994

Analogous colours

Also known as harmonious colours, analogous colours are colours that appear side by side on the colour wheel. When used together in a composition, they create subtle variations.

Harmonious colours

Complementary colours

Also known as contrasting colours, complementary colours are opposite and separated by colours on the wheel. These colours are often used together to create contrast. Colours that are direct opposites on the colour wheel can provide the strongest contrasts and draw the viewer's eye to key information within a composition. Complementary/contrasting colours can create deliberate tension in a composition, which might be required for emphasis or to create a sense of dynamic visual force. They can sometimes appear to vibrate – for example, red text on a blue background – and are deliberately used to create such an effect in some artwork.

Complementary/Contrasting colours

The colour wheel in practice

a Monochromatic (black and white)
b Primary colours
c Secondary colours
d Tertiary colours
e Analogous (harmonious) colours
f Complementary (contrasting) colours

We can look at colour from many directions, including its psychology, its symbolism and the extraordinary communicative power that colour holds. Though it is not possible to fully understand the significance and symbolism of every colour, it is essential to appreciate that colour has many facets, and to understand the influence it has in our lives.

Colour surrounds us – in language, in advertising, in fashion, and can even affect our behaviour. We quickly recognise that red means 'Stop' and green means 'Go'. When the use of colour challenges our understanding of its meaning, the message can become very confused. This can be seen in the illustration below.

To fully utilise the power of colour, designers need to understand its significance in many contexts, including that of culture. In designing for a specific audience or market, the choice of colour may be influenced by various factors. In Chinese culture, for example, red is symbolic of good luck, and at Chinese New Year you might see streets festooned with red lanterns and red decorations. In China, white is representative of death, so whereas we are accustomed to seeing brides wearing white dresses, such clothing would not be appropriate there – in fact, the colour of traditional bridal attire is red.

Just as we need an awareness of cultural sensitivities in all areas of design, including the application of colour, we also need to understand the emotional power of design elements.

Colour can elicit strong responses from an audience, even though a reaction may be quite subconscious – we are often quite unaware of the influential and persuasive effects of colour. The colours of a country's flag or the uniform worn by its athletes at the Olympic Games might encourage a sense of patriotism, which manifests itself in emotions such as pride.

As colour stimulates our emotions and senses, it can suggest a wide range of possibilities. Colour might suggest a fashionable and appealing lifestyle; it may soothe and placate, or suggest energy and dynamism.

Colours can also be described as either warm or cool. Blues, greens and purples are traditionally referred to as cool colours, while reds, oranges and yellows are termed warm. Actual temperature has nothing to do with it, but warm and cool colours can describe a 'feel' in a design composition.

▲ The two posters pictured use the same graphic elements to promote a festival, yet the use of colour helps to describe the theme and season of the events.

In environmental design, colour has been used to stimulate and sedate. Some research has shown that hues of pink can have a calming effect, and in fact a colour close to bubblegum pink was used in a US prison to subdue violent and angry prisoners. Schools often choose to paint walls in vibrant colours, such as yellow and green, colours that are said to stimulate learning and creativity.

Colour production

When producing images for print or electronic publication/screen, you will need to have an understanding of the methods of colour production. There are different colour systems that are designed for use on a computer screen and in print. It is likely that at some stage you have selected a colour on your computer screen only to have it print quite differently on an inkjet or laser printer. This is because screen colour and print colour use different systems and interchanging from one to the other will often cause colour change. Tools such as a Pantone colour swatch and colour management software assist in countering the differences.

PANTONE

Pantone, established in the 1960s, is a company that provides a universally adopted colour matching system used in the design community. Although the website is predominantly a commercial one, it also features helpful colour forecasting and colour selection tools.

Access all weblinks directly at http://nsg.nelsonnet.com.au.

RGB

RGB stands for red, green, blue. The RGB mode, on which your computer monitor is based, defines all possible colours as percentages of red, green or blue. RGB mode is used for on-screen editing or viewing of graphics.

Red

Green

Blue

ISBN 9780170349994

RGB colours are called additive colours because the RGB system involves starting with black and adding coloured light. For example, adding green light and red light gives yellow light, while red plus green plus blue light gives white light.

CMYK

Known as process colour, CMYK divides your images into four colour channels: cyan, magenta, yellow and black, which correspond to the inks used in four-colour printing. When using design software, RGB images can be converted to CMYK, but some colour change will occur. CMYK is used by commercial printers to produce full colour work.

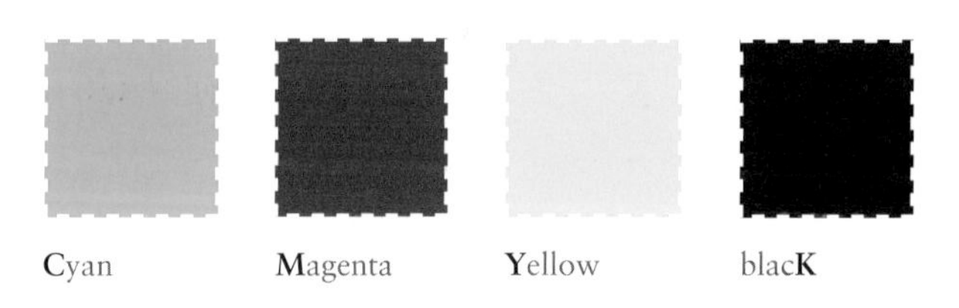

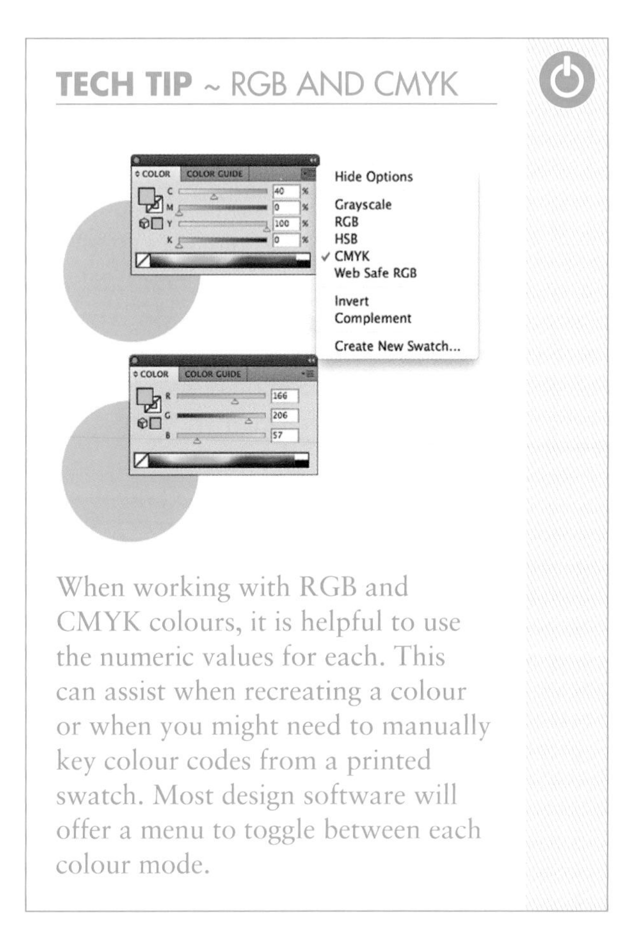

TECH TIP ~ RGB AND CMYK

When working with RGB and CMYK colours, it is helpful to use the numeric values for each. This can assist when recreating a colour or when you might need to manually key colour codes from a printed swatch. Most design software will offer a menu to toggle between each colour mode.

Spot colour

At times, it is more cost effective to use a colour that already exists in ink form and does not have to be mixed via the CMYK process. Spot colours are premixed colours produced commercially and available for professional use. When using one to three colours in a document or presentation, spot colour can be the most cost-effective approach. Pantone uses the PMS (Pantone Matching System), which is perhaps the most widely recognised spot colour matching method. Designers select the most appropriate colour from printed swatches that show hundreds of colours. They can be confident that the final print will reflect their choice, even though PMS colours often appear different when they are shown on a monitor as RGB percentages.

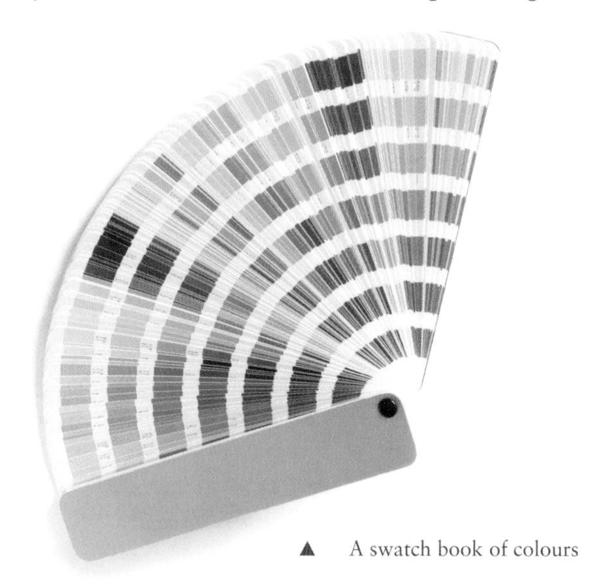

▲ A swatch book of colours

KEY WORD ~ COLOUR

Colour can be:
+ bold
+ subdued
+ vibrant
+ bright
+ subtle
+ warm
+ cool
+ primary
+ secondary
+ tertiary
+ contrasting
+ complementary
+ eye-catching
+ dominant
+ dynamic.
+ calming
+ emotive.

Use colour to:
+ define space
+ create contrast
+ create hierarchy
+ create a mood.

ISBN 9780170349994

Buro North (creators of the wayfinding signage) with contribution from Jane Resigner

▲ Colour can be used to create different zones in interior design. Colourful illustrations by Jane Resieger distinguish each level of the Royal Children's Hospital. Each floor of the hospital is identifiable by distinctive coloured images that create different visual environments on each level. The illustrations depict ground dwelling creatures on the lower levels to birds and the sky on the uppermost levels.

FORM

Form generally refers to objects that are three-dimensional in nature. We readily recognise the forms around us – from the pencil on the desk to the form of the human body. Form is often depicted visually through the application of other elements such as shape and line. Form can be rendered to enhance its three-dimensional qualities. The addition of shadows and highlights can help us to 'read' the true form of an object.

Indicia Design, Kansas City, Missouri, USA with permission of The Buckley Group L.L.C.

As you know from the physical environment that surrounds you, forms are infinitely varied and range from the geometric and constructed to the organic forms of the natural world. The representations of these forms are similarly varied and can range from the precision of an isometric engineering drawing to a loose and flowing charcoal life drawing.

Designers who work with the constructed environment – such as architects, industrial designers and interior architects – constantly experiment with our perceptions of form. Many variables impact on the design of new forms. These variables include:

+ ergonomics
+ structural constraints
+ the environment
+ fashion and trends.

ISBN 9780170349994

MARC NEWSON

Marc Newson may be Australia's most successful living designer. Newson was born in Sydney, Australia, in 1963 and studied sculpture and jewellery at Sydney College of the Arts. His design expertise spans fashion for G-Star and domestic items for Magis and Alessi to aircraft interiors, automobiles and hotels. Newson's ability to create functional yet unique forms is his trademark. His innovative use of materials is also remarkable. His website provides a comprehensive archive of his extraordinary body of work. Access all weblinks directly at http://nsg.nelsonnet.com.au.

Marc Newson, Photography by Tom Vack

▲ A design by Mark Newson

Many professionals involved in environmental design – whether it is landscape, product design or the constructed environment – are heavily influenced by the versatility of form. Take a chair, which is useful for that most basic of functions – sitting down. Yet this deceptively simple, functional object has developed, changed and evolved over the past century into a product that has challenged our ideas about form.

Form follows function

A phrase that was embraced by the modernists of the mid-20th century, 'form follows function' suggests that aesthetic considerations should come secondary to the pure functionality of a design product. Many designers believe that beautiful design is only achieved when the successful function of the design is fully realised. Contemporary (and post-modern) interpretations of this phrase are less rigid and many current designs reflect a balance between functionality and decorative elements. A good rule of thumb is to remember that no matter how attractive a design looks, if it doesn't achieve its primary purpose, it is not a successful design.

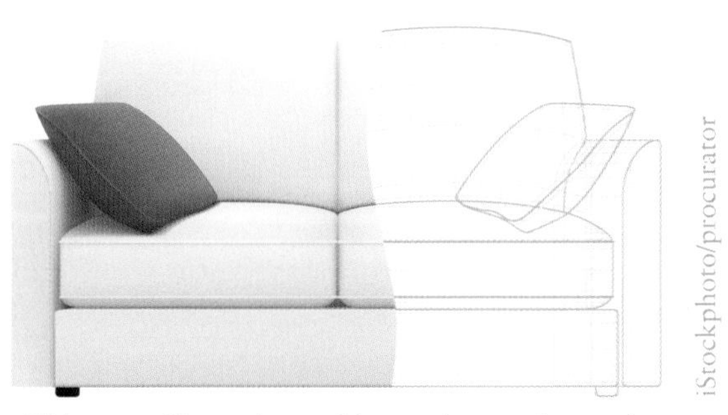

iStockphoto/procurator

▲ This vector illustration enables purchasers of a couch to understand the structural form beneath the upholstered surface. The wireframe and fully upholstered views provide contrast as well as context.

KEY WORD ~ FORM

Form can be:
+ three-dimensional
+ organic
+ geometric
+ dominant
+ subtle
+ tactile
+ solid
+ fluid
+ graceful
+ irregular
+ textured
+ natural
+ manufactured
+ modelled
+ sculpted.

Form can be used:
+ to define space
+ to create contrast
+ as a model or prototype.

ISBN 9780170349994

▲ Animi Causa 'Feel' seating system. The form of this innovative seating system can be changed in multiple ways by the user. Made from 120 soft foam balls covered with a smooth elastic fabric, the form was inspired by molecular structure.

LINE

Line is a versatile design element using only the dimensions of length and width. In technical drawings, line is integral to the representation of shape and form. Linear details – such as the outline of an orthogonal drawing and the appropriate dimension lines – are represented through lines of varying type.

The purpose or intent of a visual communication can vary through differences in the width or 'weight' of line. A fine or light line can suggest a specific technical detail or, in the context of an illustration, a sense of lightness or minimalism. Bold or heavy lines might be used for emphasis or to represent a structure within a given space.

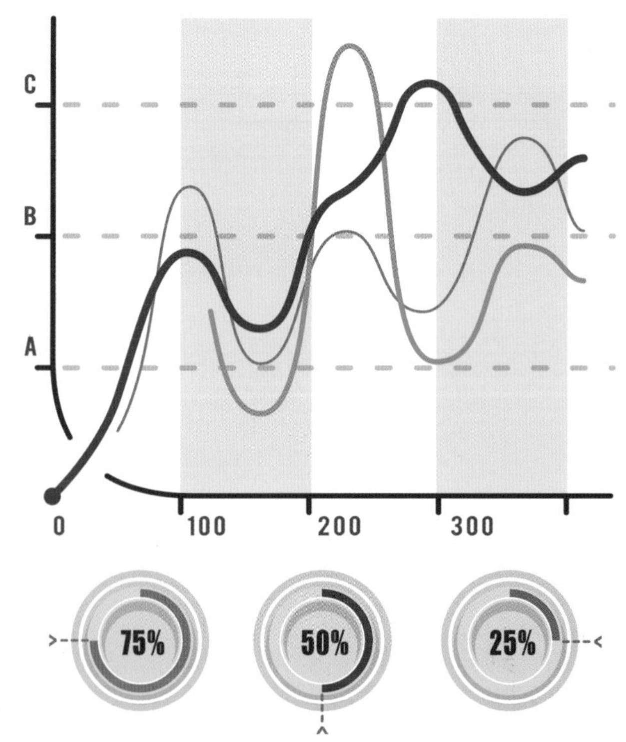

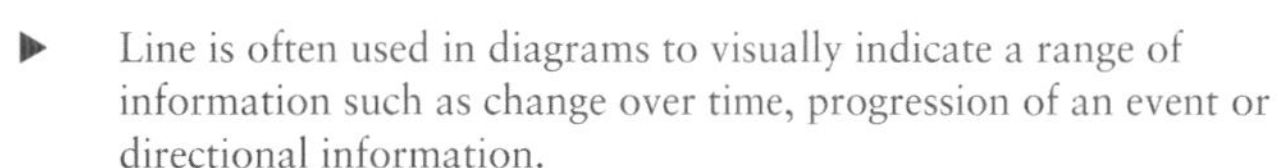

▶ Line is often used in diagrams to visually indicate a range of information such as change over time, progression of an event or directional information.

ISBN 9780170349994

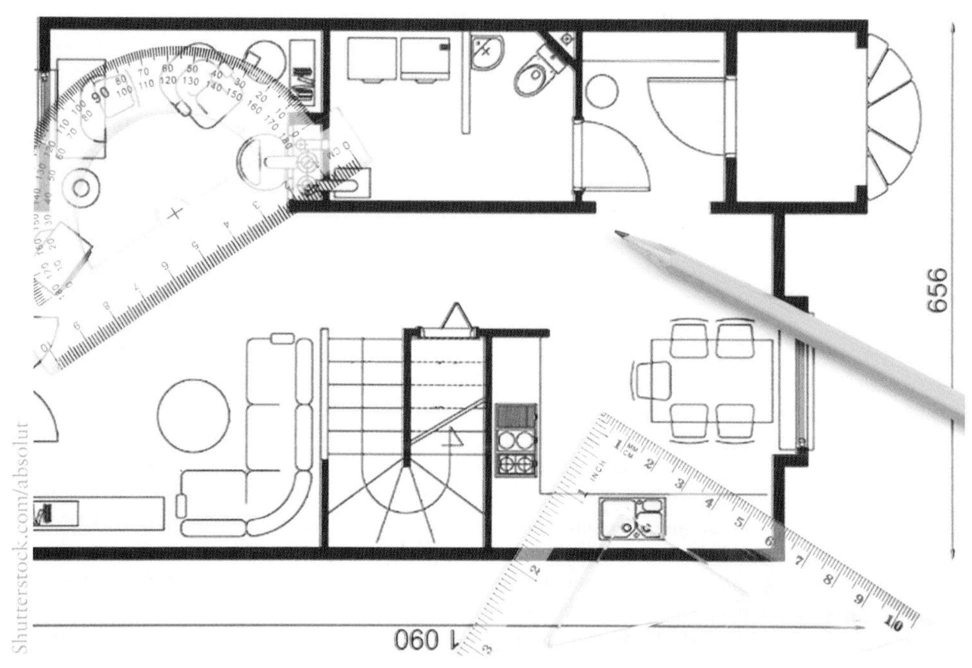

▲ Line qualities vary in architectural drawings. Varying line widths represent specific structural information.

Line can suggest direction and movement. It can draw the eye into a composition and direct it along a path. Used as a border or to define areas of a composition, line can create structure and stability.

▲ Line has been used as an illustrative device in this example. Organic lines lead the eye to the key feature of the image: the shoe.

Used as an illustration tool, line forms the basis of many popular techniques. In relief and intaglio printmaking, line is used extensively in the development of imagery. Linocuts and dry point etching, in particular, lend themselves to the application of diverse line types.

When used to render objects, line rendering techniques, such as crosshatching, create variations in tone and texture, which in turn serve to emphasise form.

▲ Line applied to fashion visualisation sketches depicts the silhouette of the garment as well as suggesting the flow and movement of the fabric and form.

▲ Examples of line use in logo design. Both logos use line as the primary element, but note that the quality of line (freehand or instrumental) makes a significant difference to the visual outcome.

KEY WORD ~ LINE

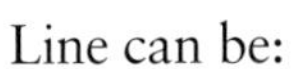

Line can be:
+ broken
+ flowing
+ bold
+ fine
+ medium
+ repeated
+ organic
+ eye-catching
+ dynamic
+ directional
+ static
+ curved
+ straight
+ sketchy
+ freehand
+ precise.

Line can be used to:
+ establish structure
+ create a pattern
+ indicate
+ direct
+ render.

Line can create:
+ contrast
+ pattern
+ formality
+ contour
+ structure.

SHAPE

Shape describes representational or abstract shapes that are two dimensional.

+ Simple geometric shapes with the dimensions of length and width include squares, triangles, rectangles and circles.
+ There are limitless irregular or abstract shapes. These might appear as natural organic shapes, or as irregular geometrically based images.
+ Shape may form the background in a composition, providing a space for the placement of other elements. Shape is used for emphasis and to draw attention to the figure in a visual communication. Shape may be the figure itself in the form of a logo or symbol.

Shape is an element that is very familiar to us, from the silhouette of the human figure to the shape of a Stop or Give Way sign on the road. Shape can inspire all kinds of reactions in a viewer; the shape of a heart or a cross might provoke an emotional response, whereas the hexagonal shape of a Stop sign demands an immediate physical response from the car driver.

Symbols are usually two dimensional and are often based on simple geometric shapes.

▲ Road signs use universally familiar shapes. It is the combination of shape and colour that communicate to road users. The yellow diamond is another common shape-and-colour combination that indicates caution.

Pictographs

A pictograph is a symbol that is based on a recognisable set of shapes or on a commonly recognised form. For example, the male/female signage used on public toilets is usually pictographic. The forms of the female and male figure are familiar and require no additional text for identification.

ISBN 9780170349994

The use of shape incorporates the application of other elements such as line or colour. Shape can be created using a range of media (such as paper, fabrics or card) and methods (such as collage or monoprinting). Other design elements, such as type, colour and line, can be used abstractly to create shapes that increase the visual interest of a composition.

▲ The Unilever logo uses shape to represent the many and varied aspects of this multinational corporation. Together the symbols express the company's core values with each unique shape also linked to the company's sustainable living ambition. Due to the diverse nature of this large business, the designer chose to apply simple shapes to showcase the specific elements that make up the Unilever brand.

SAUL BASS

Graphic designer Saul Bass was also a filmmaker and he is remembered for his striking film titles, particularly for a number of classic Hitchcock films. Bass' use of simple shapes was innovative and changed the nature of film title design forever.

Access all weblinks directly at http://nsg.nelsonnet.com.au.

KEY WORD ~ SHAPE

Shape can create:
+ hierarchy
+ pattern
+ background
+ contrast.

Shape can be:
+ two dimensional
+ solid
+ outlined
+ irregular
+ organic
+ geometric
+ defined
+ cropped
+ symmetrical
+ open
+ closed
+ free form.

▲ The Bahamas logo uses shape to depict the individual islands of this Caribbean nation. The use of scale and colour also identifies each island within the larger group.

SPACE

Space refers to the area around and between objects. It may refer to the distance between different shapes and forms within two-dimensional and three-dimensional environments. It may refer to physical spaces or to the spaces on a page or screen.

'Less is more': using white space

Attributed to Bauhaus designer Ludwig Mies van der Rohe, this statement defines much of the philosophy of the influential design movement of the 1930s. The Bauhaus rejected the decorative details and motifs seen in previous design movements. Their preference for the minimal influenced generations of designers who believed that what is left out of a composition can sometimes be as powerful as what is placed in it.

A cluttered composition can be distracting and difficult to understand. It can be important

ISBN 9780170349994

▲ Designed between 1923 and 1925 by Le Corbusier and Pierre Jeanneret, Maison La Roche in Paris is an early example of the Modernist style embraced by many architects of the era including the Bauhaus. Note the use of simple, uncluttered spaces and unadorned surfaces. These characteristics were typical of mid-century modernism.

to include detailed information in a composition but it is also important to recognise that too much information can make a viewer quickly lose interest.

White space does not necessarily mean blank white space – it may contain a colour or pattern – but it will lead the viewer's eye to the crucial information. White space can be used effectively to balance a composition. For instance, a large area of white space may balance an equally large area of text, as it will be equal in visual 'weight'.

In environmental design, space may refer to the physical characteristics of a room, building or other environment. Consideration is applied to the arrangement of objects within a given space and, as with two-dimensional applications, the organisation of a space might direct and control the way a user interacts with their environment. The arrangement of a space can also control user behaviour; for example, the thoughtful design of space can control the flow and direction of passengers through a busy airport terminal. Ultimately, the function or purpose of the environment will determine the design and arrangement of space.

▲ Balance can be created with 'white' or empty space.

RESPONSIVE ARCHITECTURE

FYI

'Recently, an emergent discipline called "responsive architecture" has begun asking how physical spaces can respond to the presence of people passing through them. Through a combination of embedded robotics and tensile materials, architects are experimenting with art installations and wall structures that bend, flex, and expand as crowds approach them. Motion sensors can be paired with climate control systems to adjust a room's temperature and ambient lighting as it fills with people.'

Source: Ethan Marcotte, 'Responsive Web Design' 2010

ISBN 9780170349994

With practice, it is possible to create striking and memorable designs using space. In combination with design principles such as hierarchy, scale and proportion, the organisation of space can influence the response of a user and achieve a variety of different design purposes.

▲ This illustration of a home interior is used to promote a new housing development. The representation of a dwelling in this way allows the viewers to imagine themselves within the environment and can assist in marketing a space that does not yet exist. In environmental design areas, the representation of space in this way enables the end user to visualise a three-dimensional space more easily than they might from a plan.

KEY WORD ~ SPACE

abc

Space can be:
+ defined
+ clean
+ expansive
+ confined
+ intuitive and responsive
+ delineated
+ minimalist
+ contemporary
+ ordered
+ flowing
+ inviting
+ targeted.

Space can be used to:
+ guide the viewer/user
+ control user/viewer behaviour
+ create a mood
+ emphasise important visual information.

TEXTURE

Texture assists in visually describing the detail of an object, and helps us to understand what an object is made from. It can also help us to recognise and understand the features of the environment in which an object exists.

▲ The textural details of these fashion designs are indicated through marker and pencil rendering techniques. The use of a texture board and the inclusion of tone helps to illustrate the characteristics of the fabrics.

Texture offers considerable challenges in illustration; it is challenging to visually represent features that we usually recognise through our sense of touch. Representing texture on a two-dimensional surface takes some practice and acute observational skills. The key to depicting texture effectively is to take into account how tonal or colour variations can affect the appearance of texture. Observe how the metallic surface of a can of soft drink, for instance, reflects light across its surface. This alters the colour of certain parts of the container in a way that emphasises its texture as well as helping to define its cylindrical form.

iStockphoto/Kirsty Pargeter

ISBN 9780170349994

Importantly, texture communicates information about the characteristics of objects. Rendered architectural illustrations might depict stone or brickwork, reflective glass surfaces and the foliage of surrounding trees. Such detail communicates information that would not be available in two-dimensional plans or three-dimensional line drawings alone.

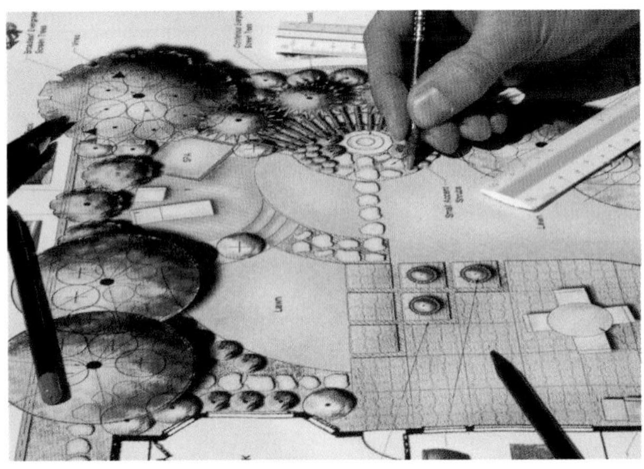

iStockphoto/Scott Feuer

Texture can be applied in logo design to reflect the nature of a company or service. Texture can be highly visually suggestive and has the power to communicate complex surface information.

◄ This logo uses a metallic, chrome texture to imply a progressive, corporate feel.

The use of texture is increasingly common in logo design as companies ensure that their corporate identity is effective in digital as well as printed formats. Prior to the Internet, very few gradients or textures were used in logo designs as they rarely reproduced well when faxed or photocopied.

TEXTURE PILOT

Texture Pilot is a visual reference library of images designed to assist designers and artists in reproducing textures and materials. The vast collection of images is arranged by detailed categories.

Access all weblinks directly at http://nsg.nelsonnet.com.au.

The texture of a product or its packaging can influence our attitude towards it as consumers. Increasingly, the pursuit of 'greener' product design and packaging has led to an increase in products packaged in recycled materials.

iStockphoto/Tom Nulens

▲ These have 'natural' textures that appeal to or encourage an environmentally aware audience.

Textures can appeal to us on a subconscious level. Humans enjoy the sense of touch; the appeal of a fluffy kitten or smooth velvet invites us to touch a surface. A soft texture may imply tenderness or luxury. Alternatively, harsh textures such as jagged edges, barbs or thorns might repel us and may even imply danger.

KEY WORD ~ TEXTURE

Texture can be:
+ smooth
+ glossy
+ matt
+ uneven
+ coarse
+ tactile
+ reflective
+ dull
+ metallic.

Texture can be used to:
+ contrast
+ emphasise
+ enhance and describe form
+ create pattern
+ communicate information about materials.

ISBN 9780170349994

Lowe GGK (Source: DEBRA Austria)

Lowe GGK (Source: DEBRA Austria)

Lowe GGK (Source: DEBRA Austria)

▲ The unexpected juxtaposition of images and textures in these images is confronting but conveys the nature of epidermolysis bullosa, a painful skin disorder. The viewer's response is to imagine the feeling of suffering experienced by those who have this genetic condition.

TONE

Tone, when applied effectively, can enhance the appearance of an object, describe three-dimensional form and provide information about the surface textures of an object.

When discussing tone, you may come across the term 'tonal scale' which refers to a series of tonal values or levels between black and white. Tone describes the play of light and shadow on an object, defining its form or shape.

Johanna Schreiner

In applying tone to an image, the light source must be taken into account. Ordinarily there will be one primary source of light, which will define the highlight and shadow areas on an object and direct the application of tone.

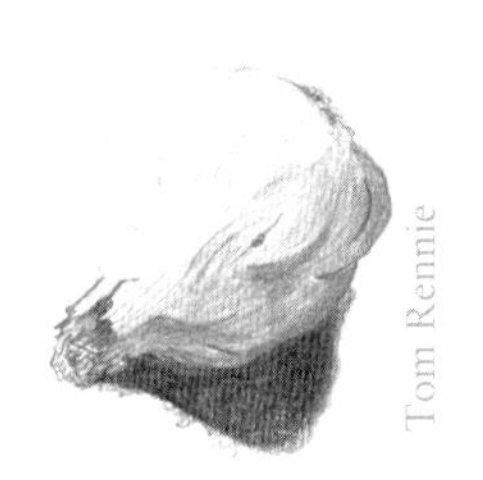
Tom Rennie

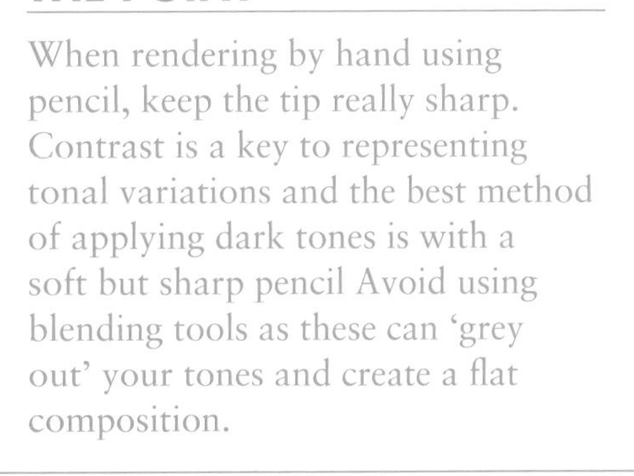
THE POINT

When rendering by hand using pencil, keep the tip really sharp. Contrast is a key to representing tonal variations and the best method of applying dark tones is with a soft but sharp pencil Avoid using blending tools as these can 'grey out' your tones and create a flat composition.

In illustrations, the application of tone can influence the mood being conveyed. Cartoonists and illustrators often use tone to emphasise a theme in an illustration. Used for emphasis, tone can create a mood that is dark – or, alternatively, a sense of lightness. Tone can be applied by a range of media and can be created through the application of different rendering techniques such as crosshatching, pencil or marker rendering and dot rendering.

KEY WORD ~ TONE

Tone can be:
+ dark
+ light
+ medium
+ subtle
+ dramatic
+ muted
+ soft
+ harsh
+ defined.

Tone can be used to:
+ render
+ contrast
+ model
+ highlight
+ emphasise
+ define
+ enhance
+ create form
+ describe texture
+ define structure.

Pinnacle South LLC

The Family Place

▲ Both of the logos above use tone to suggest depth and structure in a two-dimensional context.

8.2 ADDITIONAL DESIGN ELEMENTS

There are many additional design elements to those identified in the Queensland syllabus. Two of these that are important to build an understanding of are Type and Point. Typography is a fundamental component of many visual communications and is applied in all design areas. Point may be small by its visual nature but it is an important element that assists in communicating visual information.

POINT

Represented as a dot or other small shape, point is one of the simplest design elements and is often used as an indicator to determine or define features on a map or document. In mapping (cartography), point is used to indicate points of interest and geographical features. In diagrams or documents, point may be used as part of a bulleted list to identify a series of concepts or statements.

When used as a tonal element in dot rendering, point can convey the texture and characteristics of an object and express tonal variations effectively.

Izzie Klingels

▲ UK illustrator Izzie Klingels' incredibly detailed dot renderings appear in magazines, on music graphics and posters.

VISUAL COMPLEXITY

This extraordinary website showcases the work of designers who strive to visually represent complex concepts. For many of them the use of line is integral to depicting detailed and involved visual information. Well worth a visit!

Access all weblinks directly at http://nsg.nelsonnet.com.au.

As an element of a pattern, the repetitive use of point can create designs ranging from simple shapes that define key points within a paragraph of text to complex arrangements of dots that create an image in the style of a halftone screen or pointillist composition.

ISBN 9780170349994

▲ These fashion illustrations use point to create pattern. The use of small repetitive dots and squares suggests the qualities of fabric and decoration.

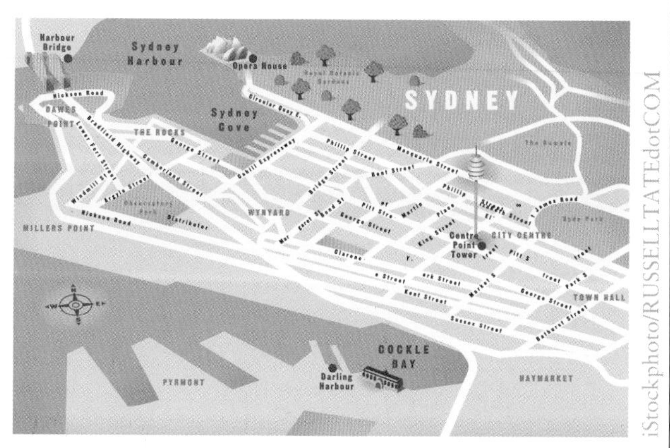

▲ Point is often used in mapping to indicate location and points of interest. This map of the Sydney Harbour area uses point to draw attention to significant sights.

KEY WORD ~ POINT

abc

Point can be used to:	Point can create:
+ differentiate	+ a map location
+ define	+ a bulleted list
+ separate	+ dot points
+ act as an indicator	+ texture
+ identify	+ pattern.
+ emphasise	
+ highlight.	

TYPE

Type is much more than just text. Type can be used as a decorative device, as the dominant visual element in a composition or as informative text. Type often forms the dominant element in contemporary graphic design. Logos that use type as their primary design device – such as Google, ANZ, IBM – have distinctive and instantly recognisable identities.

Google

▲ The Google logo is one of the most recognised logos in the world and although it is often manipulated, it retains its fundamental characteristics of colour and six letterforms. The logo was designed by Ruth Kedar in 1999. Google logo reproduced with permission.

▲ Many corporations use type alone to identify their organisations. Many typefaces are described as having their own personality, derived from their appearance, historical background and common applications. In branding a business or organisation, the selection of an appropriate typeface can be a challenging and complex process. ANZ logo reproduced with permission.

Typography communicates in many ways. From a collection of letters that form recognisable words to the arrangement of type to convey a visual message, typography is fundamental to communication. An understanding of typography, along with the language, rules and conventions that inform how it is used effectively, is essential for students of design. Of course, there are uses of type that defy all the rules yet succeed all the same; type in knowledgeable and expert hands is what makes the study of typography so fascinating.

FONT OR TYPEFACE?

Typography is defined as the art of working with type.

Typeface refers to a specific style or 'family' of type (face is also used).

Font has essentially the same meaning as typeface, and was commonly used to describe a typeface of a specific point size. Font has now became a standard term used in electronic publishing often interchangeably with 'typeface'.

Letterform or **glyph** can refer to individual type forms including symbols, numerals and icons.

There exists a long and rich history behind many of the typefaces we use today. Some type styles that we commonly see used in newspapers (and other publications that use large areas of text) date from the Renaissance. There are many type styles and an Internet search for 'fonts' will provide you with thousands of libraries of vastly different typefaces. In making decisions about the use of type, as with other elements, it is important to understand the purpose and context of type in your design.

Lorem ipsum

When preparing a mock-up of a composition, designers often use 'dummy' text rather than actual text. Dummy text leads to fewer distractions while the effectiveness of the design of a visual communication is being assessed. The most commonly used dummy text is known as 'Lorem ipsum'. This is an excerpt in Latin from a book on ethics written in 45 BCE by Cicero. It begins: 'Lorem ipsum dolor sit amet, consectetur adipisicing elit, sed do eiusmod tempor incididunt ut labore et dolore ...'

LOREM IPSUM GENERATOR

When you need to use a slab of Lorem Ipsum dummy text, visit this site to generate the specific number of paragraphs that you need.

Access all weblinks directly at http://nsg.nelsonnet.com.au.

Type essentials

Family

Typefaces or fonts exist in families. A font family might include bold, narrow, extended and italicised versions of the same typeface. Although it is possible to make many typefaces bold or italic by electronic means on the computer, using the italic or bold version from the font family reduces the risk of this 'dropping out' during the printing process.

WE ARE FAMILY!

Fact: Using too many fonts in a composition looks terrible! Try sticking to one font family and experiment with the different styles, such as bold, light and medium versions of the same family. If you must use other fonts as well, try to use no more than three in your composition. Look to the font family first and then decide if you need another font as well.

ISBN 9780170349994

Style

Some of the most common type styles are:

+ serif
+ sans serif
+ slab serif
+ script
+ decorative/graphic
+ handwritten.

Serif

A serif typeface has an extra mark at the end of the vertical and horizontal strokes of the main type. These additional features are known as serifs. Serif typefaces can improve the readability of blocks of text by leading the eye along the line of type, and are therefore used in books, newspapers and other slabs of text.

There are a number of serif styles including bracketed and non-bracketed serifs, slab serif, slur serif, wedge serif, hairline serif and rounded serif.

Bracket serif

Unbracketed serif

Slab serif

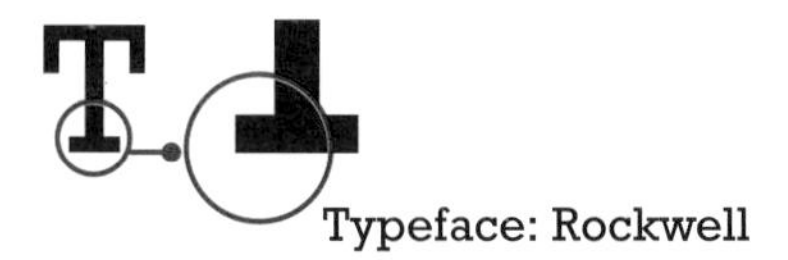

Sans serif

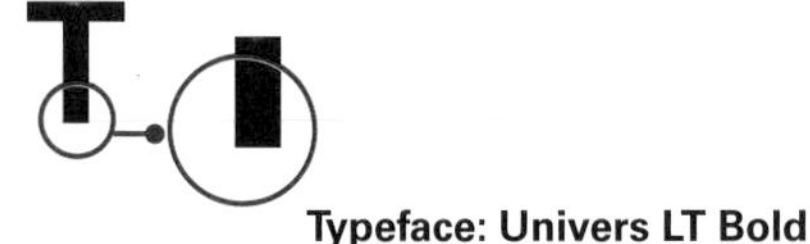

▲ Common types of serifs

When used as the dominant type in a composition, serif typefaces can sometimes appear to be traditional and conservative but do allow for easy recognition and reading.

Common examples of serif typefaces are Times New Roman, Palatino and Garamond.

Serif
Highly legible, classic and recognisable.
Use for body text as serif fonts are easy to read.
Contrasts with, and complements, sans serif type.
Commonly used in newspapers and books.

Sans serif

Sans serif typefaces lack the decorative features of serif type – 'sans' is a French word meaning 'without' – and can create a contemporary and non-traditional appearance. Sans serif typefaces can be harder for the eye to cope with in large areas of text, but work very effectively in smaller paragraphs and large headings. The simplicity of sans serif type often makes it suitable for large-scale applications such as billboards, or very small text such as footnotes in a document. Common sans serif typefaces are Arial, Futura and Helvetica.

Sans serif
Contemporary and highly legible at large sizes.
Easy to read on screen but less so in body text.
Contrasts with, and complements, serif type.
Avoid using for large blocks of text, e.g. novels.

Slab serif

Originating in the early 19th century, slab or block serif was used on promotional posters, theatrical flyers, brochures and billboard advertising. Slab serif typefaces have a geometric serif detail and can appear to be striking and solid in appearance. Traditional typewriter typefaces used a slab serif type. Examples still in use today are Courier and Rockwell.

Slab serif
Strong, solid and sturdy.
Used when legibility rather than beauty is required.
Common slab serifs include typewriter fonts.
Sometimes known as 'Egyptian' typefaces.

ISBN 9780170349994

Script

Developed from handwriting and maintaining the visual connections or 'flow' between individual letters, script typefaces imitate traditional pen handwriting. Script typefaces, such as Edwardian or Engravers Script, are often used to create a sense of history or tradition within a composition.

Script
Traditional, elegant and refined.
Use for headings, contrast and special emphasis.
Not suitable for large areas of text.

Decorative or Graphic

Decorative typefaces, also known as graphic type are often used to create visual interest and they can add humour to a composition, but they are not suitable for blocks of text. Often used in promotional advertising material, decorative type might reinforce the meaning of the type with visual elements. For example, 'Red Hot Sale' on a brochure might include flame or a melting typeface. Decorative type can grab attention or, in the case of traditional illuminated manuscripts, add beauty to a document, but should be used conservatively.

Decorative
Use with care and consideration.
Can be eye-catching and have visual impact.
Best used for titles and not body text.

Handwritten

An increasingly popular type style used in contemporary design is handwriting. The development of less structured typefaces that contain characteristics reminiscent of handwriting has become common. Like decorative type, these should only be used in an appropriate context but can add a highly contemporary edge to compositions.

Handwritten
Highly contemporary, edgy and individual.
Best used for short paragraphs or titles.
Not suitable for large areas of body text.

FORTHELOVEOFTYPE. BLOGSPOT.COM.AU

Australian typographer Gemma O'Brien is widely recognised for her handwritten type and has been known to cover her body in her own type designs. Many of Gemma's designs can be seen in popular advertising campaigns, as well as her regularly updated blog featuring personal projects and professional design briefs. Gemma uses freehand drawing to develop and refine her designs, creating engaging and whimsical work.

The language of type

When it comes to using type in practice, there are some important conventions to follow. No two typefaces are identical so it is essential to know what you are working with. There are many factors to take into account when selecting and using typefaces in your design work.

Baseline

The baseline is the imaginary line that a typeface sits upon. Some letters, such as the O in certain typefaces, may sit slightly below the baseline. When a designer needs to adjust the position of letterforms above or below the baseline, they create what is known as a 'baseline shift'.

Body text

Body text refers to the main areas of text in a document. Body text may also be called a text block. The selection of a typeface for body text is crucial and entirely defined by the context of the design. For example, newspapers and magazines often use serif type for body text as it is considered to be easier to read.

ISBN 9780170349994

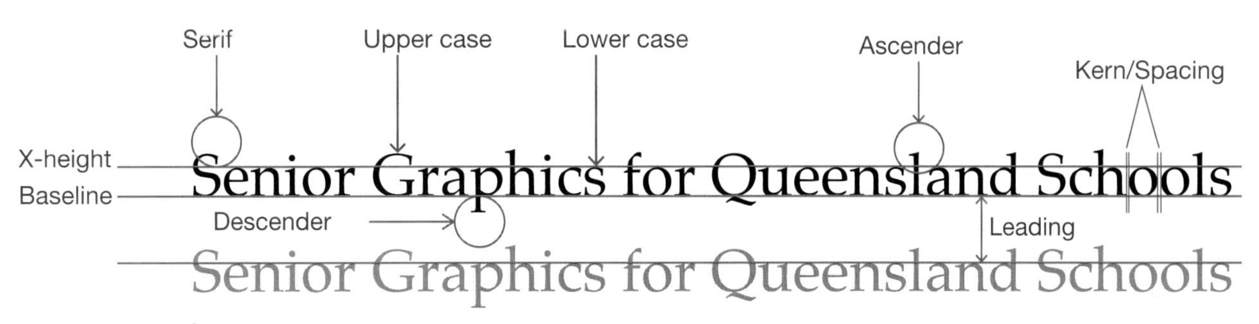

▲ Key typographic terms

Case

Upper- (majuscule) and lower- (minuscule) case letters are named as such because printers using metal type kept them in the upper and lower type boxes or cases. Upper-case letters are less legible when used in body text. Combinations of upper- and lower-case letters are often known as sentence case.

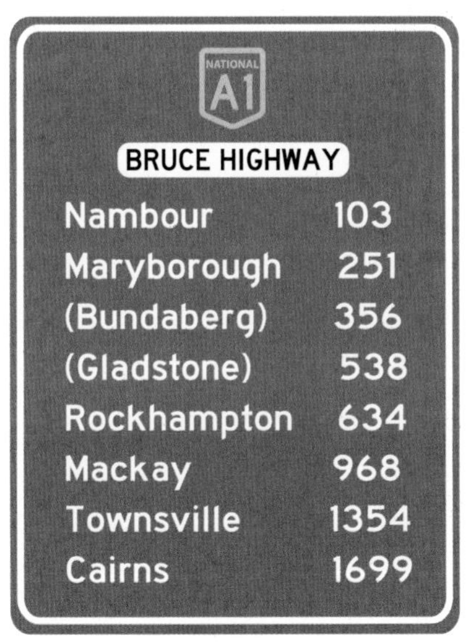

▲ Upper-case and lower-case letters are used on directional Australian road signage. The combination of letterforms is easily recognisable at speed. The typeface used in Australia is called Highway Gothic.

Kerning

Kerning refers to the space between individual letterforms. Some letterforms need to have the space adjusted when they are used together; for example, a T and L used together have larger spacing than an M and E. The type designer kerns most commercial typefaces but design software programs allow some adjustment to kerning to improve visual appearance if required.

Leading

Leading (pronounced ledding) is the distance between two lines of type. The term is derived from the strips of lead that were placed between lines of type in traditional typesetting. Leading directly affects the legibility of type and is usually set so that the eye flows easily from one line to the next. Leading is often set automatically in computer software but can be manipulated depending on the context.

Spacing or tracking

Spacing and tracking refer to the distance between all letters in a sample of text. Normal tracking leaves the spacing as the type designer intended. Negative tracking moves letterforms closer together, and positive (or open) tracking moves them apart.

Tracking
Tracking
T r a c k i n g

Type size

Points are the units of measurement used in typography. One point is 1/72 of an inch or 0.352 millimetres. Point refers to the height of the type block rather than the letter itself.

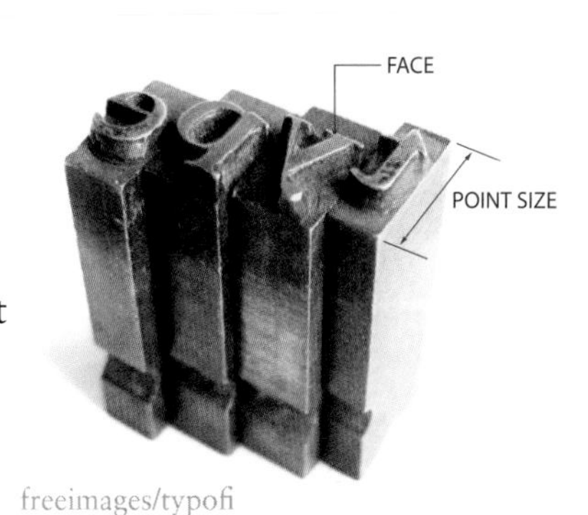

freeimages/typofi

ISBN 9780170349994

Leading is the space between lines of type.
Leading affects the legibility of text.

Type without leading or with leading that is too
close can interfere with the legibility of the text.
16 pt Myriad Pro with 12 pt leading

As can leading that is too far apart as the flow of the

text may be too difficult for the reader to follow.
16 pt Myriad Pro with 44 pt leading

Most auto leading in computer software is set 1–2 points above the point size
For example, this sentence is set in 16 point type with 18 point leading.

x-height

The x-height of individual typefaces varies widely. Here are three similar serif faces that have very different x-heights.

Type Type Type

▲ Garamond, Times New Roman and Georgia, all set at the same point size. Note that all have different x-heights. This variation should be taken into account when working with multiple typefaces in a composition.

Legibility

The purpose of typography is to communicate language. Legibility is extremely important. Type is often used for visual effect but its main purpose is to be read.

There are some effective 'rules of thumb' when considering legibility. One is to avoid using more than three typeface families in a design. Too many different typefaces can distract from the meaning of a visual message and can make a design product difficult to read.

The selection of typeface is of primary importance; the face should suit the context of the design product. It would not be suitable, for example, to set a formal document such as a financial institution's annual report in a graphic typeface such as Comic Sans. Similarly, to set an invitation to the opening of a children's play centre in a formal script would not suit the style or context of the event.

The use of kerning, tracking and leading are important factors in creating legible type. Especially important in body text, the distance between letters and lines of type will influence how easily a reader can follow the flow of words. Headlines and titles will often be set with greater spacing and leading to create impact and draw attention to the type.

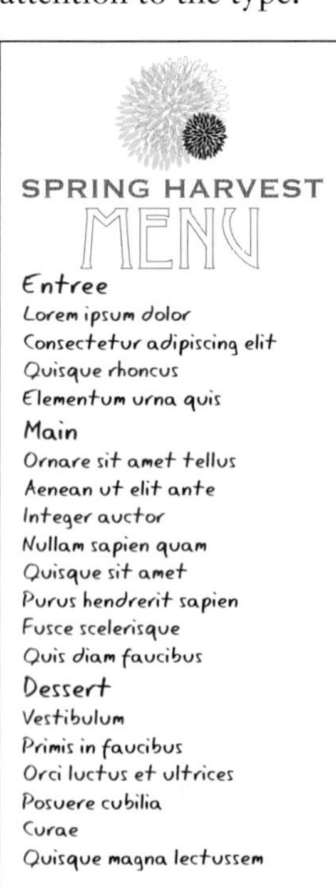

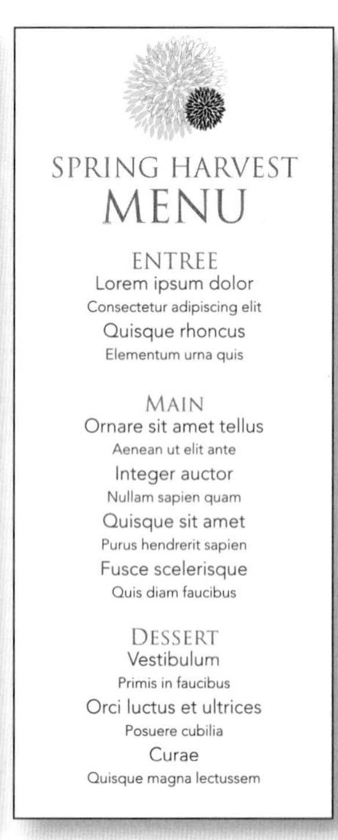

▲ Legibility. Same menu, same text, but note the difference between the two designs. The design on the right uses two typefaces (Trajan Pro and Avenir) only; adjustments to alignment, leading and point size increase its legibility as well as its aesthetic appeal.

ISBN 9780170349994

Lining and non-lining numerals

Numerals can be identified as upper case and lower case. Lining, or upper-case numerals, adhere to the baseline. Non-lining, or lower-case numerals, feature descenders that drop below the baseline. Not all typeface families carry both lining and non-lining numerals.

1234567890

1234567890

TYPOGRAPHIC POSTERS

An online showcase of contemporary typographic design, this website features designers from across the globe.

Access all weblinks directly at http://nsg.nelsonnet.com.au.

8.3 PRINCIPLES OF DESIGN

ALIGNMENT

Alignment is the placement of elements in relation to one another. When using word-processing software, you may have used the text alignment tools, which enable you to justify (align) your text to the left, right or centre of your page. These tools can give your text and images a sense of order and organisation that keeps the message clear. Alignment tools exist in all graphics software packages.

Effective use of alignment demonstrates that your composition is organised and implies that elements have relationships with other elements and images. Establishing a relationship between elements helps to lead a viewer's eye to – and through – your design. Elements and images placed without organisation will appear lost and unrelated to the composition.

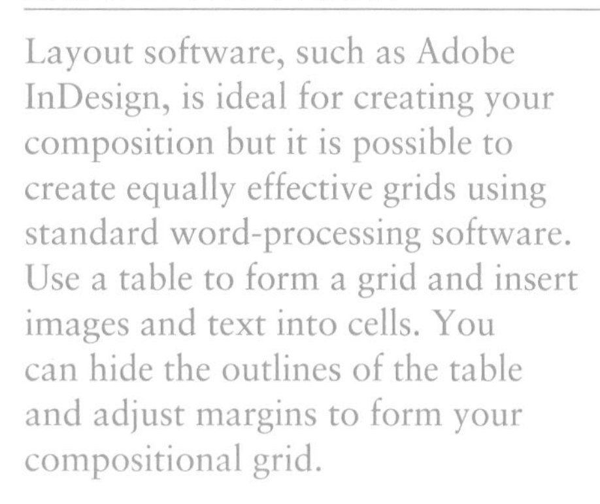

LAYOUT SOFTWARE

Layout software, such as Adobe InDesign, is ideal for creating your composition but it is possible to create equally effective grids using standard word-processing software. Use a table to form a grid and insert images and text into cells. You can hide the outlines of the table and adjust margins to form your compositional grid.

In graphic design the considered alignment of type, imagery and other visual elements form effective compositions. The various visual components of a design are often managed within a visual 'grid'. A grid is an invisible structure that supports the layout of print and digital content. A designer will use a grid to create hierarchy

Left alignment

Centre alignment

Right alignment

ISBN 9780170349994

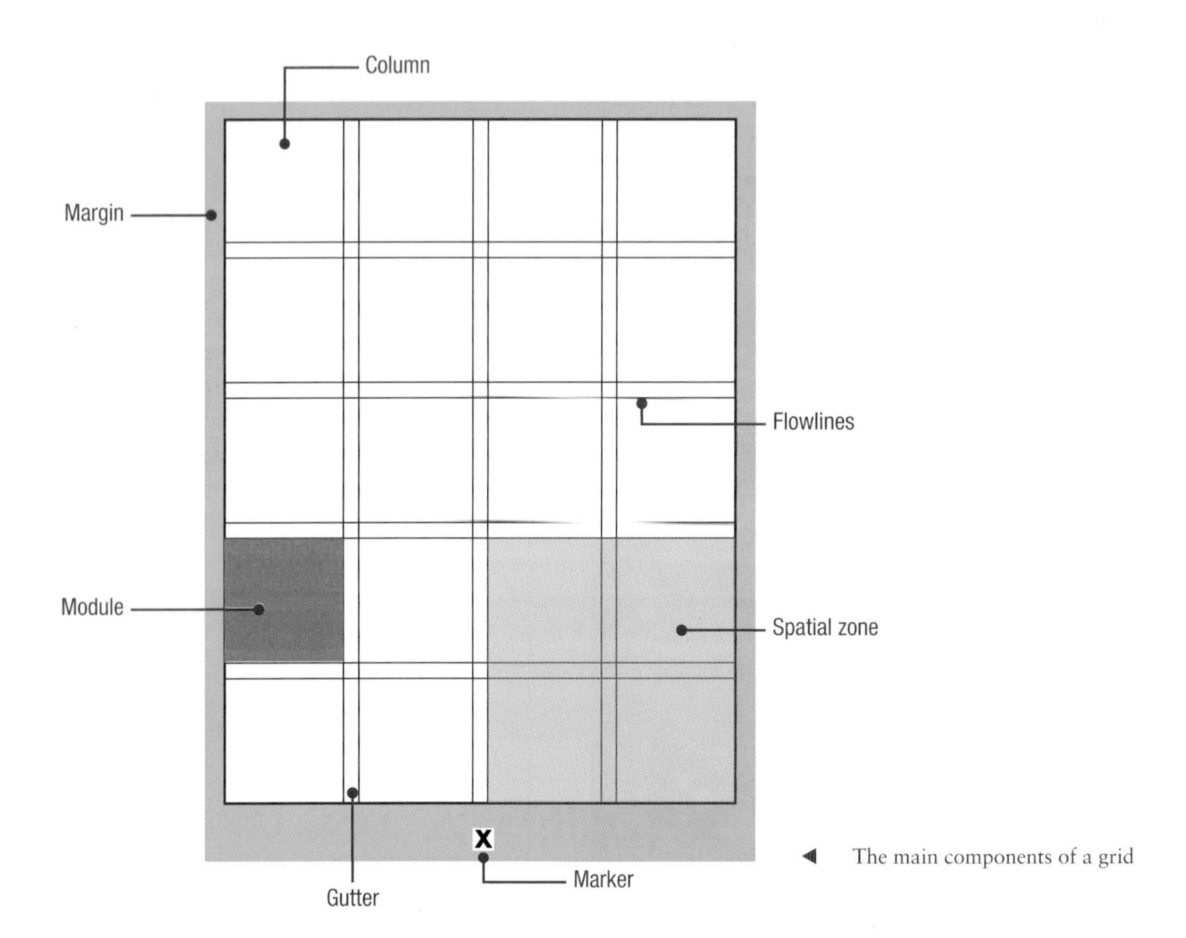

The main components of a grid

within a design composition and to delineate the placement of text and image. Grids can be seen in newspapers, magazines, web pages and even mobile devices. The grid can be a powerful tool when used well; it can draw the eye through a composition and create strong visual relationships between type and imagery.

Columns

Columns provide a sense of order in a design. They are vertical 'containers' that hold text and visual elements. The width and number of columns in a composition is established in the planning stages.

Flowlines

These are the horizontal grid lines that define areas for the placement of type and images. The combination of column and flowline creates the modules of the grid.

Gutter

The gutter refers to the spacing between columns. It is also used to describe the space between pages, near the binding, in a book or magazine.

Margins

Margins are the white space that surrounds a composition and separates the design/ artwork from the edge of the composition. Printed compositions allow enough space in the margins for the page to be cut (trimmed).

Marker

A marker is a repeating element that assists navigation on a page. It may be a page number, footer or even an icon.

Modules

Modules are the grid areas defined by the columns and flowlines. These are the spaces that may contain text or images. Multiple modules create spatial zones.

ISBN 9780170349994

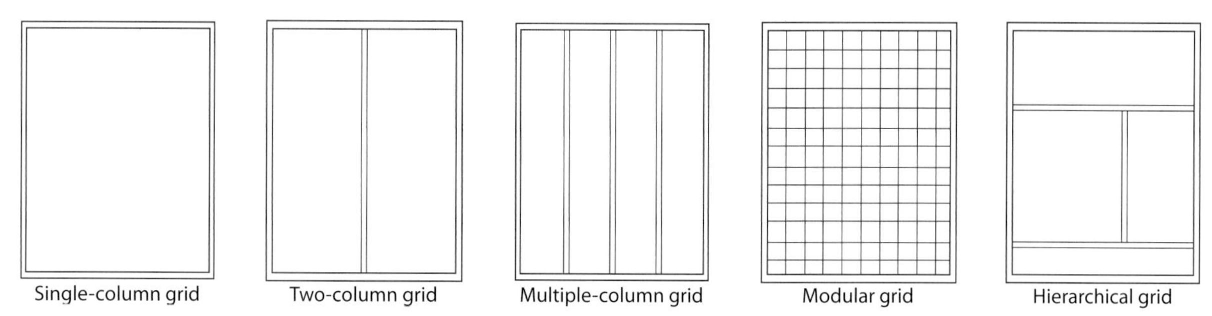

▲ Some common grid formats. The grid lines enable visual elements to be aligned effectively and assist in drawing the eye of the viewer to key information.

Interestingly, many striking designs have been created by designers 'breaking' the grid. This involves challenging the 'rules' of composition while maintaining visual balance and harmony.

Non grid-based alignment

Generally, alignment is defined in terms of a grid but more complex alignment can be achieved using other visual pathways such as angles and circles. However, when visual objects are aligned along a diagonal path they should be placed on a relative angle of more than 30 degrees as anything less is too subtle for the viewer to differentiate and may simply be seen as an unbalanced and unappealing design. The same is true for alignment along a circular path; ensure that elements are aligned clearly and do not appear randomly arranged. The use of additional design elements and principles can assist in linking visual elements together and highlighting the path a viewer needs to follow; for example, colour and proximity.

BALANCE

Balance in design establishes harmony in a composition, and harmony creates visual appeal. Whether we realise it or not, we like to see balanced compositions. It has been suggested that the appeal of balance reflects the equilibrium of the human body. Designs that are unbalanced can lack emphasis and visual appeal and may ultimately discourage us from looking, using or consuming.

In establishing harmony, balance helps to create successful designs, but don't be fooled into

▲ Web page design. The left image shows the modular grid used by the designer to create the final web page. It is possible to see how some elements use single modules while others use larger spatial zones.

thinking that harmonious design means quiet, dull and boring – quite the contrary!

Balance can be symmetrical or asymmetrical, and each style has appropriate applications. The purpose, audience and context of your design will determine the style to use.

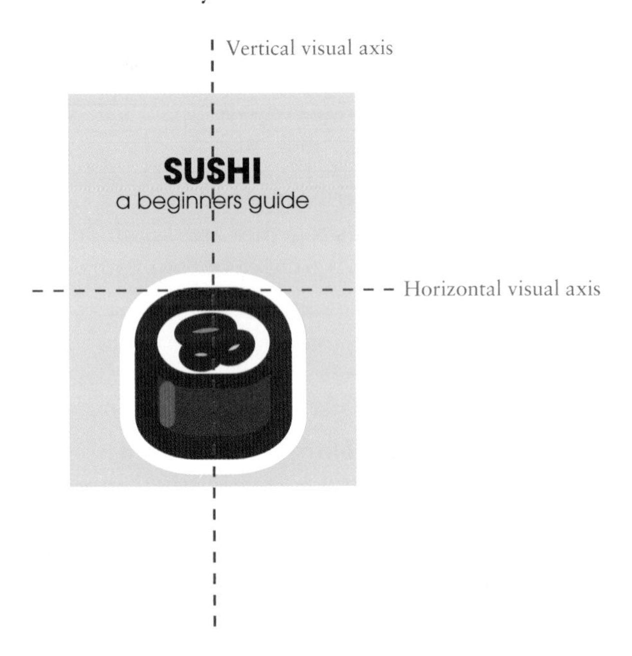

Imagine a composition that has been divided through the centre by an invisible horizontal line (axis) and an invisible vertical line. Both axes provide reference points for creating balanced and visually harmonious compositions. Although we cannot see the axes, they provide a structure that can be used to assist in planning an effective design.

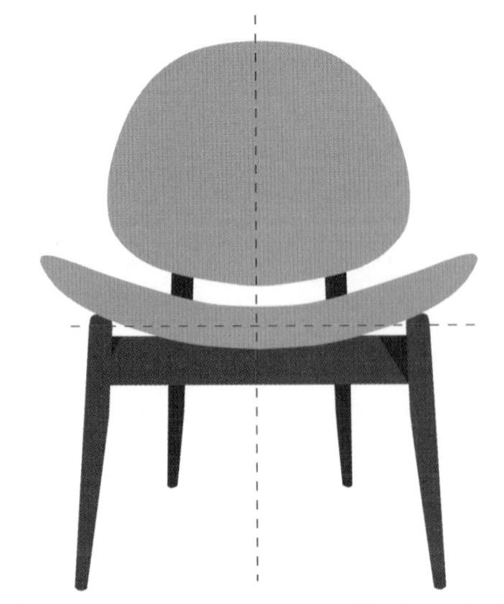

▲ Balance in reference to a vertical and horizontal axis is equally applicable to three-dimensional design.

Symmetrical balance

A composition with symmetrical balance mirrors the elements on opposite sides of the visual axis – from one side to the other.

▲ Symmetrical balance

Symmetrical composition is seen to be stable, static and passive. Such composition has a sense of regularity or conformity, which makes it suitable for a purpose that requires such characteristics. Symmetrical balance can be perceived as formal and organised in style, but it can also achieve a sense of unity between design elements, creating order and even a sense of beauty.

> **FYI**
>
> Recent studies confirm what artists have for centuries believed – that our perception of beauty in the human face is based almost entirely on symmetry. The more symmetrical a person's facial features, the more likely they are to be perceived as beautiful. Very few faces are perfectly symmetrical.

When approaching a symmetrical composition, it is possible to over-emphasise the centre and align elements in a restricted manner. It is important to be aware of the entire space you are working with.

The placement of elements in relation to an axis leads the viewer's eye into the composition. Creating a harmonious balance means that left-to-right balance and top-to-bottom balance are equally important in order to keep the viewer focused on the composition.

ISBN 9780170349994

▲ Many images appear more stable if the bottom seems slightly heavier. If the top appears too heavy, the composition may look unstable.

Sometimes our own sense of balance is refined enough to sense when the balance in a composition is wrong. With practice you will recognise when a composition is balanced and when it is not.

Asymmetrical balance

Based around a central visual axis, asymmetrical balance is characterised by an arrangement of elements that is not mirrored or equal in appearance. Asymmetrically balanced compositions appear to be more dynamic than symmetrical compositions because the placement of elements creates a sense of dynamic energy.

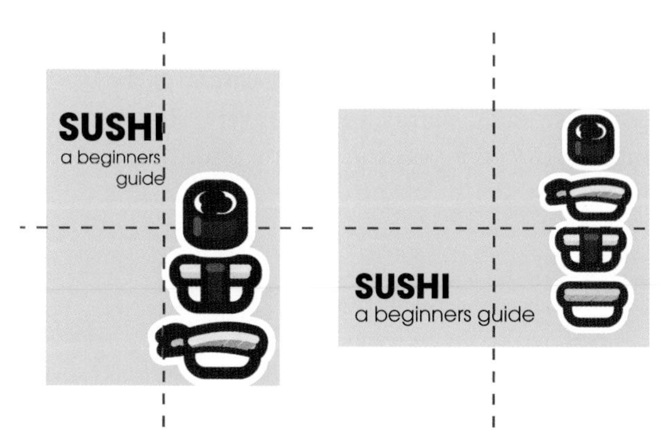

▲ Asymmetrical balance

The important thing to remember with asymmetrical balance is that the composition is still balanced! Balance is created by manipulating elements and does not have to fit the traditional 'left, right and centre' alignment approach (see Alignment on page 109). Asymmetrically balanced compositions can be created by repetition of elements and images, and creative use of scale, cropping and type.

It is even possible to use what appears to be nothing at all! The use of white space or areas that do not contain important visual details can lead to eye-catching results (see Space on page 99).

Asymmetry is often found in two-dimensional and three-dimensional designs that challenge, provoke and inspire us. Experimental designs challenge our perceptions of balance and harmony, stimulate debate and discussion, and force us to reassess our values and comfort zone.

CONSISTENCY

Consistency is important in a range of design areas as it enables users to recognise similarities in aspects such as the branding of a business or the functionality of a product. Consistency helps users understand that a system is in place, that they are within a specific environment, or that they are purchasing from a recognised and trusted brand.

There are four commonly recognised types of design consistency: aesthetic, functional, internal and external.

Aesthetic consistency

Aesthetic consistency occurs when the same visual elements are used to create a style or appearance and can most commonly be seen in identity design. Companies often have strict control over the application of their identity to ensure that it remains consistent across many areas. An identity design often involves the creation of a 'Style Guide', which outlines in detail the appearance of corporate logos and signage. For example, Facebook provides specific information about the use of their logo on web and print products including requesting that the 'f' logo not be changed in any way.

◀ Designers cannot change the font, style or colour of the Facebook logo. The logo remains consistent across all applications. Facebook logo reproduced with permission.

ISBN 9780170349994

Functional consistency

Functional consistency relates to consistency in the actions of products. For example, the user easily recognises the 'POWER ON' symbol used on most electronic devices. Functional consistency means that the end user does not have to re-learn simple functions to otherwise familiar devices. Products that conform to existing and recognised functionality are easier to use and subsequently preferred by the end user.

◄ The Power On symbol is consistently applied to many electronic devices and therefore easily recognised by the user, who does not require any new learning to use the device.

Internal consistency

Internal consistency refers to a consistent design approach within a system, space, organisation or location. For example, the signage used within Brisbane Airport uses a consistent colour palette and typeface (Zurich BT). Consistency in signage is controlled by the Brisbane Airport Corporation, which provides contractors and designers with a comprehensive signage manual outlining the application of text and colour across all applications. Ensuring that all signage is visually linked assists airport users in navigating through the terminals, car parks and surrounds.

External consistency

External consistency is harder to control as it refers to the application of consistent visuals across non-related systems, organisations, spaces and locations. For example, on entering an unfamiliar building, the location of elevator controls may differ from other premises and be difficult to locate. Australian Standards control the consistency of appearance of some systems such as fire alarms and emergency stop controls but not all systems are required to adhere to a standard. Even when the appearance of a control differs, it is the role of the designer to ensure that users can navigate, locate and operate systems successfully.

CONTRAST

Sometimes, conflict is a good thing!

Contrast is created when two very different elements are used together for visual effect. Contrast can create conflict between elements – light versus dark, bold versus fine – which leads to a visually dramatic composition.

Contrast creates a tension between elements. In fiction and film, tension heightens the interest of the reader or viewer – an increased level of tension encourages a sense of anticipation by raising the heart rate and stirring further interest in the storyline. Although visual contrast may not always make the heart race faster, it attracts attention and encourages interest in the content of the composition. Contrast, used bravely, stimulates interest in a composition that might otherwise go unnoticed.

DR4 - LIFT TOWER DIRECTIONAL

↖ Gates
82-87

dotdash

▲ Brisbane Airport Corporation (BAC) created a comprehensive manual that outlines the requirements of all airport signage. The manual ensures that there is visual consistency throughout the airport.

ISBN 9780170349994

iStockphoto/jpa1999

▲ Bold contrasts between light and dark in this illustration convey a sense of drama and menace.

Contrast is created in many different ways and with many different elements. The key to effective use of contrast is to use it boldly. Don't be afraid to take risks with contrasting elements.

Juxtaposition

Juxtaposition refers to the placement of two different elements together within a composition in a proximity that suggests a comparison. Otherwise unrelated elements are contrasted to create a strong visual relationship that communicates a message.

▲ Notice how the appearance of this composition becomes immediately more striking when contrasting elements are added.

Iveco Latin America, reproduced with permission

▲ An elephant and a truck are, ordinarily, unrelated elements, but in this vehicle advertisement, the juxtaposition of the two objects serves to emphasise the strength and load capacity of the vehicle.

ISBN 9780170349994

Contrast in type

Bold, sans serif type and script are vastly different in style but are sometimes seen together to contrast two words. A less dramatic version is the use of a bold typeface and a regular typeface from the same typographic family. Contrast is sometimes used at the beginning of a chapter in a book. You may see the title letter in a bolder and larger scale font than is used in the body of the text; this is called a drop cap and establishes the beginning of a chapter or section.

*L*orem ipsum dolor sit amet, consectetur adipiscing elit. Maecenas imperdiet orci vel dolor consectetur vitae aliquet ipsum sodales. Nam venenatis libero at metus tristique in lacinia est rutrum. Phasellus feugiat auctor felis quis bibendum. Cras congue lacus ac ligula accumsan imperdiet. Pellentesque habitant morbi tristique senectus et netus et malesuada fames ac turpis egestas. In blandit, est vel viverra pretium, eros lacus ornare leo, at blandit ligula felis laoreet ligula.

▲ The use of a drop cap in a contrasting script typeface clearly defines the beginning of this section of body text.

Create contrast
within the same type family
(Helvetica Neue LT Bold and Light)

CREATE CONTRAST
using upper and lower case
(Futura Medium)

CREATE CONTRAST
using different type styles
(Univers and Bookman Old Style)

▲ Using only one or two typefaces, a designer can easily create contrast using type. Typeface families are designed to work together but also offer opportunities for contrast.

Contrast in colour

Contrast can cause colours to virtually move around on a page, creating dazzling and dramatic effects. Some combinations of colour can create a discordant result, which may be uncomfortable to view. (Deep blue and deep red combined can have this unpleasant effect on the viewer.) However, as the human eye responds quickly to colourful stimulus, contrast using the element of colour can be a powerful tool. Colours that sit opposite one another on the colour wheel generate the greatest amount of contrast. Contrast can be generated through the application of cool and warm colours, or by one colour used at significantly different levels of intensity.

▲ The use of contrasting colour emphasises the seasonal nature of these greetings. The clever use of contrast creates the metaphorical Christmas tree in both images.

The use of contrasting colours can create optical effects. As mentioned, blue and red used at equal intensity can seem to 'fight' for the most dominant role in a composition, creating the illusion of movement.

Contrast in tone and texture

Contrasting tones can assist in defining the form of objects in rendering. The difference between a dark tone and a very light or white highlight creates a sense of an object in space. Tone provides information about the surface of a form that, without contrast, would appear flat. When applying tone to an object, it is important to use a wide range of tones and to be comfortable about applying black and using white. Use the paper itself as the lightest highlight and use a 6B or even softer pencil to create very dark shadows.

Contrast can be used for dramatic effect when combining textures. In fashion and textile design, it is possible to see clear plastics and soft fabrics incorporated together in some contemporary clothing and accessories. Clashing colours and contrasting fabrics are often used to draw attention to part of a garment.

ISBN 9780170349994

Scott Burrows

▲ Contrasting textures including wood and concrete draw attention to the angles and forms of this Southport Park project by Rothelowman Architects.

Contrast in line

Bold line, fine line, broken line, solid line …

Variations in line thickness provide subtle but effective contrasts. Segments of a text-heavy document can be separated with line, or a heading can be underlined by a bold line. In rendering form, line is often used as a single element or as part of a crosshatched pattern. Lines placed close together can create a dark tone – and when spaced further apart, can appear lighter. A rendered line can contrast with areas that contain no rendering at all to create a greater sense of form.

Contrast in space, shape and form

Contrasting shapes such as a square and a circle can be used as part of an alternating pattern to create visual interest. A pattern that contains variation and particularly strong contrasts is immediately more dominant and noticeable than a pattern that does not. Point can also be used in this way to create noticeable visual variations.

Shape is used to contrast with other elements. Organic shapes can soften strong colours, and geometric shapes can provide a contrasting ground for text and the placement of images.

Geometric forms can be used with organic abstract forms to create contrast in interior and exterior architecture. In the design of residential and commercial buildings, architects often work with landscape designers to create contrast between the natural and constructed environment; for example, the contrast between a gently sloping garden and the geometric lines of a contemporary home can create a dramatic effect.

▲ Defying logic, the large building appears to rest on a soft, red ball. In architectural design, the contrast between forms and materials, textures and colours can create striking and provocative structures.

HARMONY

Imagine that you are listening to a friend sing during a session of Karaoke, although some people have a natural gift for singing their favourite songs, many people sing off tune, off key and with no 'harmony'. That lack of harmony can be painful to listen to and the same goes for visual harmony; a lack of it can be painful to view. The human eye is instinctively attracted to harmonious design and this has an impact on our preferences for products, environments and communication designs.

A design that lacks harmony, like music, becomes 'discordant' and less attractive to the viewer/user. The use of other elements and principles of design create harmonious compositions; for example, the application of balance, whether symmetrical or asymmetrical creates visual harmony.

ISBN 9780170349994

AESTHETIC-USABILITY EFFECT

FYI

The aesthetic-usability effect is a documented phenomenon where users identify attractive (aesthetic) designs as easier to use than less attractive (less aesthetic) designs. Whether the design is, in fact easier to use or not, the user is generally drawn to a more attractive product. Research indicates that consumers have a more positive response to attractive design and even appear to tolerate malfunctions better when a device is more attractive to the eye.

▲ Two business cards for a traditional Hawaiian massage business. Note that the card on the left lacks harmony. The addition of harmonious colours, the use of effective proximity and scale and an appropriate typeface create a harmonious composition.

'Colour harmony' is a term often used in design and refers to the application of colours that work effectively together. Although we might consider harmonious colours to be easy on the eye, the term harmony in this case simply refers to how successfully colour combinations work together. Harmonious colour combinations are more attractive to the eye and can be a means of influencing the interactions and responses of users and consumers.

Supermarket shoppers spend no more than 0.03 seconds looking at a grocery item. Therefore, packaging needs to grab attention in a time frame that is literally the blink of an eye. Use of harmonious and appealing design is a vital marketing tool!

▲ This digital pattern for a webpage is created using analogous colours. The use of cool, harmonious hues of blue and green provide a pleasing visual background to more important information.

HIERARCHY

As we grow up, we become familiar with the concept of hierarchy. If you are a youngest child and were forced to sit in the middle seat in the car, or were the last person to have your opinion heard, you may have been painfully aware of family hierarchy. Hierarchy is the establishment of an order of importance. Just like a 'pecking order' within a family, there is a hierarchy within a composition.

THE NEWS

PRICE OF TEA IN CHINA
SKYROCKETS

Price of a cuppa to soar

Lorem ipsum dolor sit amet, consectetuer adipiscing elit. Nullam malesuada lectus. Praesent enim neque, pellentesque pulvinar, laoreet non, tristique in, tellus. Quisque et neque. Vestibulum ante ipsum primis in faucibus orci luctus et ultrices posuere cubilia Curae; Aliquam eget ipsum in enim luctus rhoncus.

Lorem ipsum dolor sit amet, consectetuer adipiscing elit. Nullam malesuada lectus. Praesent enim neque, pellentesque pulvinar, laoreet non, tristique in, tellus. Quisque et neque. Vestibulum ante ipsum primis in faucibus orci luctus et ultrices posuere cubilia Curae; Aliquam eget ipsum in enim luctus rhoncus. Pellentesque et libero.

ISBN 9780170349994

In producing a composition, it is essential to understand the purpose for which it will be used. This will influence the arrangement of the most important elements. In learning about hierarchy, the front page of a newspaper is a great place to start. Every day, the newspaper will feature a masthead, main headline, subheadings, photographs and text. We quickly recognise that the headline is the most dominant element – the type is usually bold and much larger than the subheadings or body of text. Hierarchy, in this case, is established through scale. The second element in the hierarchy may be the photograph, followed by (or equal to) the masthead, the text and other material.

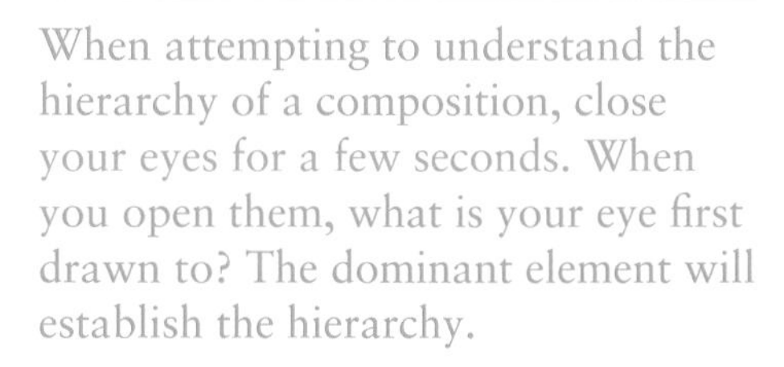

THE EYE HAS IT!

When attempting to understand the hierarchy of a composition, close your eyes for a few seconds. When you open them, what is your eye first drawn to? The dominant element will establish the hierarchy.

Hierarchy can be established in many ways. The use of scale, as shown above, is only one method. Dominant colours, shapes and textures can also draw the eye to the most important aspects of a composition. However, poor use of hierarchy can distract from the message and meaning of a design, so it is essential to control the dominance of elements in a composition.

PROPORTION

Proportion in visual communication is about relationships – relationships between the scale of parts of an object or a composition. Proportion, properly used, creates balance and in turn provides visual harmony – an essential in the creation of pleasing visual communications.

How do objects relate to one another within a composition? A chair that has a small seat and an oversized back support may be uncomfortable and ugly. That doesn't mean that an overly large back support is not a feasible design option; its elements simply need to be in proportion to one another to become an effective design. Likewise, in two-dimensional design, proportion is essential to creating successful designs.

▲ Note the proportional differences between the two images of the teacup and saucer.

▲ No established visual harmony

▲ Hierarchy established through scale

▲ Hierarchy established through scale, pattern and colour

ISBN 9780170349994

In the first image, the proportions of shape and line are incorrect. This interferes with the believability and the attractiveness of the illustration. If the viewer becomes distracted by inconsistencies such as poor proportion, it is likely that they will miss the message of the visual communication.

FYI

During the Renaissance, artists were concerned with the pursuit of visual harmony and beauty in drawing, painting and architecture. This pursuit of aesthetic perfection led to the development of complex geometric systems of proportion. Artists and architects used systems such as the 'golden ratio', 'golden section' or 'divine proportion' devised by the ancient Greeks, which defined a clear visual order as a geometric equation.

Similarly, the 'harmonic ratios' of the Renaissance established a visual balance in objects by establishing that the proportions within the form matched the overall proportion of the form as a whole.

GOLDEN RATIO

This online tool allows the user to calculate the golden ratio of a series of measurements; a handy tool when creating layouts for web and print pages.

Access all weblinks directly at http://nsg.nelsonnet.com.au.

Proportion relates to the comparison of different elements within a composition. Relationships are important in any composition; they indicate which elements relate to one another and lead the eye through information in the most effective manner.

We are naturally attuned to proportion, and intuitively understand when something is 'out of proportion'. Particularly important in observational drawing, proportion helps us depict realistic representations of objects. Renaissance artists, such as Leonardo da Vinci, researched the 'ideal' proportions of the human body, and da Vinci, among others, established a scale to guide artists in the depiction of the idealised classical male figure.

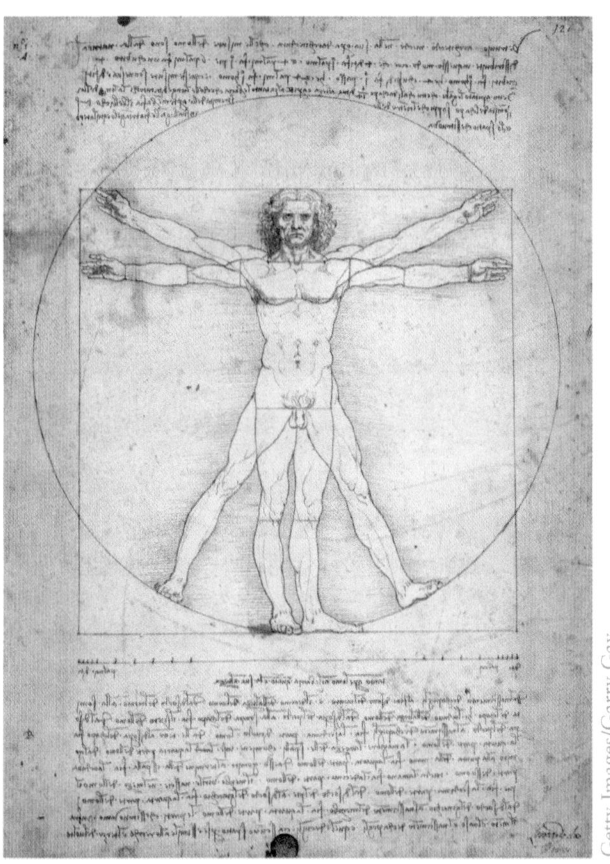

Getty Images/Garry Gay

▲ Leonardo da Vinci created this drawing of average human (male) proportions based on the writing of the ancient Roman architect Vitruvius.

Much later, in the 1940s, French artist and architect Le Corbusier established a scale of proportion between the human body and architectural design called the 'modulor system'.

'Playing' with proportions can lead to innovative and creative design solutions, so don't be afraid to experiment in your own work. Cartoonists commonly use exaggerated differences in scale and proportion to draw attention to a humorous concept.

ISBN 9780170349994

iStockphoto/Magdalena Tworkowska

▲ Exaggerated proportions are used in caricature and cartoons.

Fashion illustrators distort the proportions of the human figure to emphasise features such as legs and necks. These exaggerated proportion elongate the human figure to a significant degree. Traditionally, this distortion has been seen as beneficial to the presentation of design concepts.

When drawing from direct observation, it is essential to establish the relative proportions of physical details in order to produce the most authentic representation. Illustrating a design within a given context assists in establishing realistic proportions. For instance, the inclusion of a human figure in architectural drawings helps a viewer comprehend relative proportions.

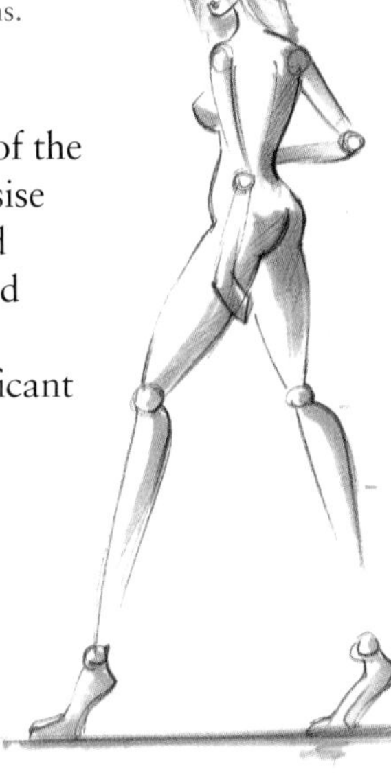

▲ Fashion illustrators often apply the '9 Heads' technique where the human body is divided into the equivalent of nine head lengths. In reality the 'average' human body is approximately seven head lengths long.

▶ The figure provides a clear context for the function of the handbag as well as providing a visual reference for the proportion of the bag to a human figure.

PROXIMITY

GESTALT THEORY

FYI

During the 1920s a group of German psychologists developed theories of visual perception based on how the human eye groups visual elements into a whole. 'Gestalt' is translated as 'whole' and reflects the theory that, when images are arranged in a certain way, our brain groups them together into a cohesive concept. The Gestalt principles most commonly applied in design are Proximity, Similarity, Continuation, Closure and Figure/Ground. The application of Gestalt theory is important when creating designs that are coherent and visually effective.

The principle of proximity is a Gestalt Principle of Perception that states that objects placed close to one another are perceived to be related. Objects that sit close to one another establish a clear visual relationship while objects that are separated or that sit far apart within a space or composition are perceived as having little or no visual relationship.

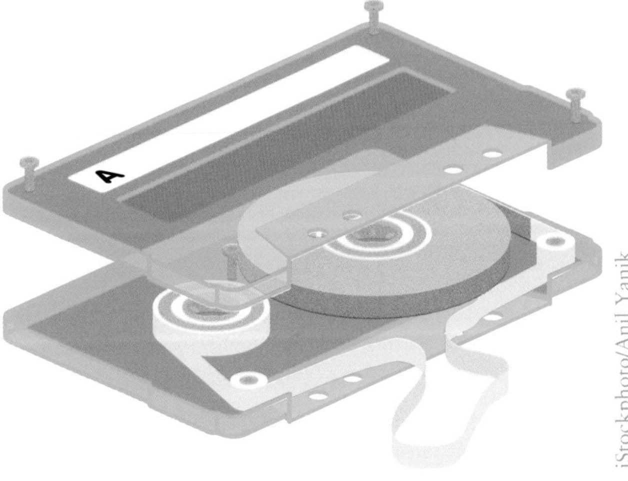

iStockphoto/Anil Yanik

▲ In this exploded view of a cassette tape, the proximity of the parts enables the human eye to visualise the object as a whole. Too far apart and it becomes more difficult to visualise the object in its original state.

ISBN 9780170349994

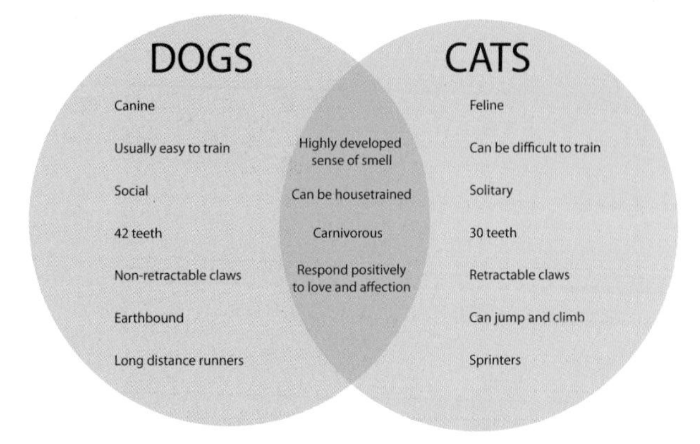

▲ Venn diagrams are a good example of the use of proximity to illustrate sets of overlapping facts, concepts and ideas. Where the circles intersect, concepts are shared.

When creating an effective space, product or composition, it is important to consider the connectedness of elements. For instance, in the design of a car, the proximity of essential functions to the driver, such as the indicator controls, means that the driving experience is logical and efficient.

Shutterstock.com/Yuri Samsonov

▲ Note the proximity of similar functions. All audio controls sit in the one area of the steering wheel while all cruise control functions are located together on the opposite side. Once familiar with the functionality and location of controls, the proximity of related buttons helps the driver to utilise the functions without taking their attention from the road. The buttons also sit in close proximity to the driver's hands.

The grouping of elements can assist in simplifying a design; the viewer may require less time to understand the visual relationships within the design. Conversely, separate elements may encourage a viewer to spend more time within a space or composition. The application of proximity depends entirely on the original purpose of the design itself. Take care to understand the needs of the audience and the purpose of the design when applying the design principle of proximity.

REPETITION

Repetition refers to the use of the same or similar visual elements repeatedly within a composition. Repetition is most commonly seen in the creation of visual pattern. Created from shapes or combinations of shapes, the repetition of visual elements can be seen in many design areas including environmental design, product design and graphic design.

Patterns that use elements over and over are repetitive. They may be simple arrangements of lines, shapes or images, but their common characteristic is that they repeat the same sequence of imagery.

Repetitive patterns create a sense of unity and establish clear relationships within a composition. The power of repetitive patterns lies in their consistency. The repetition of elements may be as basic as a bulleted list in a document, or as complex as the structure of enlarged snowflakes or a Byzantine tile mosaic.

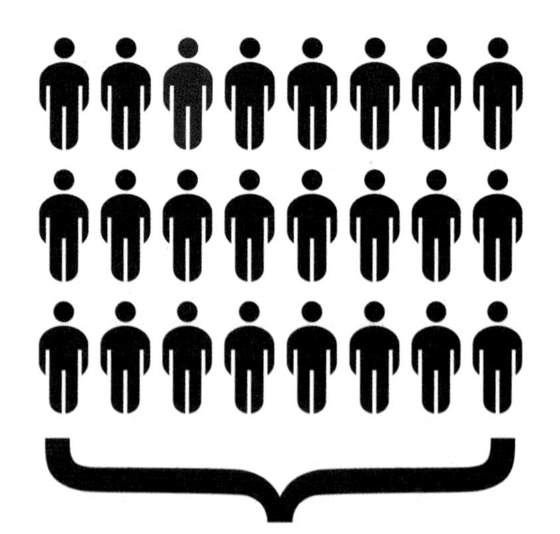

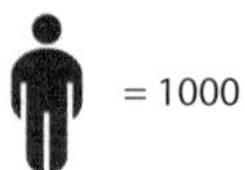

▲ In this extract from an 'infographic' or 'statistical diagram', pattern has been used to convey a numeric value. The repetition of elements can assist the viewer to visualise a complex concept such as quantity.

ISBN 9780170349994

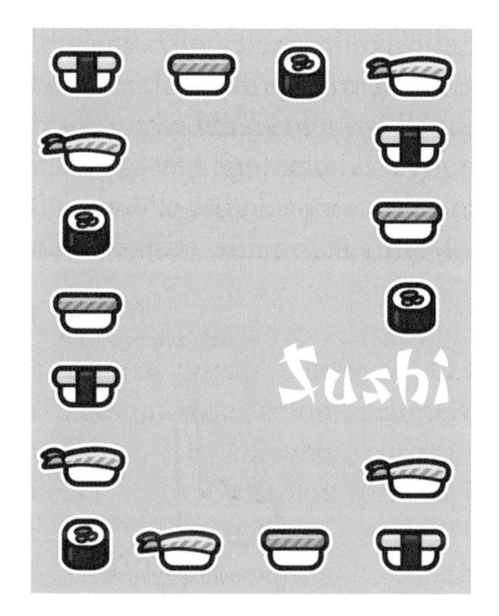

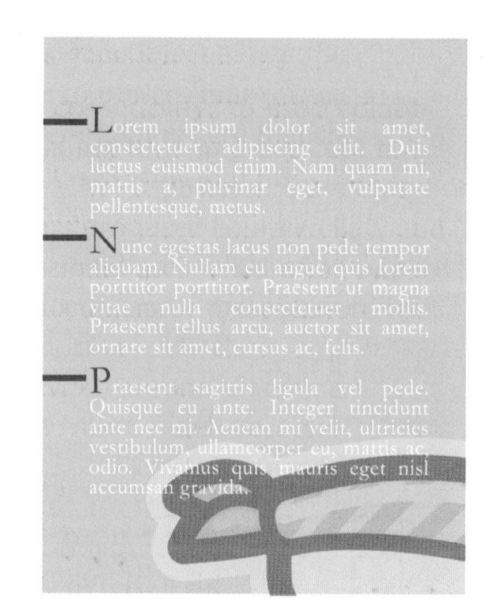

▲ Repetition of elements can be used in many different ways: in borders, to distinguish lists of information and as a visual device.

Repetitive patterns can create a sense of rhythm in a composition, adding movement to the elements. Repeating an arrangement of shapes in a manner that is dynamic adds energy and visual interest. Creative use of figure-ground can allow for the construction of patterns that are visually ambiguous and optically intriguing.

Repeated elements are also seen in patterns that alternate. Such patterns may consist of several different elements used in a changing sequence. Alternating patterns can be created using any visual element and can add visual variety and dynamism to a composition.

Textile designers commonly use alternating patterns that display variations in colour, line and shape. Although designers who work with fabric and textiles may focus on the purpose of the material – such as its application to an individual item of clothing or a handbag – they also have a keen sense of how pattern will appear on a larger scale.

Like repetitive pattern, alternating pattern creates a sense of order, but in a very different way. The variation of a pattern that alternates a range of elements conveys innate energy and life.

Many patterns occur in nature and in the constructed environment. These can be a great source of inspiration in design and may trigger ideas for two-dimensional and three-dimensional design concepts.

iStockphoto/AlterYourReality

iStockphoto/millsrymer

iStockphoto/Brainsil

▲ Patterns occur in nature and the constructed environment.

ISBN 9780170349994

Pattern is used in many areas of design. Digital designers use repeating patterns to create wallpapers and backgrounds for computer operating systems and web pages.

Textile and interior designers use patterns in fabrics and surface decoration. Patterns may alter as fashions and trends change but they are integral to many areas of design. Pattern designers may use traditional techniques and materials in the creation of pattern such as screen-printing, drawing and dyeing. However, in fashion, many commercial fabrics have their patterns developed and refined entirely by computer.

◀ When designing repeating patterns, ensure there is a link between elements on opposite sides of the original image to create a seamless pattern. The image can then be repeated like a set of tiles.

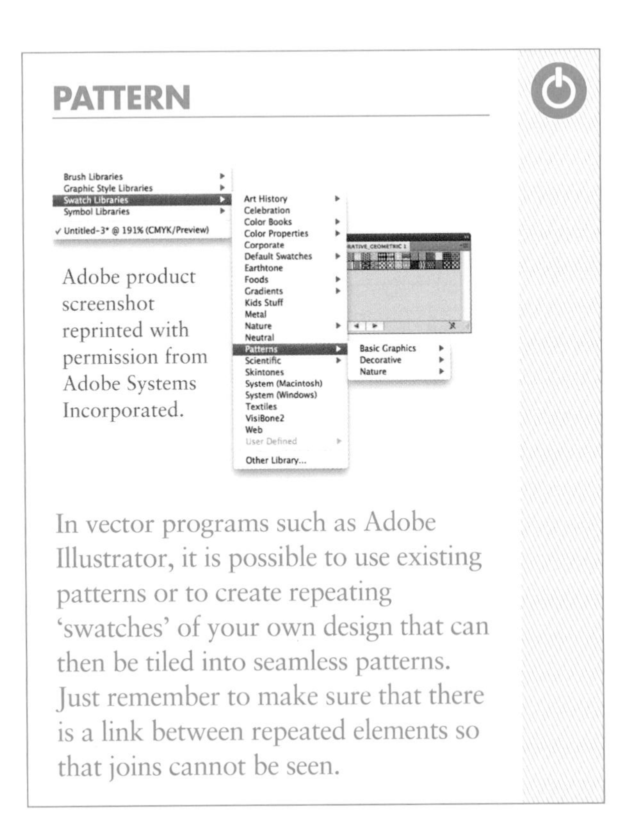

PATTERN

Adobe product screenshot reprinted with permission from Adobe Systems Incorporated.

In vector programs such as Adobe Illustrator, it is possible to use existing patterns or to create repeating 'swatches' of your own design that can then be tiled into seamless patterns. Just remember to make sure that there is a link between repeated elements so that joins cannot be seen.

SCALE

Scale concerns the size of elements within a composition. Scale exists because of relationships between different elements in a hierarchy. Scale, which we can also refer to as size, assists the viewer to make sense of depth, distance and proportions in a composition. Scale enables us to make comparisons between elements or objects, and helps to add meaning. A statistical diagram, for example, uses differences in scale to depict differences in the data.

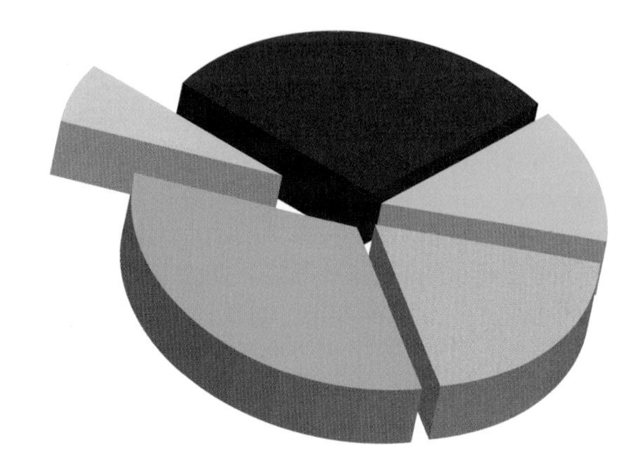

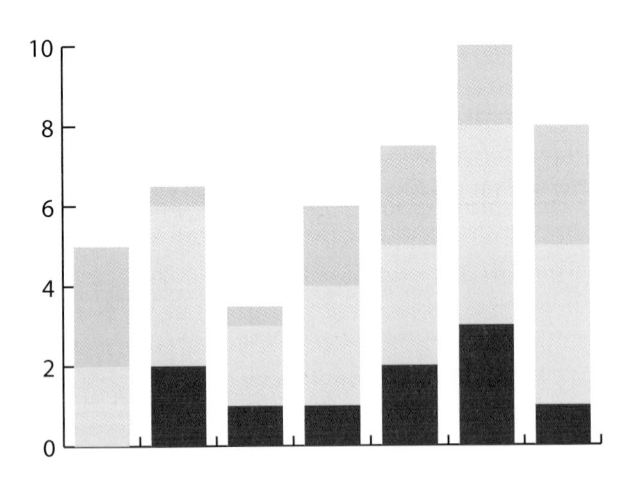

Scale is used to add realism to an illustration. The placement of familiar images, such as the human body, can provide a context in which to depict a product or construction. Architectural illustrations often use trees, foliage, cars and figures to suggest the realistic scale of a construction.

ISBN 9780170349994

Scale can be exaggerated and used to create dynamic contrasts in visual communications. Using elements that play with our innate sense of scale can reinforce the meaning or theme of a composition. Scale and proportion often work together for visual impact. Adjustments to scale can often affect proportions in an illustration, construction or composition.

Through our understanding of instrumental drawings, such as orthogonal drawing, scale is a familiar tool. Scale is used to illustrate details that cannot be represented at their actual size. Scale is applied to complex information such as maps to assist our comprehension of large amounts of information.

▲ Scale model of a student project

Stuart Robinson

Scale models are used in many areas of three-dimensional design to evaluate a concept. The opportunity to observe a design concept in three-dimensional form is valuable, as the strengths and weaknesses of the concept are more clearly identifiable. A model enables a client or other interested party to view a realistic representation of a product that may otherwise be difficult to visualise from a two-dimensional form. Scale models are used in many areas of design, such as automotive design, architecture, design of theatre sets, and product design.

The identification of scale can be used in illustration and layout to highlight important visual information.

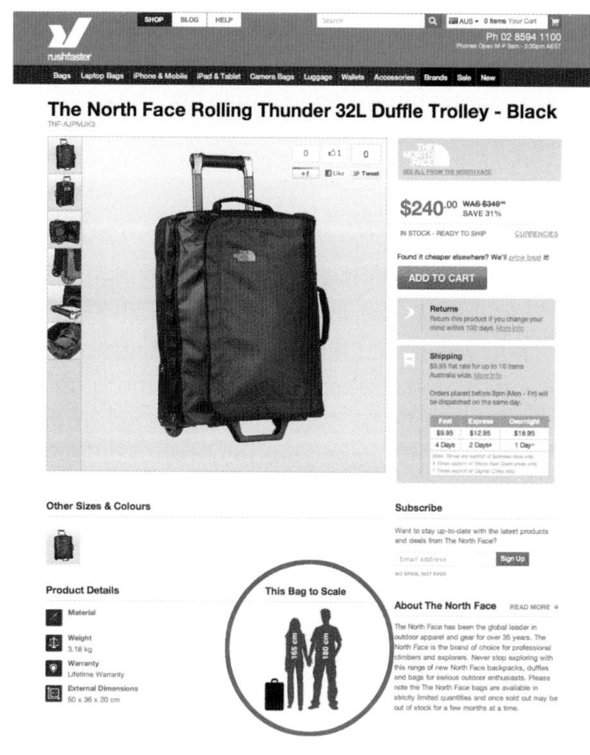

▲ Rushfaster sell bags and luggage online. To assist purchasers with decision making, a scale is used to help visualise the size of each bag.

Rushfaster.com.au

UNITY

Unity ties a design together. Unity can be achieved by making use of many of the design elements and design principles covered in this chapter.

▲ The wide public recognition of the Nike brand is achieved through consistent application of the logo on all Nike products. A unified brand becomes familiar and easily recognised.

David Paul Morris/Bloomberg via Getty Images

ISBN 9780170349994

A unified design is a design where clear links can be identified between visual elements. Consistency is important in achieving unity and many brands require that their corporate identity, colours and typefaces be applied consistently across a range of carriers. From signage to packaging, transport and visual communications, consistent use of brand elements ensures that the values of a brand are reinforced to consumers in a unified way.

Johanna Schreiner

▲ This student designed packaging for a spice company named '1602'. She used pattern to unify disparate but related products under the same branding.

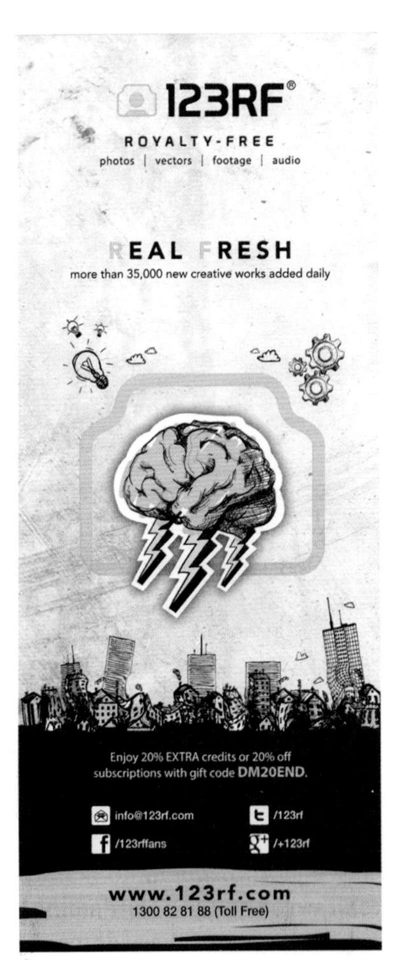

Image courtesy of 123RF.com

▲ These three print advertisements for the stock image company 123RF appeared on different pages of *Desktop* magazine. Despite being placed at irregular intervals throughout the magazine, the artwork retains unity due to the application of a range of design elements and principles. The advertisement use consistent application of COLOUR, the brand logo and SHAPE to retain UNITY across each page.

ISBN 9780170349994

CHAPTER 9
DESIGN TECHNOLOGIES

'What is the most treasured and well-used piece of equipment in your studio? My head.'
Alan Fletcher

The term 'design technologies' may conjure up the latest graphics software but it also refers to the fundamental tools that are used in design. There are essential tools that designers use and it is valuable to have an understanding of their role whether they are used in manual drawing tasks or digital tasks. Often, traditional design tools are reflected in the names of tools used in design software.

Tools and technologies vary with each design area and, increasingly all areas of design rely heavily on digital media for design and production. However, drawing is still an important tool for all designers and is invariably applied during brainstorming and the generation of design ideas early in the design process.

9.1 DRAWING TECHNOLOGIES

PENCILS

Pencils are your most important piece of equipment in drawing tasks. Traditional drawing pencils are made of graphite encased in a wooden body and are still the most inexpensive tools for putting your ideas onto paper. The 'lead' in your 'greylead' pencil is not actually lead; it is graphite, and virtually harmless. Some students and designers prefer to use solid graphite pencils, which are contained within a plastic rather than wooden casing.

Pencils are graded according to the softness or hardness of the graphite inside.

+ B = Black. B series pencils are softer than others, and are usually used in sketching, illustration and rendering.
+ H = Hard. H series pencils feature hard graphite and are often used for line work, technical drawing and drafting.
+ HB = Hard/Black. HB pencils are equally hard and soft.
+ F = Fine. F series pencils are similar in quality to HB, but slightly harder.

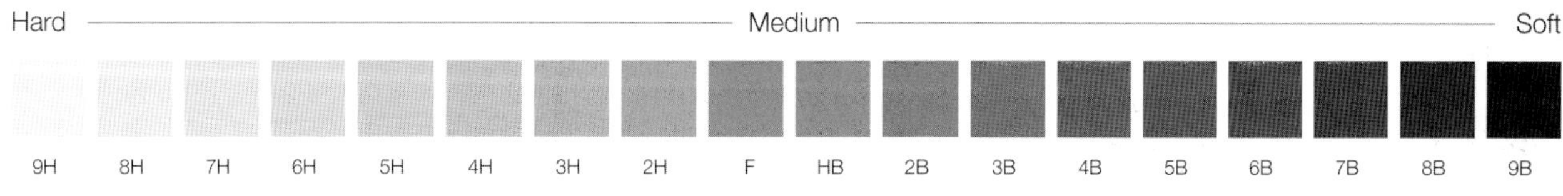

▲ Pencils are graded by the softness or hardness of the graphite inside.

ISBN 9780170349994

ERASING

When using pencils, ensure that you have a clean vinyl eraser handy. In technical drawing an erasing shield, which is a template that allows for accurate erasing, can be helpful and assist you in maintaining a clean drawing surface.

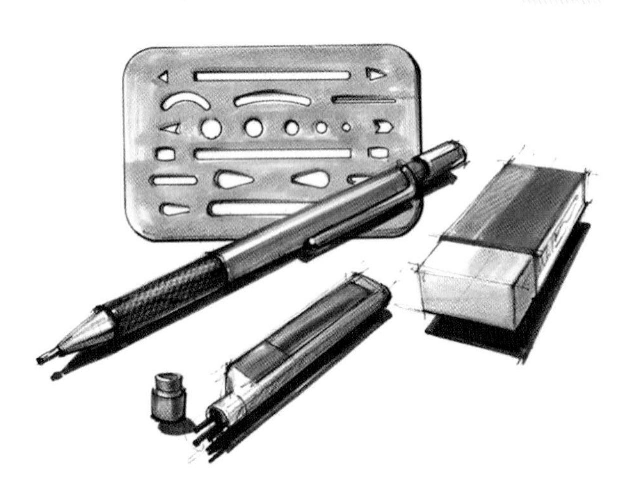

MICRO PENCILS

Also known as mechanical pencils, micro pencils contain graphite of a specific width and grade and never require sharpening. The graphite feeds through the plastic casing of the pencil and maintains a consistent line width. Graphite refills are available in a range of widths and grades, which makes them ideal for technical drawing.

COLOURED PENCILS

The range of coloured pencils on the market is enormous – everything from small tins of 12 pencils to wooden boxes with 100 or more pencils. Generally, a pack of 24 or 36 good quality pencils, with leads soft enough for effective rendering, is adequate.

PAPER

The International Organization for Standardization (ISO) established a system of standard paper sizes known as the ISO 'A' series.

The series is based on a large paper sheet with an exact size of 841 mm × 1189 mm. This sheet is known as size A0. Other sizes include:

+ A1 – 594 mm × 841 mm, or half A0
+ A2 – 420 mm × 594 mm, or half A1
+ A3 – 297 mm × 420 mm, or half A2
+ A4 – 210 mm × 297 mm, or half A3

Papers are graded by weight. It is possible to recognise their grade and suitability for a specific purpose by identifying the grams per square metre (gsm) standard. The higher the gsm number, the heavier the paper.

+ Newsprint: At 52 gsm, newsprint is a very lightweight paper that is used for preliminary sketches (and as tablecloths in trendy cafes!).
+ Bulky news: This is a heavier version of newsprint, with some texture. It is about 80 gsm.
+ Cartridge paper: Cartridge can range anywhere between 110 gsm and 200 gsm, and is often a crisp white colour. A visual diary or sketchbook will contain cartridge of a medium weight. Cartridge is a good multipurpose paper, but markers or felt pens may 'bleed' into the grain when used with this paper.
+ Bleedproof paper and fineliner paper: These useful papers are coated to give a smooth surface that inhibits the 'bleeding' of markers and fineliners into the paper grain.
+ Special coated papers: These papers are designed specifically for use in colour inkjet printers. They are coated with a choice of gloss, satin or matt finish, which allows the ink to sit on the paper rather than be absorbed into the grain.

ISBN 9780170349994

PENS

Technical pens are a popular choice with many designers. Ink-filled technical pens are refillable and available in a range of nib widths to allow precise line work. The most common nib widths are 0.25 mm, 0.5 mm and 0.7 mm. Technical pens are more durable than fibre tip pens such as fineliners, but they require considerable maintenance.

Technical pens: line thickness in mm

Fineliners and other fibre tip pens are a cost-effective alternative to technical pens, and their deep black ink enables work to be easily copied or scanned. Unlike technical ink pens, fineliners have a flexible felt tip.

MARKERS

Water-based markers are commonly used by graphic, industrial and product designers, who find that this marker allows quick, effective colour coverage and ease in rendering a surface. Most markers come with both a fine tip and a broad tip and can be refilled with pigmented inks. It is possible to purchase markers in sets or as individual pens. If you decide to invest in markers, it is recommended that you buy a product that can be refilled.

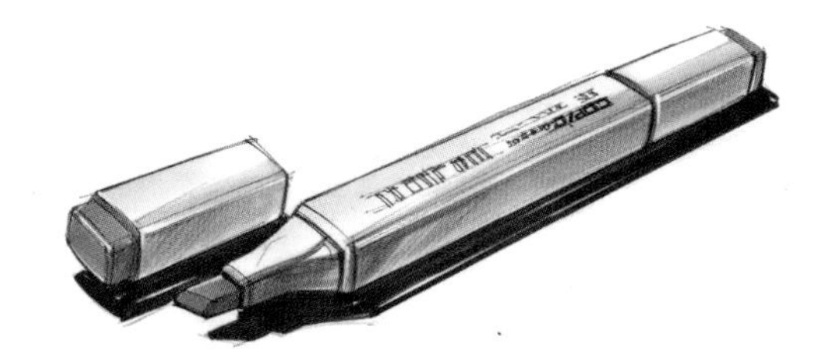

Although the use of markers may take some practice and they can be somewhat expensive, they are a very effective tool for rendering and illustration. Bleedproof paper should be used when working with markers.

DRAWING EQUIPMENT

Although you may complete most of your technical drawings using CADD software, you may be required to manually draw some images. If so, you may need to familiarise yourself with the following equipment.

Set squares

Set squares are designed to be used with a T-square (or straightedge) to create vertical lines or lines at a required angle. The most common set squares used: 30°/60° and 45°.

T-square

The T-square (or straightedge) provides a horizontal base for the creation of true horizontal lines when used on a drawing board. The T-square supports the set square and templates.

ISBN 9780170349994

Curves and templates

Curves, such as flexible curves, snail curves and French curves, enable you to draw curved lines or create irregular, curved shapes with accuracy.

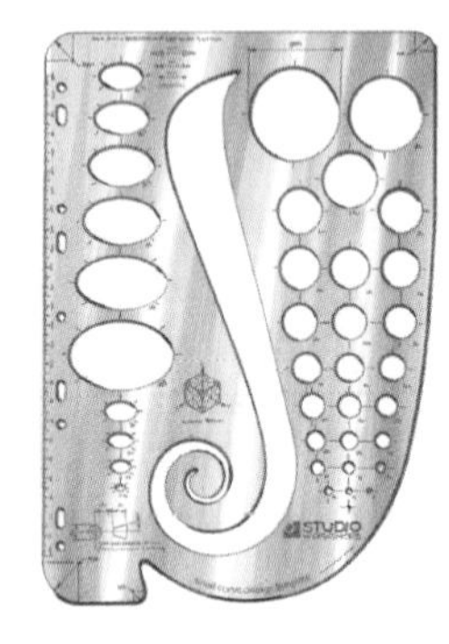

Templates provide machine-cut shapes of set sizes to assist in constructing instrumental drawings. They include circle templates, ellipse templates and symbol stencils.

Texture board

Texture boards are a helpful tool when rendering with pencils or pastel. Offering a range of textures on both sides, they can be used to suggest surface detail when rendering objects and environments.

USE THE BEVEL

Notice that most set squares, templates and plastic rulers have a profiled or bevelled edge. This edge will stop the ink from ink pens, fine liners or markers from bleeding onto the surface. If necessary turn your instrument over to utilise the bevelled edge for cleaner work.

Drawing board

When drawing manually, the drawing board is the support base for your drawing equipment and may be a simple, melamine coated board or a sophisticated drafting setup. Drawing boards feature a 'true' straight edge for the T-square to rest against, enabling you to work with accuracy. Paper can be attached to the board with board clips or masking tape.

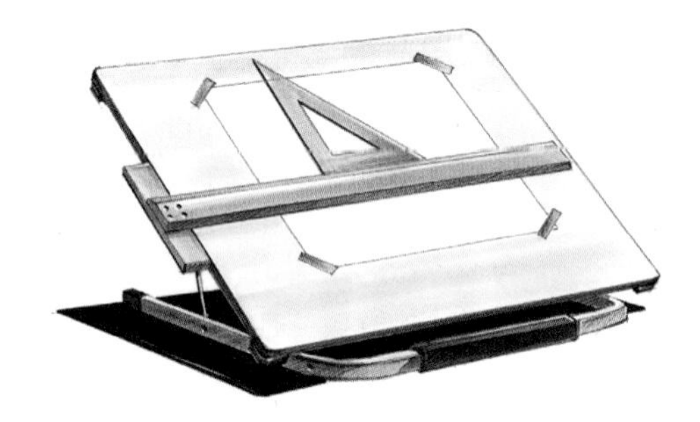

9.2 DIGITAL TECHNOLOGIES

CADD software used on desktops, digital tablets, other portable devices, drawing tablets and touch screens provide scope to create some extraordinary digital designs. Digital production is an integral part of the Senior Graphics syllabus.

Setting yourself up with the latest PC or Mac computer can be an expensive exercise, especially if you buy the latest design software. The computer, no matter what the brand, is a tool, and should be seen as just that. Like your pencils and sketchbook, the computer can help you to work through the requirements of the study and assist you in creating successful final graphical presentations. However, the computer should not be seen as the only answer in design – the use of computers is one part of the solution to your design problem, not the solution itself.

That said, the computer offers you many opportunities for experimentation and visual risk-taking, and can provide you with the chance to depict the images that you have previously visualised and sketched. CADD packages, such as the Autodesk suite, enable you to create and manipulate complex designs into two-dimensional

ISBN 9780170349994

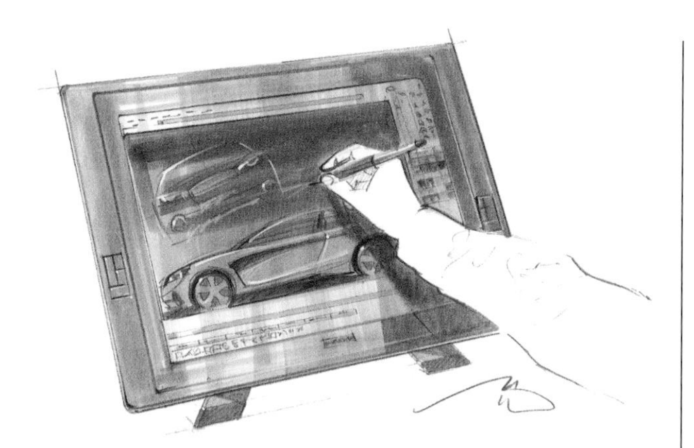

▲ Technologies such as very high resolution touch screens enable designers to draw directly onto the screen surface.

and three-dimensional presentations. Three-dimensional imagery can be animated to present a working view of a final design. Other software packages, such as the Adobe suite, also provide boundless opportunities to turn ideas into high-quality final presentations. Software such as Adobe InDesign enables you to combine vector, bitmap and text together.

Many schools offer access to high-end programs such as the Autodesk and Adobe suites. If you do not have access at school, there are many similar programs – and even some freeware programs – available on the Internet. There are many levels of software available, just as there are variations in the hardware that runs it. Remember though, it is how it is used that defines the value of any software package.

WORKING WITH DIGITAL IMAGES

There are so many different varieties and versions of digital imaging software available that it isn't possible to discuss the functions of each one in this book. The two main types of digital design software you are most likely to use are vector-based and raster-based. Many bitmap and vector-based programs can be used together to create documents and presentations using both image methods.

Vector images are mathematically defined images that consist of lines and curves. Formed in programs such as Adobe Illustrator, AutoCAD and Autodesk Inventor, vector images are sometimes known as object-oriented images. This is due to the ability to move and manipulate entire lines, shapes and curves independently of other image elements. Vector images are not affected by resolution and can be resized with minimal loss of image quality. Common uses for vector images are assembly drawings, logos or symbols, illustrations and diagrams.

▲ Vector illustration of an apple – note no pixelation

Raster images (also known as bitmap images) use a grid (or raster) of small squares of data known as pixels to create images. The term 'pixel' is based on the words 'picture' and 'element', and refers to the smallest element of visual information on a computer monitor. Unlike the shapes, lines and curves of vector images, bitmap images can be edited pixel by pixel, or in groups of pixels. Arguably, the most popular pixel-based editing program is Adobe Photoshop.

▲ Raster (bitmap) image of an apple – note the pixels that create the image

ISBN 9780170349994

Many raster and vector-based programs can be used together to create documents and presentations using both image methods. Programs such as Adobe InDesign support both vector and raster images.

Anti-aliasing

Pixels are squares of digital information, and it can be difficult to create smooth curves or rounded edges with square elements. Consequently, many bitmap software programs feature an option called anti-alias. Anti-aliasing creates an illusion of smoothness by adding small graduated areas of colour similar to that of the surrounding pixels. This tricks the eye into seeing pixels that are smooth, and is helpful when applying text in bitmap (raster) programs.

Resolution

The higher the resolution of an image is, the higher the quality. Image resolution is commonly referred to in pixels per inch (ppi), which indicates the number of pixels of information that are represented within an inch of image space. Artwork that will be printed, either professionally or on an inkjet printer, needs to be of a high resolution. In many cases resolution higher than 300 ppi is not generally required for final presentations. An image scanned at 300 ppi can comfortably be enlarged to at least three times its size without losing any sharpness or clarity. An image created in Photoshop at 300 ppi with an image size of 210 mm × 297 mm, for instance, can be enlarged to 630 mm × 900 mm with minimal loss of quality.

Images that are to be used mainly for web-based or screen-based purposes, such as web pages and digital presentations, have resolutions of 72 or 75 ppi. Images scanned or created at 72 ppi are not suitable for printing because they lack the sharpness and detail of higher-resolution images. However, the advantage of low-resolution images is that they are fast to download and view online, or send via email, because of their smaller file size.

THE DIFFERENCE BETWEEN PPI AND DPI

FYI

PPI (pixels per inch) refers to pixels within an image and is related to screen resolution. For example, a 300 ppi image contains 300 pixels in each inch of image size. It is the preferred term when referring to the quality of an image.

DPI (dots per inch) is related to the printing of images using a printer. Generally, a printer uses four or more coloured inks to create images. Each pixel of the screen image is created by a series of tiny ink dots. A 1200 dpi printer, for example, will print 1200 dots of ink per inch of image. The higher the dpi, the better the print quality; however, the printer will use more ink and the print will take longer to execute.

Note that images created at a low resolution, such as 72 ppi, cannot have pixels 'added' later to create a higher-resolution image, as the 'extra' digital data simply doesn't exist. This is why it is vitally important to establish the end purpose and subsequently the resolution of your images before you start.

Image types

You will come across a number of common image types in your studies and in your use of the Internet. Each of the types has one or more distinct applications. Your choice of image type will be determined by the intended use of the image itself.

COLOUR MANAGEMENT

There are several colour modes that apply to digital images and, as with resolution, it is helpful to know the purpose of the artwork in order to identify the most appropriate mode of colour.

ISBN 9780170349994

Image type	Use	Description
TIFF **Tagged Image File Format**	Print	One of the main formats for preparing graphics for print and publication. TIFF files retain the maximum amount of visual information and consequently maintain a large file size.
JPEG **Joint Photographic Experts Group**	Web (and print if high quality)	A popular web graphics format. Although JPEG files are compressed (which causes some loss of quality) they are a good format for high-quality images that need to be emailed or published electronically. JPEG files are often smaller in size and this can be adjusted by altering the quality in a program such as Photoshop or Image Ready.
GIF **Graphics Interchange Format**	Web	A standard format for web graphics such as icons, toolbars and thumbnail images. GIF files contain minimal visual information and exist in a highly compressed format so their file size is very small.
EPS **Encapsulated PostScript**	Print	A file type preferred by commercial printing companies that print publications on very high-resolution printers. PostScript is a vector language used in desktop publishing software, such as InDesign and QuarkXPress, and it treats all images, including fonts, as objects rather than bitmaps.
PDF **Portable Document Format**	Print and web	A proprietary Adobe format that is created and/or read in Adobe Acrobat. A PDF file enables a document to be compressed, sent, viewed and printed electronically without any loss of the original format, fonts or graphics. Many commercial printers prefer files submitted as PDF format.
BMP **Bitmap**	Print and web	A file format for saving high-quality bitmap images, originally designed for Microsoft Windows systems. A BMP file is similar in type and file size to a TIFF.
Proprietary file types	Print and web	All graphics software programs use their own proprietary file formats. In Photoshop, the default file format is PSD, which allows you to keep all layers, selections and channels intact. Illustrator uses the default file format AI; CADD uses DWG; Flash uses SWF; and so on. A good tip is to always save a version of your work in the proprietary format so that you can return to it at a later stage and make any necessary changes.

CMYK: Cyan Magenta Yellow Black

If you are creating images and presentations for commercial print production, you will need to save your images in CMYK mode.

Many home inkjet printers also use four separate ink cartridges to produce image colour. RGB images can be converted to CMYK but some colour change is likely to occur. CMYK is a subtractive colour mode and is also referred to as four colour and process colour.

RGB: Red Green Blue

RGB colour is used for screen and digital designs where professional printing is not the desired final outcome. Varying amounts of red, green and blue light create RGB colour. RGB mode is used on digital devices such as computer monitors, laptops and mobile devices. RGB is an additive colour mode.

Planned use	Suggested resolution	Suggested colour mode	Suggested file format
Artwork for full-colour printing	300+ ppi*	CMYK or PMS/ Spot colours	TIFF, EPS or PDF (or large, high-quality JPEG)
Black and white for printing	150+ ppi	Greyscale	TIFF, EPS or PDF (or large, high-quality JPEG)
Simple web graphics such as icons and buttons	72–75 ppi	Indexed colour or RGB	GIF
Detailed, photo-realistic graphics such as large images for use on the Internet	72–75 ppi	RGB	JPEG
Graphics for digital presentations such as PowerPoint, video, etc.	72 ppi	RGB	JPEG

*pixels per inch

Web-safe colours

There are 216 RGB colours that are considered 'web-safe'. This means that colours used on a website (or other presentation designed for viewing on a computer monitor) are guaranteed to appear as intended.

As computer users have monitors adjusted to different colour settings, the web-safe colours provide for the most basic setting. You can use a myriad of RGB colour combinations, but there is no guarantee that every user will view them in the same way, due to differences in device quality. Each of the web-safe colours is identified by a hexadecimal code; for example, a bright red may have an RGB value of R255 G0 B0 and a hexadecimal code of #FF0000. This code is used in hypertext markup language (HTML), the code used to create web content, to identify colour.

Making the right choices

It is important to know the context in which your digital work will be used, so that you can plan ahead. There is nothing worse than spending hours on an image, only to find that the resolution is not good enough to print or that your colour mode is incorrect.

Be aware of image sizes when downloading images from the Internet to use in your design work. Make sure they are large enough to print them if you wish to. It is possible to adjust your image search to 'large' to find images of better quality. When downloading images, remember to attribute the source of your images.

9.3 PRINTING TECHNOLOGIES

INKJET AND LASER PRINTING

The two main types of 2D printers that you are likely to have access to are inkjet printers and laser printers. Inkjet printers precisely place tiny drops of coloured ink onto paper to create an image. The inkjet printer is known as a non-impact printer because it sprays drops of ink onto the paper and does not touch the paper itself. The laser printer is also a non-impact printer and uses toner (or dry ink), static electricity and high temperatures to position the image on paper.

Inkjet printers are the most popular means of printing colour images for school and home users. They range from small A4 printers to very large format printers. It is possible to obtain high-quality prints without the expense of professional printing processes. Inkjet printers generally use four or more colour cartridges, one for each of the CMYK colours cyan, magenta, yellow and black, or hue variations of these colours; for example, light magenta, light yellow. Different percentages

ISBN 9780170349994

of each of the four colours can produce seemingly endless variations in colour. As your computer screen operates in RGB mode, it is helpful to have access to a colour swatch so that you can more closely identify the printed colour. Some swatches contain the CMYK percentages for each colour to assist in generating the appropriate colour in your chosen software package.

Colours you see on the screen are often different when printed. Colour always appears brighter on the screen. Commercial products that calibrate your monitor are available to ensure greater parity between your screen and the printed page. These are placed on the monitor itself and ensure that your colours remain true. A less expensive alternative is to test print your colours as you go, or use a colour palette swatch.

When using an inkjet printer, the variables that may affect your work are print resolution and paper quality.

PRINT RESOLUTION

The number of dots per inch or dpi value of an inkjet printer will affect the clarity of the print. This value refers to the number of dots of ink placed on the paper within each inch of image space. A printer that produces 1440 × 720 dpi will print a higher quality image than a printer that prints 300 × 300 dpi.

Laser printers can also vary in quality and some printers produce a clearer image than others. Often, laser printers do not produce the subtle tonal variations that can be seen in inkjet prints, and colour laser prints can appear flat. However, for reproduction of single colour letterform, logos and developmental work, the laser printer is quite suitable.

PAPER QUALITY

There are many specialty papers available for use in inkjet printers. To produce high-quality results, inkjet papers are treated with a vivid white coating, preventing ink from bleeding into the fibres of the paper and preserving the integrity of ink colour. When traditional 90 gsm papers, such as photocopy paper, are used the ink is absorbed into the surface fibre, which reduces not only the sharpness of the print but also the intensity of the colour.

Special coated papers are available in different surface types, similar to photographic papers, including gloss, satin and matt. It is also possible to print onto iron-on transfers, transparency sheets and adhesive films.

Like photocopiers, laser printers use toner rather than ink, which does not bleed, although it can smudge in heavily toned areas. Specialty papers are also available for laser printers and include transparencies, varnished papers (gloss, matt and satin) and some recycled and textured varieties in a range of colours.

The choice of paper will be determined by the purpose of the print and the suitability of the surface type for the task.

OFFSET PRINTING

Offset printing is a process used by professional printers. A digital file is converted to a series of 'plates', which are coated with ink. The plates travel through a complex printing press, which transfers the inked image on the plate to the chosen 'substrate' or printing surface. Offset printing originates with digital files that have been prepared in line with the printer's specifications. A professional printer will indicate what is required in a printable file. Generally, the specifications include the colour mode (usually CMYK), the need for outlined type (to ensure type looks as intended), bleed (to extend colour, image or type to the printed edge) and trim marks (where the paper, card or other stock is cut) as well as the stock on which the product will be printed.

The professional printer can also coat the printed stock with a varnish to create a matte or gloss appearance and provide binding of books and magazines. The complexity of offset printing means that clear communication between designer and printer is essential to ensure a successful outcome.

ISBN 9780170349994

THREE-DIMENSIONAL PRINTING

In recent years, 3D printing has become an affordable option for at-home printing and student work. According to Google Trends, searches for information about 3D printing increased tenfold in 2013. Searches ranged from 3D printing of food and houses to jewellery and hats. 3D printers range in scale from small printers created from kits for home or school use to large, oven-sized printers used in university settings and commercial enterprises.

Known as 'additive' printing, 3D printers add layers of a selected material (usually a fibre composite or plastic) over and over to form a three-dimensional representation of a CADD drawing. Often used for prototyping and the construction of models, 3D printing is developing rapidly into a competitive commercial manufacturing process. Clothing, footwear, artworks and products with working parts can be created with 3D printing.

3D printers create forms using an STL file (stereolithography file) that describes the surface geometry of an object created in CADD software. To create the print, the printer builds cross-sections of the object, which correspond to cross-sections in the original STL file.

Like 2D printing methods, the thickness of layers in a three-dimensional print is described in terms of dots per inch (dpi). Typical layer thicknesses are approximately 250 dpi but vary according to the capabilities of the printer. Due to the complexity of the 3D process, printing can take long periods of time, ranging from a few hours to a few days.

In the Bond film *Skyfall*, many props were created using 3D print technology. The helicopter that attacks Skyfall Lodge at the conclusion of the film was a 1:3 model created entirely by 3D printing. Each part of the vehicle was 3D printed, painted and assembled.

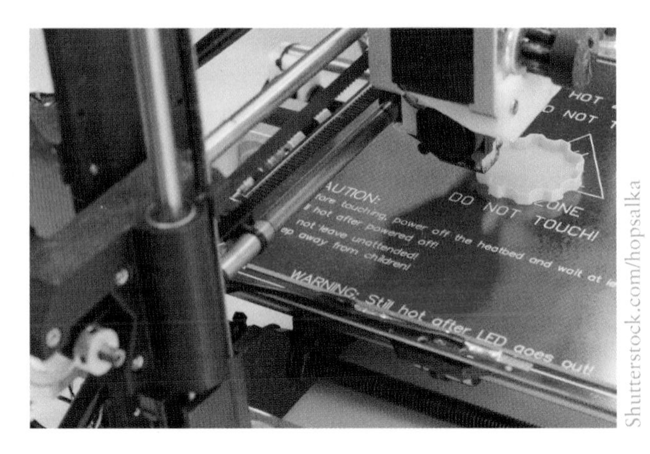

Shutterstock.com/hopsalka

Shutterstock.com/hopsalka

Shutterstock.com/hopsalka

▲ 3D printers create objects through the layering of plastic filament. The printer converts digital information supplied by a CADD program and layers multiple strands of materials to build up a three-dimensional form.

ISBN 9780170349994

CHAPTER 10
LEGAL RESPONSIBILITIES

'Design is creativity with strategy.'
Rob Curedale

As in any professional area, designers need to consider the legal and ethical issues that affect their field. As professionals they have responsibilities towards their clients, users and the wider community.

These issues include:

+ copyright
+ intellectual property
+ standards and safety
+ image manipulation
+ cultural sensitivities.

Legal issues are set; they are issues that are defined by law and cannot be breached without serious consequences. Legal regulations apply to designs created and sold in Australia and might include safety regulations and standards that designers are required to adhere to. Ethical concerns are less concrete; they may range from 'Can I work for this client?' or 'Is this a conflict of interest?' to 'Does this design negatively impact those who will see it or use it?'

Designers in all design fields take legal issues into account in their work. An architect may be required by council by-laws to consider the impact a design could have on the community and make aesthetic judgements with those in mind. The same architect has an obligation to create a safe construction for the client and there are legal implications if safety is compromised. All designers are faced with issues of copyright and attribution; both in their own work and when using the work of others. Decisions about the use of materials from other sources may be constrained by copyright, costs and licensing restrictions. Likewise, a designer may seek to protect their work under copyright laws.

10.1 LEGAL RESPONSIBILITIES

COPYRIGHT

Copyright is designed to protect the products created by writers, designers, artists, composers, filmmakers and other creative professionals. In Australia, copyright is automatically granted to a product once it is put into 'material form' such as being drawn or written down. The owner of the copyright has the right to show, publish or perform the work in the public realm and can prevent others from reproducing the work without explicit permission. The copyright owner may sell the rights to that work or 'assign copyright' to another party.

Copyright protects:

+ artistic works – paintings, photographs, maps, graphics, cartoons, charts, diagrams and illustrations
+ literary works – novels, textbooks, poems, song lyrics, newspaper articles, computer software, computer games
+ musical works – melodies, song music, advertising jingles, film scores
+ dramatic works – plays, screenplays and choreography
+ films and moving images – feature films, short films, documentaries, television programs, interactive games, television advertisements, music videos and video podcasts
+ sound recordings – MP3 files, CDs, DVDs, vinyl and tape recordings, podcasts
+ broadcasts – pay and free-to-air television and radio.

Copyright logo. Even when the logo is not present, copyright still exists under Australian law.

Copyright does not protect techniques, concepts or ideas but it does protect their tangible physical representation. An idea for a chair design, for example, is not copyright; however, copyright law covers the sketches, technical drawings, models and final design product.

The owner of copyright may be separate from the owner of the designed item. An individual may own the chair mentioned above yet the copyright to the chair design remains in the hands of the original copyright owner, who may be the designer or manufacturer.

Many designers use sourced imagery in their work. Copyright images and photographs may be used in publications, websites and other public domains only with the permission of the copyright owner.

THE COPYRIGHT TIMELINE

Just because a work appears online does not mean that it is out of copyright. For artistic, literary, musical and dramatic works, the period of copyright protection runs from the time of creation until 70 years after the death of the creator. Films, sound recordings and broadcasts are protected for 70 years from the end of the year in which the work was released. Sometimes it takes some searching and research to identify whether creative products are in or out of copyright. Once out of copyright, a work is considered to be 'in the public domain' and can be used freely.

ISBN 9780170349994

COPYRIGHT

The website of the Australian Copyright Council. The council supports the copyright needs of creative professionals in Australia. The site offers information organised by profession and features detailed information about all aspects of Australian copyright law. Council members include representatives from the peak professional associations for Australian writers, musicians, photographers, visual artists, journalists, filmmakers and architects.

Access all weblinks directly at http://nsg.nelsonnet.com.au.

INTELLECTUAL PROPERTY

Intellectual property (IP) is literally, 'the property of your mind' and refers to the creative production of a new invention, brand, design or artistic creation. For example, in designing a new product, intellectual property rights enable the designer to legally protect their design idea from copies and imitations. Unlike copyright, IP is not automatically recognised and a product, design idea or concept must be registered for a given period of time at a cost. Using the Australian Government's Intellectual Property Organisation, designers can apply for patents, trademarks and registered designs to protect the originality of their work and ideas. IP Australia takes care of four distinct types of intellectual property: patents, trademarks, designs and plant breeders, rights. The following table outlines the relevant types of intellectual property that is usually registered.

What is protected	Type of IP right	What the IP provides
Inventions	Patents	The owner has the exclusive right to use, sell or license the invention. Patents also allow the owner to stop others from manufacturing, using, copying and/or selling the device or process.
Letters, numbers, words, colours, a phrase, sound, smell, logo, shape, picture, aspect of packaging or any combination of these	Trade marks	A trade mark identifies the particular goods or services of a trader as distinct from those of other traders. The owner has the exclusive right to use, sell or license the trade mark.
The way a product looks or a design on a manufactured product	Designs	The visual appearance of a manufactured product is protected, but not the way it works. The owner has the exclusive right to use, sell or license the registered design.
Art, literature, music, film, broadcasts and computer programs	Copyright (Automatically applied)	The owner's original expression of ideas is protected, though not the ideas themselves. The owner has the exclusive right to use, sell or license the copyright work.

▲ Table adapted from www.ipaustralia.gov.au/understanding-intellectual-property/how-to-use-ip/what-can-you-protect/

ISBN 9780170349994

IP rights are legally enforceable in Australia and protect many designers from copying and misuse of their designs. Areas of design that benefit from IP protections include:

+ architecture
+ digital media
+ exhibition design and display
+ fashion design
+ furniture design
+ graphic design
+ industrial design
+ interior design
+ jewellery design
+ landscape design
+ television, film and set design
+ textile design

When creating work for a client, the ultimate owner of the design (the intellectual property) is usually established in the contract at the beginning of the design process. In the majority of cases, the contract between client and designer will state that all intellectual property generated becomes the property of the client in return for design fees.

Although a 'normal' contract would assign all IP rights to the client, a designer might negotiate at the beginning with the client over what will be assigned – and under what payment terms – and then write this agreement into the contract. For example, one option is to agree to assign the IP rights to only the final, selected idea, retaining rights to any other design ideas.

INTELLECTUAL PROPERTY RIGHTS

The website of IP Australia provides comprehensive information about all areas of Intellectual Property. The site explains the key differences between copyright and IP rights and offers visual examples of each IP category: Patents, Trade Marks, Designs and Plant Breeder's Rights.

Access all weblinks directly at http://nsg.nelsonnet.com.au.

USE AND MANIPULATION OF IMAGES

Print media often sees the use of images that have been altered and the use of software, such as Photoshop, is often used to retouch and alter the appearance of individuals.

Shutterstock.com/Elena Rudyk

▲ Image of woman's face before and after Photoshop retouching.

There have been many well-documented cases of celebrities seeking compensation from publishers for overt and exaggerated manipulation of their images. Although there is no specific legislation in Australia that protects people from having their likeness altered without their consent, there are areas of law that offer some protection. Defamation law offers recourse if an individual believes that the alterations to their image cause ridicule, contempt or a loss of reputation while Consumer Law protects against deceptive or misleading interpretations of an image. It is important that designers are aware of relevant legislation that pertains to image use.

STANDARDS

Australian Standards are documented requirements that designers and manufacturers must follow. Standards set out specifications and procedures that ensure that products, services

ISBN 9780170349994

and systems are safe, reliable and consistent. The documents use consistent terminology that defines levels of safety and the quality that products are required to meet. Although Standards documents are not legal documents they are a requirement in Australian design and manufacture and they can be mandated by Parliament as compulsory.

Standards can be published in a range of formats and may include:

+ Australian Standards®
+ international standards and joint standards
+ codes
+ specifications
+ handbooks
+ guidelines.

The Australian Standards cover many industries including:

+ agriculture, forestry, fishing and food
+ manufacturing and processing
+ building and construction
+ electrotechnology and energy
+ water and waste services
+ transport and logistics
+ health and community services
+ consumer products, services and safety
+ education and training services
+ communications, information technology and e-commerce services
+ public safety; public administration; business and management.

Cultural sensitivities

Although it is not a legal consideration, cultural sensitivities are a factor that designers need to consider. In a global marketplace many designs are sold and used in countries around the world. What is appropriate in one market may not work in another due to cultural or religious sensitivities. For example, the 'thumbs up' symbol has positive connotations in Western countries but has less clear meanings in some Middle Eastern societies. A designer may not be aware of the distribution of a design, therefore if a design is likely to be used in varied markets it is incumbent on the designer to ensure that the product is not culturally offensive or misleading. Such considerations reinforce the importance of building a solid understanding of the end user (see Chapter 7) during the research phase of the design process.

ETHICS

Ethics in Graphic Design is a US forum for the discussion of ethical issues in graphic design. Issues range from social responsibilities in design to privacy and copyright.

Access all weblinks directly at http://nsg.nelsonnet.com.au.

It is a good idea to apply appropriate legal and ethical habits early. Begin by using simple steps in your work to meet recommended best practices in design.

10.2 COPYRIGHT FOR STUDENTS

The rules for using copyright-protected materials in education are slightly different. Under Australian law, schools have expanded rights to use copyright materials without seeking permission from the copyright owner as long as content remains within the classroom. This doesn't mean open slather for schools. There are still parameters set as to the amount of copyright works that may be copied, displayed and reproduced but the rules make the use of copyright materials for educational purposes much more flexible.

Sources being used by Senior Graphics students in Queensland should always be acknowledged.

When using images, the original source of the image should be acknowledged in an annotation that records the original author or copyright owner. If the owner cannot be identified and the image has been sourced from an online location, note the web address or use a screen capture to identify the source. This is called attribution.

If student work is to be displayed publicly, there must be clear acknowledgement and attribution of any content used that has not been created by the student.

TIPS FOR STUDENTS USING THE WORK OF OTHERS

1 Most importantly, always identify the source when using the work of others.
2 You are entitled to use a 'fair' amount of work from other sources for 'research and study' without gaining permission from the copyright owner; this is known as fair dealing. Fair dealing requires that the work is used only for research, criticism, satire and parody, or reporting of news. It is likely that most work used in Senior Graphics will fall under the 'research and study' area.
3 You are entitled to use the work of others when you have express permission from the copyright owner to do so. You should have evidence of the permission.
4 You are entitled to use work with a Creative Commons licence that allows use by others.

SMART COPYING

This official website is designed for teachers and students at Australian schools and TAFEs. Educational use of resources entails different copyright requirements and these are clearly outlined here.
Australian Attorney-General
The official website of the Australian Attorney-General offers information about current copyright law.

Access all weblinks directly at http://nsg.nelsonnet.com.au.

CREATIVE COMMONS

▲ Creative Commons logo

Creative Commons is an international non-profit organisation that provides free licences to copyright owners to allow others to legally share, reuse and 'remix' their material. Creative Commons was created in direct response to the increasing accessibility of materials via the Internet and a perceived lack of control that creators have in the digital domain. A Creative Commons licence is identified by a series of symbols, which indicate the context in which the author of the work is prepared to allow others to use the work. When a creator releases their work under a Creative Commons licence, it is made clear what the user can and cannot do with the work. All Creative Commons licences allow works to be used for educational purposes. Teachers and students can copy, share and often modify a Creative Commons work without seeking permission from the work's creator.

How to attribute a Creative Commons work

Include the following:
+ the author name
+ the title of the work
+ the URL where the work was located
+ the type of Creative Commons licence attached to the work
+ any copyright notice attached to the work.

CREATIVE COMMONS

This is the website of Creative Commons Australia. The organisation supports Creative Commons in Australia and administers the Australian Creative Commons licences. The website features detailed information about licences including fact sheets and case studies.

Access all weblinks directly at http://nsg.nelsonnet.com.au.

ISBN 9780170349994

Creative Commons licence types

Symbol				
Meaning and letter code	Attribution BY	Non-commercial NC	Non-derivative works ND	Share alike SA
Description	This applies to every Creative Commons work. Whenever a work is copied or redistributed under a Creative Commons licence, the original creator (and any other nominated parties) must be credited and the source linked to.	Allows others to copy, distribute and perform the work for non-commercial purposes only.	Allows others to distribute, display and perform verbatim copies of the work. The work may not be adapted or changed in any way.	Allows others to remix, adapt and build on the work, but only if they distribute the derivative works under the same licence terms that govern the original work.

ATTRIBUTION OF RESEARCH

Source attribution is an important skill to learn and there are guides to assist you in correctly documenting your sources. The Harvard system of referencing materials is probably the most widely used.

HARVARD REFERENCING GENERATOR

The Harvard Referencing Generator is an online tool that can help you to acknowledge your sources easily and quickly. Type in the URL, book title or magazine and the generator will produce a correct citation for you to copy and paste.

Access all weblinks directly at http://nsg.nelsonnet.com.au.

Generally the following should be included when attributing non-original content.

+ The name of the work (if available)
+ The author and/or copyright owner's name (Usually this should be the surname followed by initial but some web content may give you only first name so use what is available or attribute to Anon. if no author can be identified.)
+ The URL of the work if found online
+ The origin of the work if found in a secondary publication (The name of the publication should be identified along with its date of publication.)
+ The date of the work
+ The date of access or download

ISBN 9780170349994

Research	Source	Attribution	Example
Images	Online, e.g. Google Images	Avoid annotations that simply state 'Google Images' or another search engine. Navigate to the source site of the image and copy the URL. The attribution should state: + the owner (if known) + the URL of the image in <___> + the date it was accessed/downloaded.	Image by Jones, J available from <www.greatbuildings.com/image_033> [13 May 2014]
Images	Print sources	You should include the publication title and date if it originates from a print source. The attribution should state: + the owner + the title image and/or the article from which it was sourced + the name of the print publication + the date of publication.	Image by Mavis Davis 'Australia's ten best photographers', *Design Journal*, Issue 12, 2014 (When annotating an attribution in handwriting, you can emphasise the source by an underline rather than italics, e.g. Design Journal)
Stock images	Online stock photo site	Free stock photo sites will vary in the level of attribution required but most will require the name of the owner of the work. The attribution should state: + the owner (if known) + the ID (#) number of the image + the URL of the image or stock site in <___> + the date it was accessed/downloaded.	Image copyright Dani007 #0436721 <http://sxc.hu> [4 June 2014]
Text	Digital sources, e.g. blogs, Wikipedia articles etc.	Quotes or references directly taken from online sources, such as blogs, must identify the author and origin. The attribution should state: + the author + the title of the blog post or article + the date of the blog post or article (in square brackets) + the URL of the blog in < ___> + the access or download date (in square brackets).	Feagins, L, Interview with Illustrator Dawn Tan [12 September 2011] <thedesignfiles.net> [30 June 2014]

(Continued)

ISBN 9780170349994

(*Continued*)

Research	Source	Attribution	Example
Text	Books	When using research from a secondary source such as a book, you must also add the publisher name and location of publication. The attribution should state: + the author + the date of publication + the title of the book + publisher + location.	Martin, B, & Hanington, B, 2012, *Universal Methods of Design*, Rockport Publishers, USA (The title should be italicised if using a computer but can be underlined when handwritten.)
Text	Magazines and journals	The article name must be mentioned and depicted in single quotation marks. The attribution should state: + the author + the date of publication + the title of the article + the name of the publication + issue or volume number and date.	Banham, Stephen, 2012, 'The Typeface: Newman', *Desktop Magazine*, No. 279 (The name of the magazine should be italicised if using a computer but can be underlined when handwritten.)
Creative Commons materials	Refer to Creative Commons in this chapter to read information about online content that is covered by the Creative Commons Licence structure.		

CHAPTER 11

DESIGN STRATEGIES AND PROJECT MANAGEMENT

'I work on (design ideas) everywhere – all the time. I've always got a pocket full of paper.'

Paul Smith

11.1 DESIGN STRATEGIES

In tackling a design problem, design strategies are methods of approach that stimulate and encourage the flow of creative ideas. They are usually practical tasks that can be applied at any stage of the design process to assist in inspiring concepts and design directions. Design strategies encourage creative and critical thinking, essential thinking tools in a designer's toolbox.

Popular design strategies include:

+ brainstorming techniques
+ SCAMPER
+ graphic organisers, such as concept maps.

Design strategies are important to designers in all areas of design as they often lead to innovative and creative thinking. For students, they are critical tools in generating concepts to explore and evaluate throughout the design process.

BRAINSTORMING TECHNIQUES

Brainstorming is the application of small, stimulating tasks that tap into the imaginative resources of your mind. These are designed to exercise your brain or to help expand on an idea that needs a bit of a push.

Brainstorming is all about quantity rather than quality and the aim is to generate lots of ideas that can be sorted through to identify potentially relevant design directions. Brainstorming is focused on possibilities and the creative thinking that it inspires is open, non-judgemental and imaginative.

Although many brainstorming techniques are word based, they don't have to be. All of the examples in this chapter can also be used with sketches alone or in combination with text.

ISBN 9780170349994

OPEN IDEO

Initiated by the renowned design studio, Open IDEO is an open platform for creative individuals from around the world to collaborate on innovative design ideas through brainstorming and discussion.
Access all weblinks directly at http://nsg.nelsonnet.com.au.

WORD LISTS

Olivia Rose

▲ This student used a word list to identify potential directions when working on a design brief.

Making a list of words and phrases associated with the design problem is a good starting point. It requires you to write as many words as possible that are related directly or indirectly to the contents of the design brief. The idea of a word list is to forget about what might be relevant or irrelevant at this stage and just get as many words down as possible. When you think you have enough words, pick out the most promising ones and develop them further. By taking one concept from the first list and expanding it into a sublist, the ideas become more specific and can lead towards initial sketches.

ISBN 9780170349994

Starting template for a word list

Where will the design be located/used? List all the possible locations and uses.	Who uses or sees the design? List every person or group that may come into contact with the design.	What does the design need to do/achieve? List all the possible functions that the design could achieve.	What features might the design include? Be creative in listing every possible feature; practical or not. There are no right or wrong answers.	How will the design be distinctive? What could be done to the design to ensure it stands out?

RANDOM WORDS

A popular brainstorming tool involves selecting a random noun from a dictionary or thesaurus and finding a link with your theme or topic. In the design of a corporate logo; for instance, the random word 'bird' might be used to trigger ideas for a high-flying, progressive identity.

The subsequent brainstorming may be as follows:

Bird → bird flight → streamlined bird flying upwards

Bird → bird's-eye view → overview of area below

Bird → flock of birds → images of birds in the distance showing future planning/thinking.

Although an unusual method, the random word technique can be helpful when ideas do not flow. The method requires an open mind but the results will be fresh and can invigorate stale ideas.

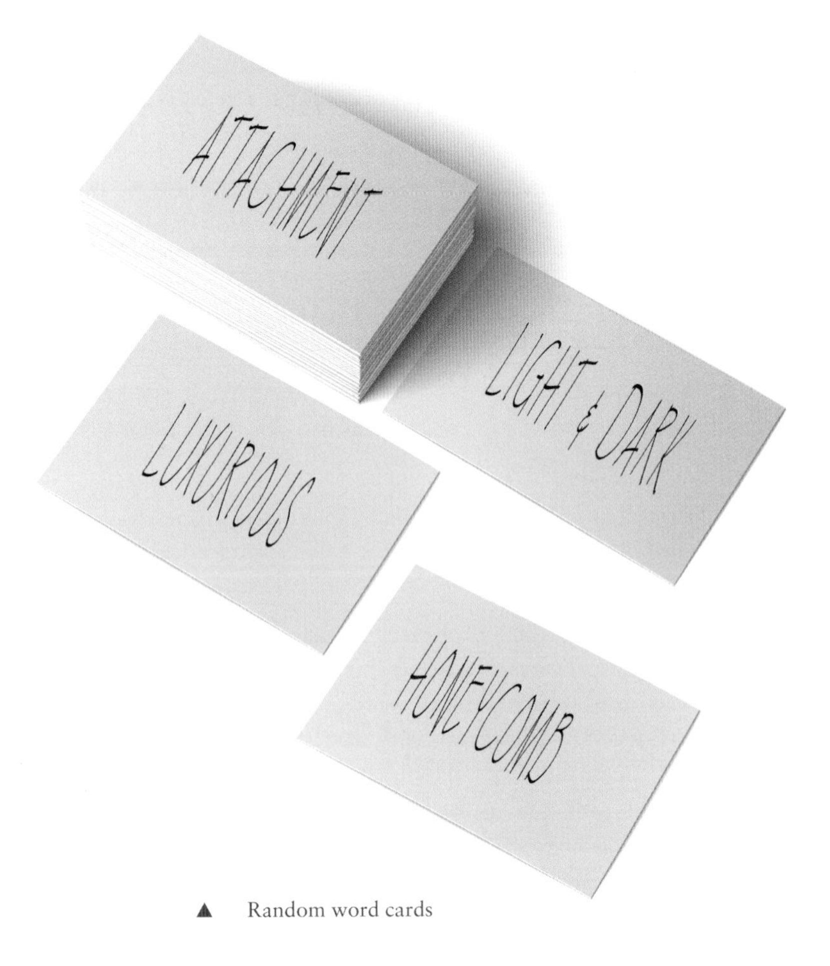

▲ Random word cards

ISBN 9780170349994

DEVELOP A BRAINSTORMING TOOLKIT

Create a set of 20 or more cards with random words placed on each card. Use them to stimulate ideas when you get stuck or need a new design direction. Add to the collection as you think of more helpful or inspiring words.

PRODUCT PERSONALITY

A unique approach to stimulating the direction of a design, particularly if it has become stale or stuck is to create a personality for the product or space in question. Imagine your design as a person and give it a creative name that captures its personality; for example, 'Greta' or 'Marvin' or 'Tiffany'. Think about the design as a person and describe its personality traits; is it conservative? Outgoing? Dynamic? It may sound silly to do so, but attaching a persona to a product design can help to make decisions about what characteristics, design elements and features might be applied!

SCAMPER

SCAMPER is a checklist that helps you to think of changes you can make to an existing idea in order to create a new one. Use these changes either as direct suggestions or as starting points for new ideas or concepts. SCAMPER is particularly useful in product design but it can be applied to any concept that requires creative development.

You may not use all aspects of SCAMPER in every design – make use of the parts that are relevant and inspiring. In using the tool consider your application of Design Factors such as elements and principles of design, materials, sustainable practices. Give thought to the use and incorporation of different graphical representations.

SCAMPER stands for:

S – Substitute
C – Combine
A – Adapt
M – Modify
P – Put to another use
E – Eliminate
R – Reverse

Key questions using SCAMPER

SCAMPER	Actions
Substitute	Consider replacing all or part of your design with alternative options.
Combine	Create something new by combining parts of the design or introducing new combinations.
Adapt	Think about how the use of function of the design could be changed to suit a different purpose or set of circumstances.
Modify	Consider radical change to all or part of the design. Think about the distortion of some aspects.
Put to another use	Think about how the design could be used in another way. Could an aspect be sourced from another design?
Eliminate	Reflect on what could be removed from the design. 'Less is more', or is it?
Reverse	Consider completely 'flipping' one or more aspects of the design – physically or conceptually.

ISBN 9780170349994

Sample SCAMPER template

SCAMPER elements	Key questions to ask	What are the possible results in your design?
Substitute	What if I swap this for that and see what happens? Who else could find this appealing or useful? What other materials, design factors could I use instead? What happens if I substitute the shape, texture, form or colour?	
Combine	What elements or principles of design can be combined? What graphical representations could be combined?	
Adapt or Add	What part of the concept can I change? What if I were to use parts of other design elements or principles? What if I reuse aspects of my design in other ways or other places?	
Modify or Magnify or Minimise	What happens if part of the concept is expanded, exaggerated, minimised or changed? What is the effect of altering proportions and relationships in the design?	
Put to another use	What other function or use can my concept be applied to? Can another design feature from another product be used in my idea?	
Eliminate or Erase	What can be removed from my concept? What can be understated or streamlined? What happens to the design if parts are taken away?	
Reverse or Rearrange	What is the opposite of what I am currently doing? What if I did it the other way round? What if I reverse the elements or the way it is used? What happens if I mix up the design?	

GRAPHIC ORGANISERS

Graphic organisers are visual methods of arranging information and are designed to sort creative responses to a design problem. Used with brainstormed ideas, graphic organisers are a means to arranging a large number of early creative ideas into categories and sub-categories that can be evaluated for design directions. Graphic organisers utilise critical thinking skills that assess the effectiveness, suitability and possible directions of design concepts.

There are many graphic organisers available to you. A quick Internet search will present hundreds of alternative organisational tools to assist your design development. The key is to find tools that are the most productive and helpful to your design process.

CONCEPT MAPS

Concept maps are visual tools used to create thematic structures. Concept maps are quick and easy to review – just a glance will help to organise and identify connections and relationships between ideas. Compared with conventional notes, a concept map engages more of the brain in the process of connecting facts and ideas.

A complete concept map will have main topic lines radiating in all directions from the central subject. Subtopics, themes and ideas will branch off these, like branches and twigs from the trunk of a tree. You do not need to worry about the structure produced – this will evolve of its own accord.

ISBN 9780170349994

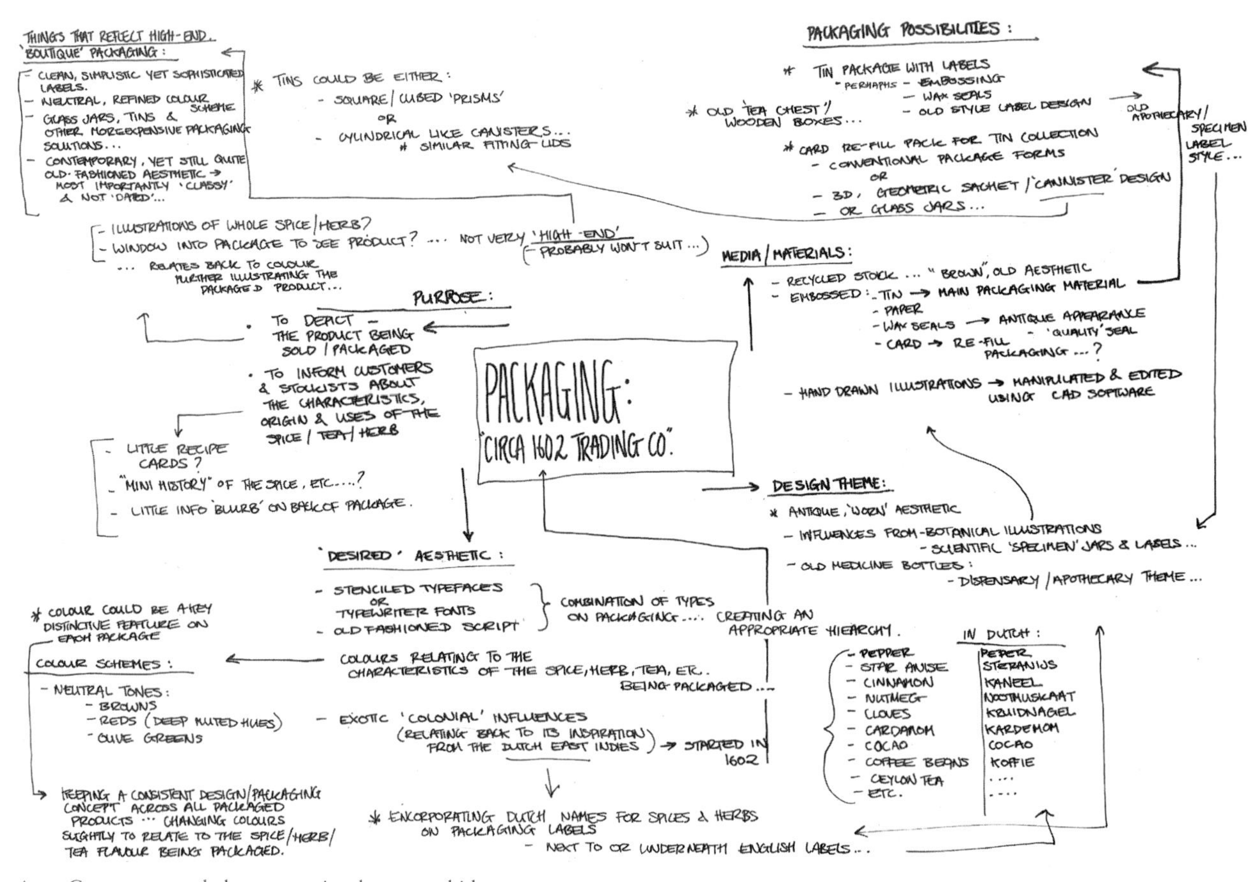

Johanna Schreiner

▲ Concept maps help to organise themes and ideas.

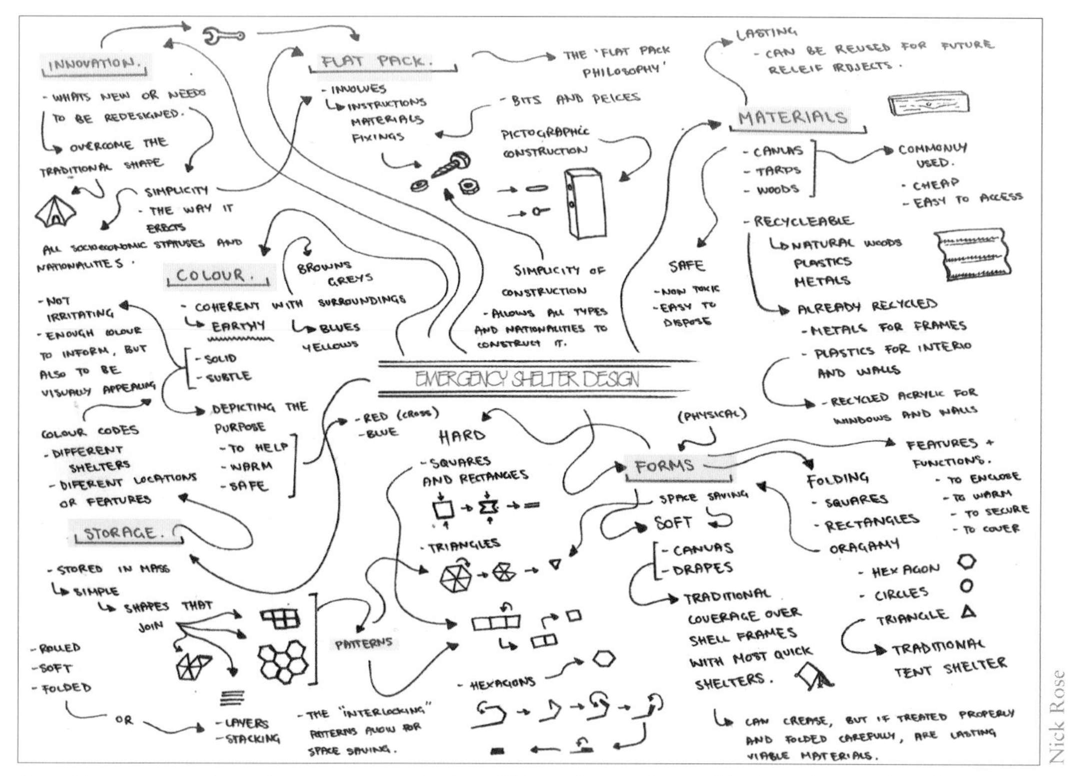

Nick Rose

▲ Concept maps do not have to use words alone. This student used a combination of text and small images.

ISBN 9780170349994

Sample concept map template

This template is a starting point but you can create your own concept map using any questions to form topic lines. For example, you may decide to use the Design Factors as your topic lines such as 'Sustainability', 'elements of design', 'principles of design' etc.

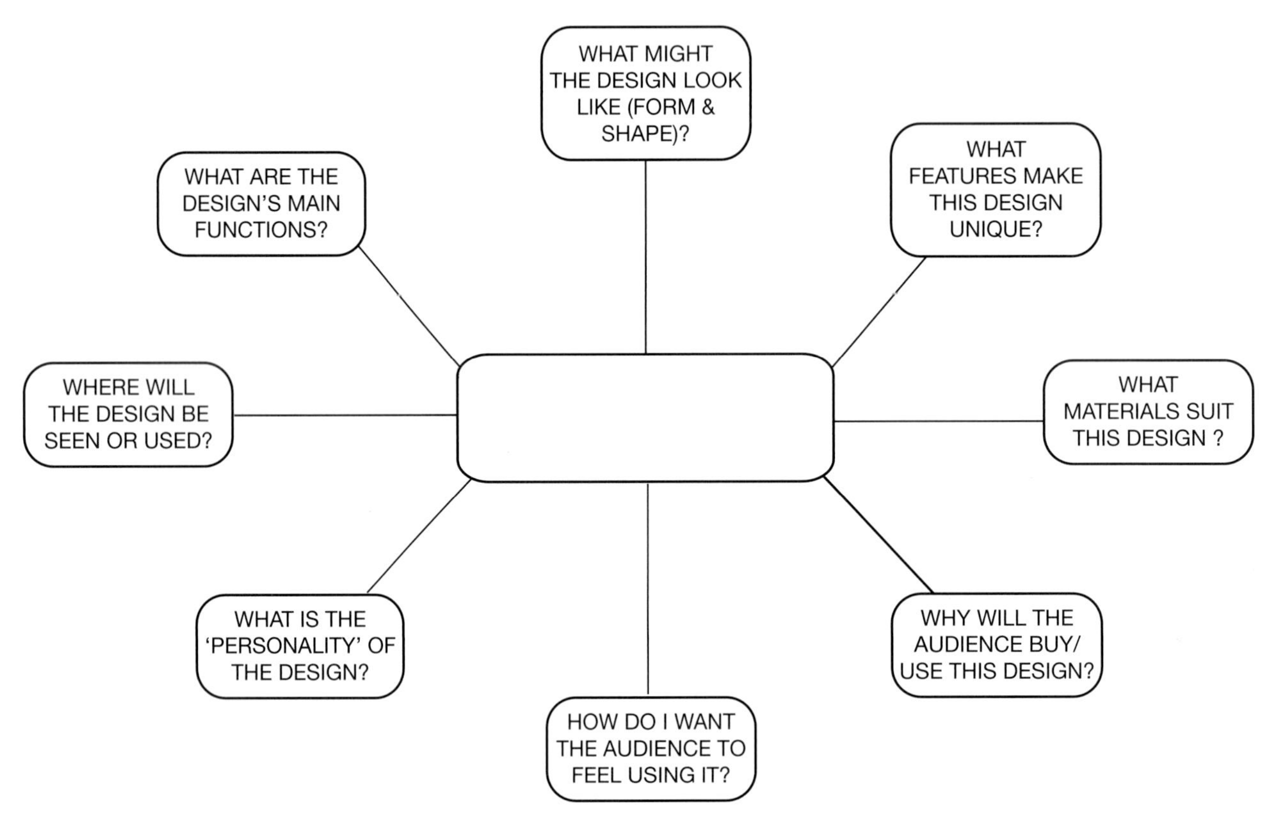

▲ Concept map template

AFFINITY DIAGRAM

An affinity diagram is a method of organisation often used by creative teams and project managers to sort brainstormed ideas into groups of related concepts. Brainstormed ideas are written onto cards or added to chart software, then the relationships between the ideas are identified and grouped into like areas. Ideas are sorted into groups that have an 'affinity' with, or relationship to one another.

Affinity diagrams are highly visual and can be created manually using coloured post-it notes or digitally using chart creation software. Colour coding is an integral part of the process as it helps to identify similar themes and ideas. Once formed into lists or groups, each idea can be prioritised to the top or bottom of the list. Ideas and concepts can be moved, removed and reorganised until a strong design direction is identified.

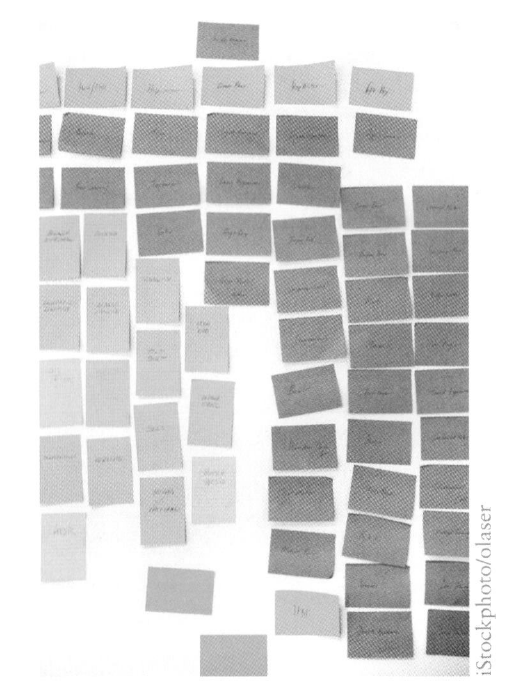

iStockphoto/olaser

ISBN 9780170349994

11.2 PROJECT MANAGEMENT

Planning for a design project is as important to the success of a design as the final product. It doesn't matter how good a final design is if the final deadline is not met. Understanding how long the design process will take including the production of final graphical presentations is essential to ensuring successful outcomes.

There are a number of tools available to designers and students alike that assist in managing time and productivity during the design process.

Many large design studios have project managers who organise scheduling and deadlines to ensure that projects meet their completion dates. In a school setting, it is invariably the teacher who sets deadlines but it is up to individual students to manage their time and output to meet the due dates.

Part B of this book introduced the design process, which lies at the core of your design work in Senior Graphics.

In planning your design project, it is possible to break the design process down into steps that may help you to schedule each step of the process.

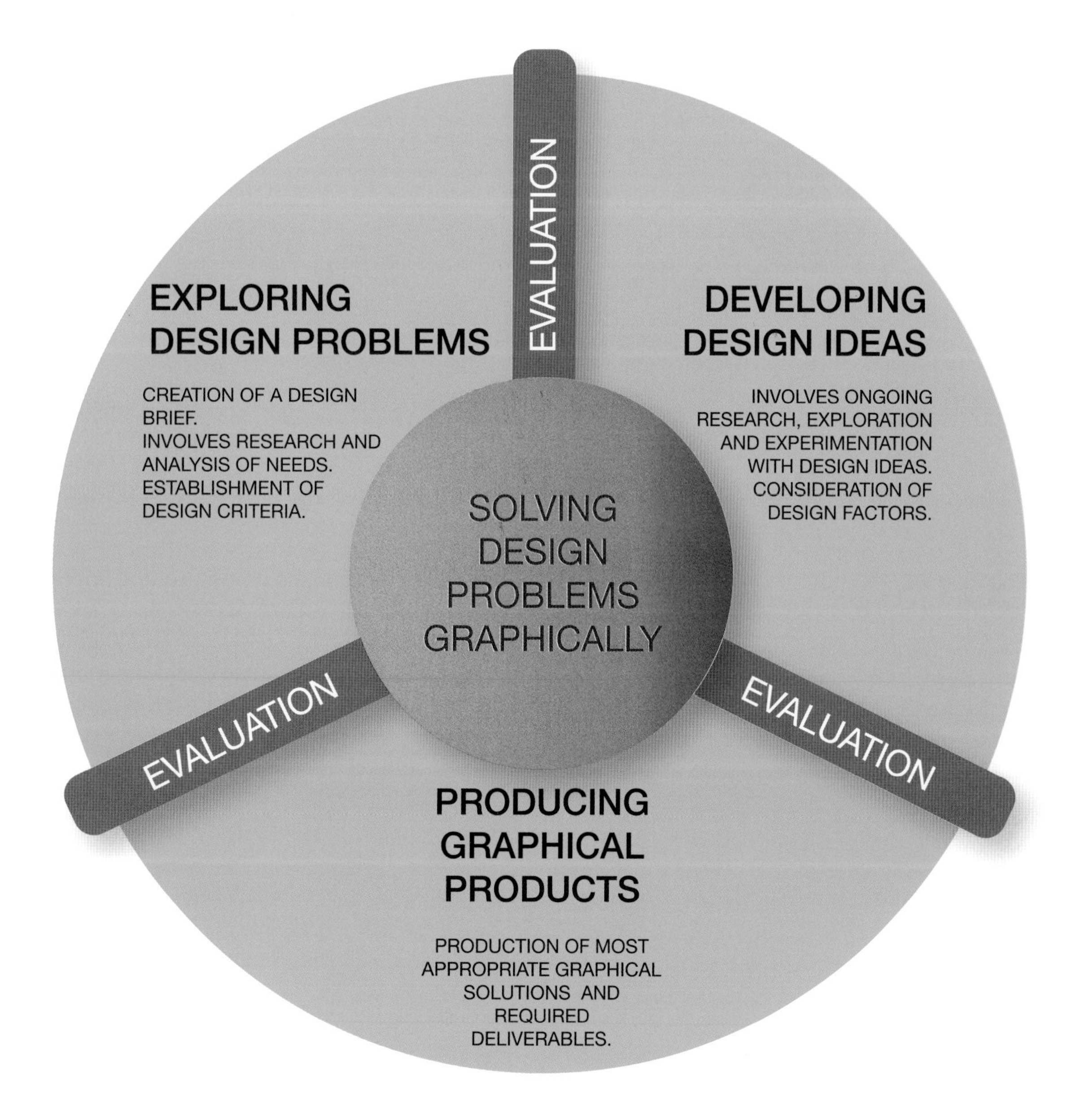

ISBN 9780170349994

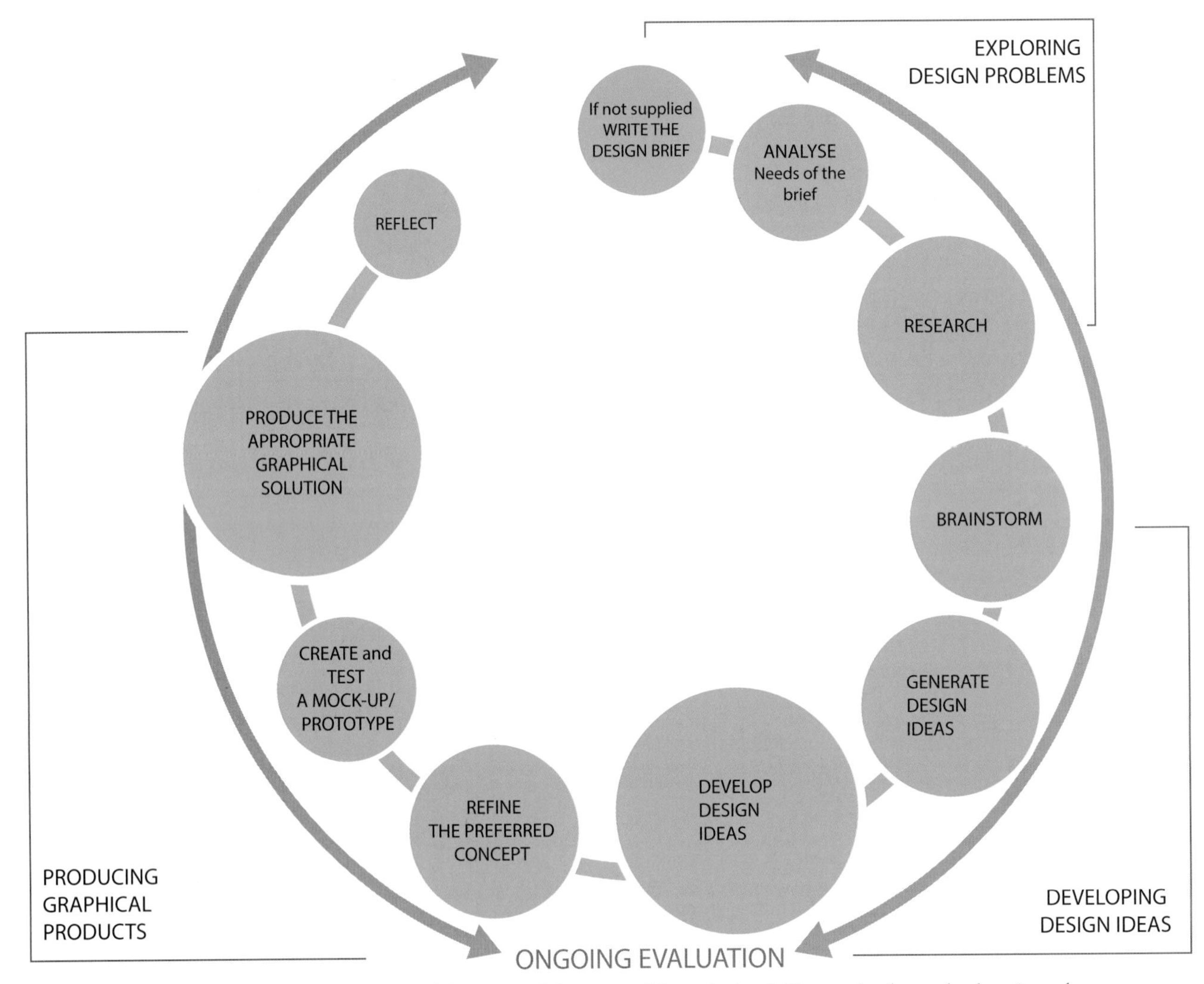

▲ Identify each step of the design process and be aware of the scope of the task ahead. Know what has to be done in each step.

TO-DO LISTS

To-do dists are prioritised lists of all the tasks that you need to undertake in your project. They list everything that needs to be done, with the most important tor most urgent asks at the top of the list, and the least important or urgent tasks at the bottom.

To-do lists can be daunting so it is important to organise them and mark off tasks when they are completed. Break tasks down into achievable steps rather than large, overwhelming jobs. Used with action items, a to-do list is an essential tool and ticking the completion boxes can provide a sense of control over the process.

ISBN 9780170349994

Sample to-do list

To Do	Completed	Action items
Interview member of target audience. Write list of questions. Formulate and send email. Gather data and analyse for helpful information.		Send email to interviewee.
Research existing designs. Collect images and sources. Annotate – identify connections with my design.		Read Chapter 5.
Develop concept map.		Read Chapter 11.
Become familiar with AutoCAD software in preparation for final production of graphical representations.		Speak to teacher about access to software.

ACTION ITEMS

An action item is a task that needs to be completed in order to reach a major goal. Unlike to-do lists, which feature diverse tasks, action items generally relate to a single objective. Action items can be used in combination with a to-do list at any stage of the design process and can assist in propelling the project forward as each task is ticked off the list.

SCHEDULING

A calendar or diary is the simplest form of scheduling available. There are many software programs available to assist with scheduling from the calendar on your phone to a spreadsheet in Microsoft Excel. All offer means of visualising the length of tasks.

Gantt charts are one of the most common scheduling tools applied in project management. They are a combination of bar graph and timeline that indicates what tasks need to be done and their due date.

In addition to stating what and when, a Gantt chart also visualises the length of the task, where the task overlaps with other tasks and the start and end of the entire project. Arrows and lines show where tasks are connected to one another. Gantt charts are usually created using specific software.

The key to an effective Gantt chart is first identifying the key tasks that need to be completed. Use colour coding to distinguish each task and be realistic about the amount of time each task is likely to take.

TASK	TERM 1							
	WEEK 1	WEEK 2	WEEK 3	WEEK 4	WEEK 5	WEEK 6	WEEK 7	WEEK 8
PLANNING								
WRITING THE DESIGN BRIEF								
RESEARCH								
BRAINSTORMING								
GENERATING IDEAS								

▲ Simple Gantt chart. The Gantt chart is a visual tool that enables the viewer to see what tasks are to be completed and when each task begins and ends.

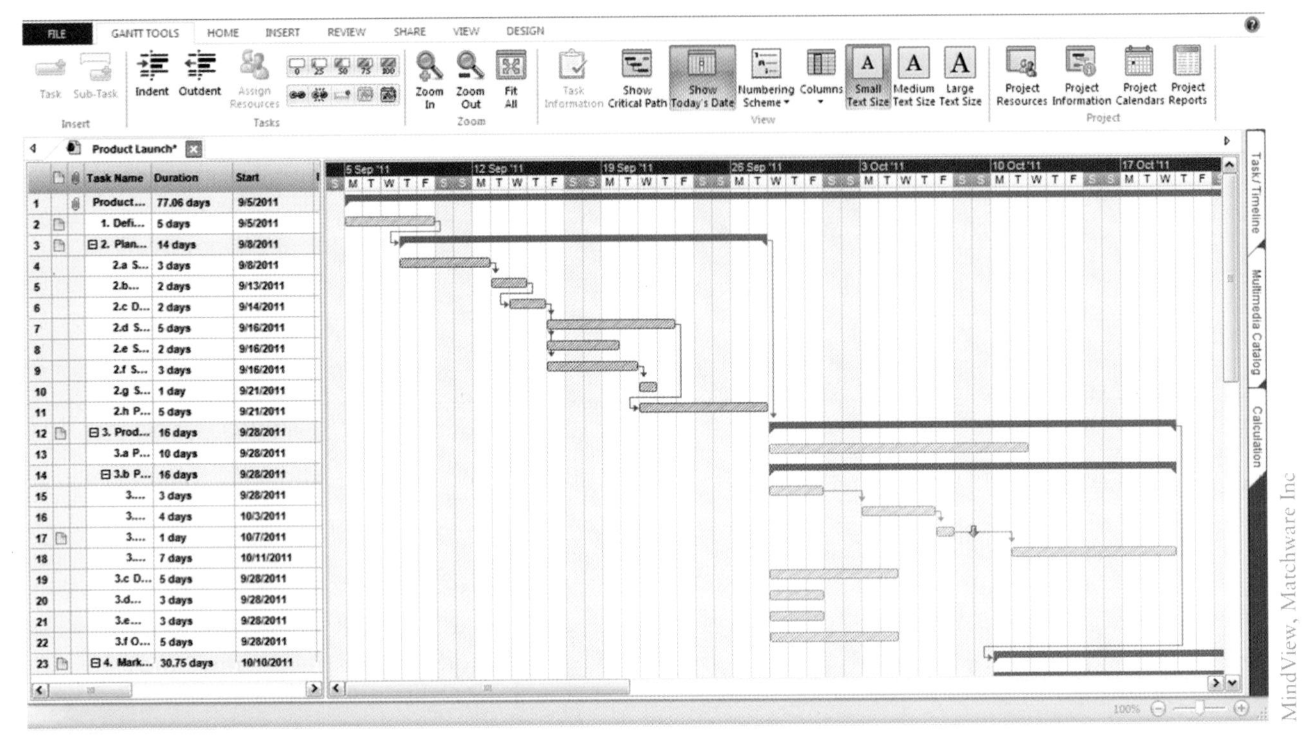

▲ Complex Gantt chart created with MindView software

MindView, Matchware Inc

TIPS FOR STAYING ORGANISED

1 Write down the due date.
It is essential to understand just how much time you have. Keep deadlines on hand in your phone or diary.

2 Make lists.
Prepare a list of the design criteria that you must include as specified by your teacher. Use of to-do lists is recommended to mark off completed tasks.

3 When given a choice, select a design task that gives you scope to build skills.
Challenging yourself will help you to stay focused. When faced with a choice, select a design task that provides you with the opportunity to build your skills and showcase your creative abilities. Don't stay in your comfort zone – you'll get bored quickly.

4 Select a task that interests you.
Given that you may be working on a task for a number of weeks, it is important that you select a design task that will maintain your interest and enthusiasm over that time. Challenge yourself, but take on a task that is achievable within the time frame.

5 Be organised.
If you are working on paper, maintain a well-organised workbook. If you are working entirely on screen, keep your folder structure and desktop organised. Make notes of the sources of images that you use. Back-up your computer files often.
Organisation goes for your space as well; make sure you have plenty of room to work without distraction. Surround yourself with creative materials that inspire and motivate you.

ISBN 9780170349994

CHAPTER 12
SUSTAINABILITY

'Providing a real way to deal with the urgent needs of environmental issues that is accessible to everyone is the new fight for the democratization of design.'

Philippe Starck

'Sustainability is about ensuring that the wellbeing of society – the combination of community liveability, environmental sustainability and economic prosperity – is maintained or improved over time.' (Department of the Environment, Commonwealth of Australia, 2012)

The things that we design, create and consume contribute to environmental impacts, which mean that sustainability is a consideration in all aspects of design and manufacturing. Although issues of sustainability, including recycling, waste minimisation and sustainable practices, are identified in the 21st century, that hasn't always been the case.

Developments in manufacturing during the Western Industrial Revolution of the 17th to 19th centuries saw the rapid growth of factories, mines and cities, changing existing economic and cultural structures.

Prior to the Industrial Revolution, manufacturing existed on a local scale and many communities were focused around fishing, farming and cottage industries. Unsurprisingly, sustainability factors remained relevant, if not applied, as the success of a community often depended on the management of natural resources such as water or soil quality. Issues such as deforestation, soil salinity and water pollution could adversely affect agrarian communities whose livelihoods depended on the production of crops and animals.

Technological advances in Europe and the Americas between 1650 and 1850 saw increased use of fossil fuels. Coal, in particular, was used to power engines and for the generation of electricity. In hand with such advancements came improvements in health, sanitation and transportation, which in turn led to significant population growth.

Within industrialised countries, cities expanded to meet the needs of an increasing population.

ISBN 9780170349994

Densely populated urban areas grew to meet the demands of people attracted to the facilities, services and employment opportunities offered by large cities. The march of technological advancement and the related consumption of resources was seen by many as inevitable progress and continued well into the 20th century.

Science, Industry & Business Library, The New York Public Library, Astor, Lenox and Tilden Foundations

The impact of industrial developments influenced art and design in a range of ways. Between 1850 and 1914, the Arts and Craft Movement became prominent in the UK, Australia and the USA. The movement eschewed the industrialised mass production techniques that had evolved over the past 200 years in favour of a handcrafted and artisan-based approach.

Designers of the Arts and Craft movement including William Morris, Charles Rennie Mackintosh, Frank Lloyd Wright and Alexander Knox looked to the natural environment for inspiration and their work often uses visual motifs directly sourced from flora. Hand-made and carefully crafted, the work was often manufactured using slow, traditional techniques, which meant many pieces were often rare and expensive.

Getty Images/Bridgeman Art Library

▲ Morris wallpaper from the Victoria and Albert collection

Other designers embraced technological advancements and used materials that were previously considered industrial in furniture design, architecture and interior architecture. Le Corbusier applied steel to both furniture and domestic interiors. The Constructivists, the Bauhaus and deStijl, used forms, textures and materials in their designs that reflected an industrial and manufactured aesthetic.

High backed chair, c.1897 (dark stained oak with rush seat & pierced oval back rails) (b/w photo), Mackintosh, Charles Rennie (1868–1928)/Private Collection/Bridgeman Images

◀ 'Argyle' chair by Charles Rennie Mackintosh, 1898, Powerhouse collection

ISBN 9780170349994

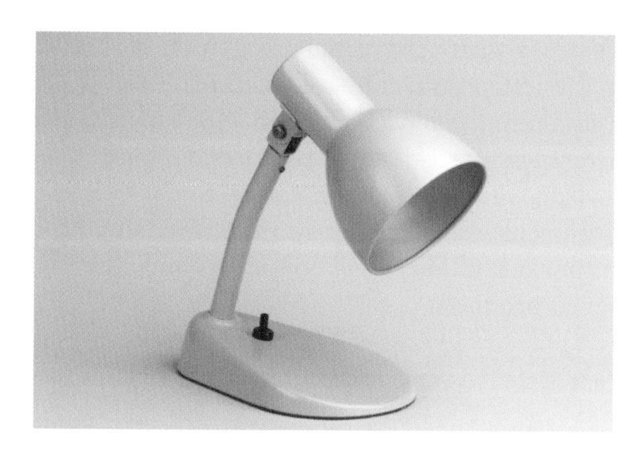

▲ Marianne Brandt (1893–1983): Kandem Bedside Table Lamp, 1931. New York, Museum of Modern Art (MoMA). Lacquered steel h. 9 1/4' (23.5 cm), base w. 7 1/4' (18.4 cm) Phyllis B. Lambert Fund. Acc.n.: 191.1958 © 2012. Digital image, The Museum of Modern Art, New York/Scala, Florence

In the 1950s and 1960s Charles and Ray Eames also adapted developments in technology to their designs. Curved wood and moulded plastic features in the furniture of the Eames partnership. Many of the techniques applied by the Eames' evolved from manufacturing advancements stemming from the Second World War. History shows that technologies that develop during wartime are often, ultimately applied in design.

Post-1945 lightweight, flexible and durable materials, such as plastics, were manufactured more readily and cheaply, providing scope for the refinement of many products. Domestic appliances made of lightweight materials became more portable, cheaper to purchase and easier to use, creating vast markets in products that were cheap and eventually, disposable. Issues of waste, pollution and increasing landfill needs for discarded products became prominent in the late 1960s and into the 1970s.

V&A Images, Victoria and Albert Museum

▲ Emerson Patriot Radio, designed by Norman Bel Geddes, mid-20th century

▲ Charles EAMES (designer) United States 1907–78, Ray EAMES (designer) United States 1916–88, HERMAN MILLER, Michigan (licensee) United States est. 1923, HERMAN MILLER (AUST.) PTY LTD, Melbourne (manufacturer) Australia est. 1962 Lounge chair 670 and Ottoman 671 (1956) (designed), (1972) (manufactured) leather, plywood, aluminium, nylon, zip, (other materials) (1) 85.2 × 86.2 × 83.3 cm (lounge chair) (2) 42.7 × 65.6 × 54.6 cm (ottoman) Purchased, 1972 (D80.1-2-1972)

ISBN 9780170349994

Following a significant global energy crisis in the early 1970s, Victor Papanek published a book called *Design for the Real World*, which outlined the responsibility of the design community in developing and using sustainable materials and reducing environmental impacts through design. In the 1980s, environmental concerns gained public momentum and companies began to use 'green awareness' to distinguish their products within the marketplace. During the 1990s and 2000s, awareness of greenhouse gas emissions, climate change and deforestation meant that consumers began to demand alternative and recyclable materials.

Volvo Car Group

▲ Increased consumer awareness of greenhouse gas emissions and the rising costs of fossil fuels has seen large growth in the design and manufacturer of hybrid vehicles.

Greater awareness of the impacts of industrialisation, including the use of fossil fuels, the consequences of urban population growth, and environmental and climate change, has evolved over the past 30 years. Access to bodies of information and images via traditional and social media mean that consumers can make informed decisions when building a home, purchasing a product or accessing energy resources.

In 2012, the Australian Government outlined its vision for a sustainable Australia and settled on three distinct indicators of sustainability: social, environmental and economic. The notion that sustainability reaches into most aspects of contemporary life is shared among academics, governments and organisations globally; it is a complex and evolving issue.

Along with governmental recognition of the need for action on sustainability, individuals and communities have developed a high level of awareness of sustainable practices. Greater knowledge of the human impact on the environment has led to significant and identifiable shifts in expectations from consumers and users of designed products.

KEY POINTS ~ SUSTAINABLE FACTORS

Social sustainability

Targets for social sustainability aim to ensure the long-term availability and access to education, health and employment, which provide both personal and community security and wellbeing.

Environmental sustainability

Environmental sustainability aims to ensure that the natural environment is monitored, protected and managed and refers to climate, land and ecosystems, water, waste and natural resources.

Economic sustainability

Economic sustainability applies to the maintenance, development and management of factors including wealth and income, transport and infrastructure, productivity and business innovation.

Design, by its creative and innovative nature, is primed to respond to change and, in many areas, has adapted to the demands of sustainable practices. Design professionals have played a major role in bringing the design of energy efficient homes, the application of renewable resources in product design and the use of sustainable materials to the mainstream. Design has been used to build public awareness of issues relevant to the achievement of sustainable targets and graphic design has played a major part in the communication of these ideas.

ISBN 9780170349994

The minimisation of waste and the application of lifecycle principles in design are issues that face professionals in the industrial design, built environment design and graphic design fields. One of the implications of designing products for mass consumption is the increased waste due to high production volumes. The choices made by designers about materials, packaging, energy use, waste and recyclability may reflect their personal and professional values but may also be influenced by the expectations of both client and end user. The use of ethical, sustainable and appropriate materials is increasingly an expectation of informed consumers. In all fields of design, professionals consider the long-term impact of their decisions in the earliest stages of the design process. Cost considerations may constrain the selection of materials used in a design, but advances in 'green' and 'eco-friendly' materials development may now offer wider choices.

TranSglass recycled glass designed by Tord Boontje and Emma Woffenden for Artecnica

▲ tranSglass® recycled glass by Emma Woffenden and Tord Boontje, 2006

In some areas of design, the use of sustainable products and the application of energy efficient and environmentally friendly practices are mandated by law. Standards set by government bodies, including Standards Australia, may specify that the design of products and particularly that of buildings must have strict environmental codes applied. Built environment designers are often required to respond to constraints related to sustainability issues.

12.1 LIFECYCLE ASSESSMENT

Lifecycle assessment (LCA) is a technique used in industrial design for assessing the environmental impacts of a product across its entire lifecycle. The lifecycle references the period of time from extraction of base materials, throughout the design and manufacturing process, to the use and final disposal. The life of a product is also referred to as cradle-to-grave design; designs that are fully recyclable at their end of life are referred to as cradle-to-cradle designs. Observations of a product's lifecycle can tell a researcher or a designer about the burdens a product or service place on the environment.

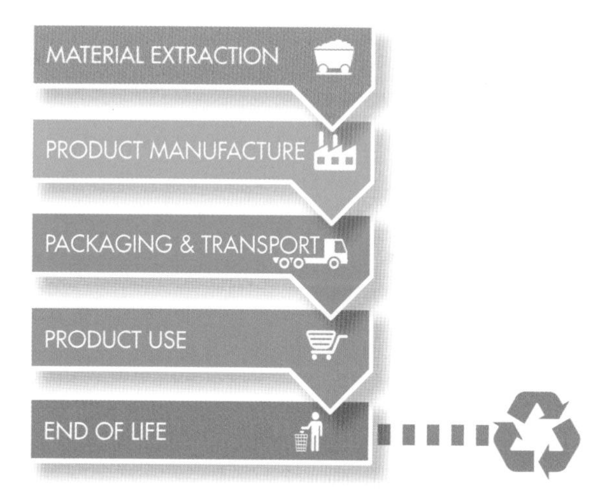

STEPS OF LIFECYCLE ASSESSMENT

Materials extraction and processing

This refers to the energy and impact of sourcing the raw materials for a design. It may relate to products such as wood (from forestry), minerals (from mining) or materials sourced to create composite materials (production of concrete, fibreglass etc.).

ISBN 9780170349994

Methods to minimise the impact of this stage might include:

+ using minimal materials
+ using renewable materials
+ avoiding materials that deplete natural resources
+ using recycled and recyclable materials
+ avoiding materials that are toxic or hazardous
+ avoiding materials that deplete ozone or create greenhouse gases.

Product manufacture

This is the phase of the LCA where the designed product is produced using the sourced materials. Manufacturing processes vary and may impact the environment and economy in a range of ways. Waste from the manufacturing process, energy expenditure and the manufacturing location are some of the issues that may be addressed when analysing this stage of the process.

Methods to minimise the impact of this stage might include:

+ using a limited variety of materials
+ avoiding waste
+ reducing the number of separate components or assembly steps in the manufacturing process
+ simplifying functionality
+ selecting low-impact materials and manufacturing processes.

Packaging and transport

Once the product has been manufactured, distribution becomes the next priority. Given the global nature of much manufacturing, very large distances often need to be covered. In the design of some products, consumers have become increasingly aware of the 'mileage cost' of products that are transported by air and sea.

Methods to minimise the impact of this stage might include:

+ using lightweight packaging to save energy in transportation
+ creating recyclable or reusable packaging
+ maximising efficiency of packaging
+ selecting an efficient transportation system.

Product use

This stage of the LCA process addresses how designed products can be best prepared for their use by the target audience. There is little that can be controlled once the product has been purchased or distributed so it is important that the design of the product itself ensures the most efficient, durable and functional use when in the hands of the end user.

Methods to ensure the sustainability of this stage might include:

+ energy efficient designs
+ durable designs and lasting materials
+ functional and purposeful designs
+ designed for a range of conditions.

End of life

Ideally, there should be minimal waste at the end of life stage of the LCA. A sustainable design will have addressed the circumstances of end of life in the design phase, well before manufacture.

The most common end-of-life options are:

+ extension of life through durability (extension of life may also require designers to think ahead and plan for changes in trends, styles and colours)
+ reuse (refilling or exchange systems)
+ recycling
+ disassembly
+ degradability
+ safe disposal.

Designers are encouraged to analyse the lifecycle of their design as an integral part of their design process.

This is becoming a more common practice in all areas of design. Three steps are recommended in a lifecycle assessment:

1 List the processes and materials required for design and manufacture.
2 Evaluate the potential impacts associated with the identified processes and materials.
3 Analyse the results to make informed decision about the impacts of the proposed design.

ISBN 9780170349994

		Potential impacts	Proposed strategies to minimise impacts
What materials will be required by the product?			
What processes will be involved in the design and manufacture of the product?			
What processes will be involved in transport and delivery of the product?			
What is the impact of the use of the product?			
Evaluation			

12.2 STRATEGIES FOR SUSTAINABILITY

RECYCLABILITY

Consideration of the recyclable nature of materials is important in product design. The more parts of the product that can be recycled or can be manufactured from recycled materials, the more the impact of the product on the environment is reduced. The reuse of products for their original purpose is known as recycling (e.g. excess building bricks used for paving or on a new construction); the conversion of recycled products into a new product for a new purpose (e.g. plastic drink bottles into non-slip mats) is called down-cycling.

MULTIFUNCTIONAL DESIGN

Products that have more than one function can increase the value for the owner. A product designed for one purpose could be reused for a different function. For example, packaging might be able to be reused as a different product such as storage. If a product has multiple functions, fewer products are needed. Consumers may view products that have many functions as having higher value and retain them for longer periods of time.

Suck UK Ltd, www.suck.uk.com

Suck UK Ltd, www.suck.uk.com

▲ The bottle light is a fully rechargeable LED light that can repurpose empty bottles into long-lasting lights. See more at suck.co.uk.

▲ The Current table by Dutch designer Marjan Van Aubel incorporates solar cells that enable charging of devices. The solar desk functions as both a working space and a charger for electronic gadgets.

Materials selection

An effective sustainable approach is to select non-toxic, recyclable and low impact materials in the first place. All materials have a Materials Safety Data Sheet (MSDS) that can be used to assess suitability and impacts. Hazardous and toxic substances are best avoided but if they are to be used, identifying appropriate means of disposal at end of life is essential.

Qubies

▲ Locally designed product Qubies uses non-toxic materials to ensure that the storage of food is safe for consumers. The materials selection is free from hazardous substances, which makes it appealing to its target market – mothers and families.

Durability

Designing quality products that are built to last means less waste is produced in the long term. Quality products are more likely to be reused, repaired, resold or recycled. The longer a product lasts, the less likely it is to be replaced. Designing for longevity requires careful selection of materials and the use of processes that ensure solid and enduring construction. The appearance of a design can communicate its longevity; design that appears durable and long lasting helps to reinforce its durability. Extending the life of a design may reduce consumption and waste.

Fairfax Syndication/The Age/Gary Medlicott

▲ Shipping containers are durable and long-lasting products that have been adapted for a range of uses including housing.

Fairfax Syndication/The Age/Wayne Taylor

▲ The structural integrity and consistent form of shipping containers mean that multistorey, multifunction dwellings can be created quickly and creatively. This dwelling makes use of existing features such as the container doors.

ISBN 9780170349994

Dematerialisation

Reducing the amount of material used in a design is a straightforward way of minimising environmental impact. Similarly, reductions in the weight, volume and size of product can lead to less waste at end of life. An example is housing design that uses a minimal range of materials.

HAY, Denmark

▲ The Nobody chair by Hay design is moulded from one piece of material. Using a technique borrowed from the car industry, the chair is manufactured from two layers of thermopressed PET felt – a 100% recyclable material made from used water and soda bottles. The chair is stabilised without the use of an internal frame, plastic, screws, glue or other reinforcements.

Efficiency

Products, particularly those that use electricity, need to be efficient in their use of energy. Creating 'standby' or 'sleep' functions reduces power use. Designs might encourage users to turn off or unplug by using automatic switches or timers. Cost-aware consumers are often attracted to energy-efficient products.

Fenner School of Environment and Society, The Australian National University

▲ The Frank Fenner Building at the Australian National University (ANU) was the first building in the Australian Capital Territory to achieve both 6 Star Green Star Office Design and As Built v3 ratings from the Green Building Council of Australia (GBCA).

Disassembly

Products that are designed to be easily taken apart at their end of life can make the recycling process more efficient. Consideration in the early stages of a design about the structure and functionality of a product can predict the speed and ease of taking it apart for recycling.

Studio David Graas

▲ The cardboard lounge designed by David Graas comes flat packed for assembly. If disposal or storage is required the design can be completely disassembled.

Degradability

Materials that can be broken down organically though compositing are known as degradable materials and can be used in some products. These include paper, cardboards and starch-based plastics. Glass, metals and petrochemical-based plastics take many years to degrade and are best recycled or reused.

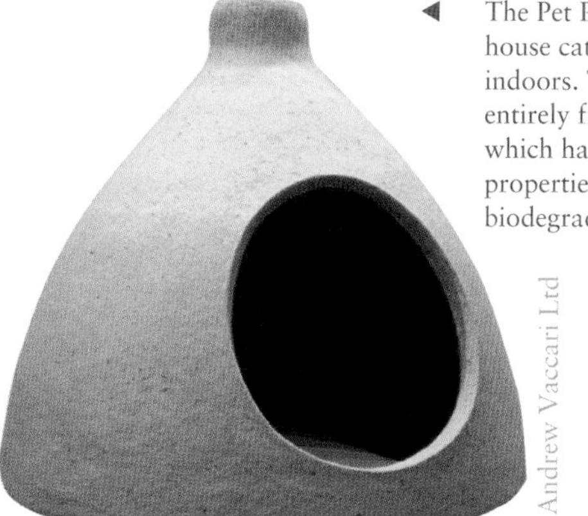

Andrew Vaccari Ltd

◀ The Pet Pod is designed to house cats and small dogs indoors. The pod is made entirely from papier maché, which has excellent insulation properties and is completely biodegradable.

ISBN 9780170349994

Product responsibility

Increasingly companies are encouraged to take full responsibility for their products. In recent years and, in some cases due to consumer or government pressure, companies have introduced policies designed to address issues with the end of life of products. Recycling schemes and returns procedures enable products to be remanufactured or recycled.

▲ Global technology company HP states that addressing the end of life of its products is a priority and central to its efforts to decrease environmental impacts across its value chain. HP started a hardware recycling program in 1987 and an HP LaserJet cartridge recycling program in 1991. Today, HP provides product take-back programs in 70 countries and territories, including in developing regions where the company says it is raising recycling infrastructure standards. 'After a customer returns his or her hardware product, our priority is to determine the best recovery solution for it. When equipment has resale value, we prefer to refurbish it and resell it, the option with the lowest environmental impact. We work with a range of reuse and recycling vendors to ensure environmentally responsible options for processing HP products at end of life and commission third-party audits to track the performance of our global recycling network.'

RESOURCES FOR SUSTAINABLE DESIGN

There are many web resources available that outline strategies that designers can undertake for more sustainable and efficient design products.

Design can change

United Nations D4S: Design for Sustainability

Treehugger

Information about the latest contemporary and sustainable designs in many design areas.

Access all weblinks directly at http://nsg.nelsonnet.com.au.

In encouraging sustainable design practices, governments aspire to provide secure housing, safe communities, clean water, employment through industry and manufacture, waste minimisation and a raft of other benefits for present and future citizens. It is not always an easy undertaking and pressures of industrial progress, economic policies and social expectations can hinder developments in sustainable practices.

Designers and students of design can apply their knowledge of sustainable practice at all stages of the design process. In applying lifecycle assessments, awareness of social, economic and environmental impacts are made clear and visible.

ACTIVITY

Observe a range of products from one room of your home such as the kitchen or bathroom. Analyse how each may be categorised under the strategies for sustainability, recyclability, multifunctional design, degradability, dematerialisation, disassembly, etc. Do they fulfil any of the strategies?

Select one or more of the analysed products and redesign it to address multiple sustainability strategies.

ISBN 9780170349994

CHAPTER 13
REPRESENTING MATERIALS

'Many of my structures are not meant to be permanent. I've always been interested in using material in a new way.'

Shigeru Ban

Applying materials to objects helps a designer create a sense of realism and enables the viewer to recognise the characteristics of a design. The inclusion of textural details, tone, colour and pattern helps to define the features and forms of objects and spaces. Designers in all design areas apply the rendering of materials to some aspects of their work. Whether executed by hand using sketches or refined in CADD software, the application of materials is key to communicating the specific qualities of a design.

Observe and familiarise yourself with the textures around you. Your clothing, the table, the carpet and flooring all reveal different textural qualities. Once you begin to observe and practise drawing the textures around you, you will see how it becomes easier to create the appearance of materials.

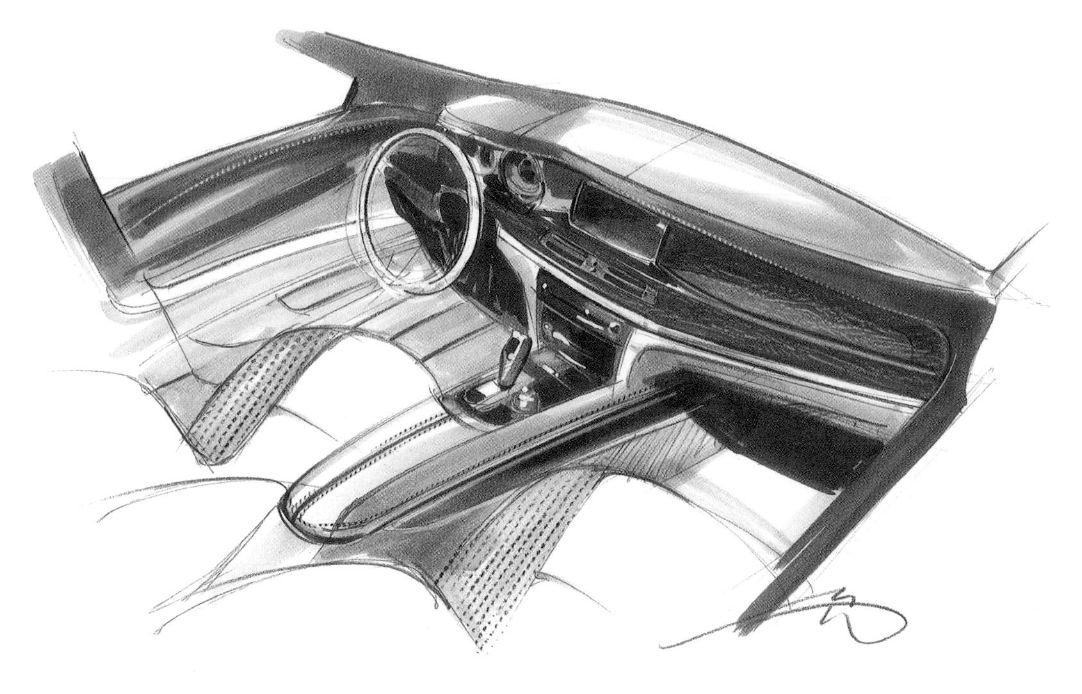

ISBN 9780170349994

13.1 NATURAL TEXTURES

Textures that occur in nature are rarely uniform and have characteristics that are not found in human-made materials. Natural textures such as wood include the grain of the timber, knots and other imperfections due to age, damage and weather.

When using natural textures it is important to include details that give an authentic appearance. However, as important as it is to incorporate texture realistically, you should judge just how much detail is required. Too much detail can detract from the purpose of the drawing.

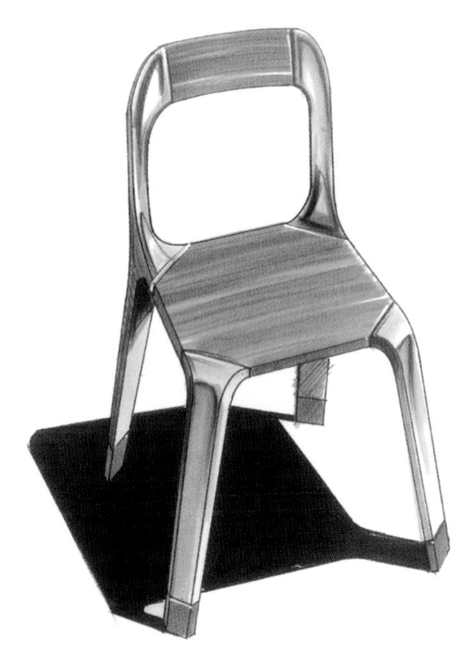

▲ Irregular lines can create the unique character of wood grain. Avoid being too uniform in your representation.

It is not always necessary to fill an area with textural detail, as a section of texture can often convey enough information. Stone, grass and foliage are commonly depicted in illustrations, and small sections of texture can communicate the characteristics of a given area rather than filling space with detail. When drawing trees avoid representing every leaf and branch as this can cause the image to appear contrived.

TIP

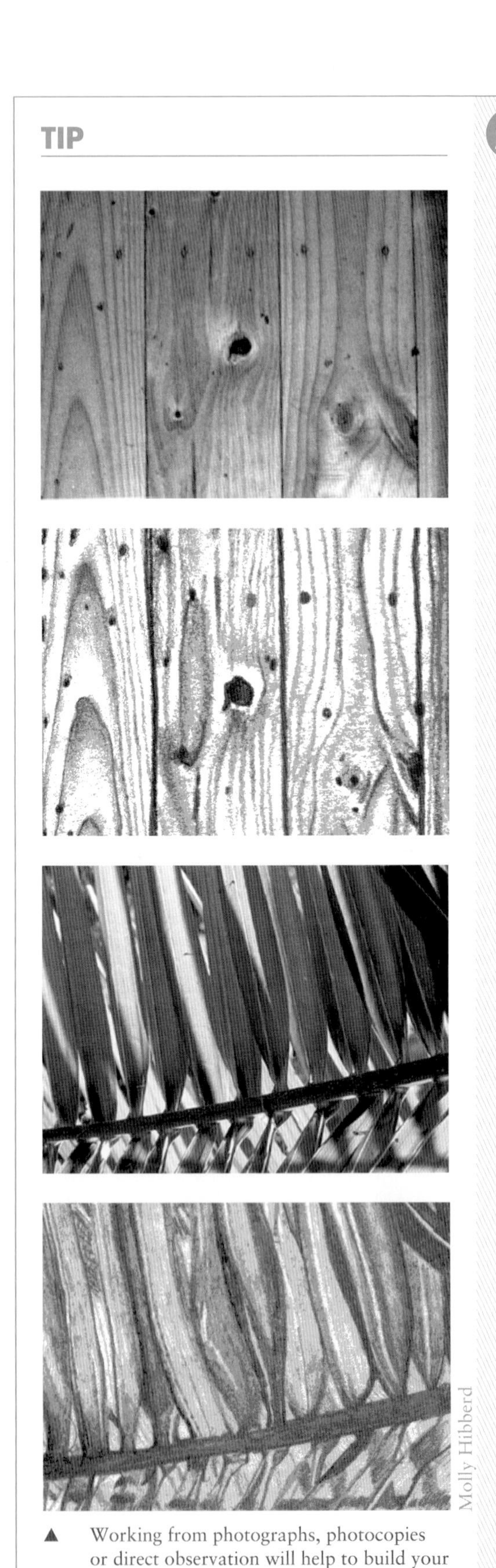

Molly Hibberd

▲ Working from photographs, photocopies or direct observation will help to build your skills and confidence in rendering effective texture by hand.

ISBN 9780170349994

By using line and tone it is possible to suggest the textural qualities of natural forms through a stylised representation. The way you choose to represent natural forms depends on the purpose of your drawing – a cartoon has a very different purpose to an artist's impression of a landscape design.

13.2 TEXTILES

The diversity of available fabrics makes for equally diverse methods of illustration. In the fashion and furniture design areas, the representation of fabric texture becomes very important.

▲ Leather handbag. Note the inclusion of some grain and a form that suggests stiffness.

When representing fabrics, such as woven cloth, you should observe the direction of the threads. Woven fabrics – whether created by hand or by a machine – have threads that travel in two different directions. Called the weave and the weft, these threads will reflect light differently. It is rarely necessary (or advisable) to show every thread but it is important to appreciate that fabric is not flat and mono-directional, and to convey this in the rendering.

▶ Fashion designers use style sketches to identify the basic form of a garment and use rendering techniques, such as crosshatching, to identify the texture.

Textile illustrations convey the qualities of fabric through freehand drawing and rendering. They are used to show not only the form of a garment or household item, such as a cushion of throw, but also the characteristics of its fabrics. Rendering in fashion drawings shows not only the appearance of the fabric but also its physical characteristics such as reflective or transparent qualities, the richness or texture of the material, or the layering of separate fabrics in one garment. Fashion drawings, by their nature, tend to be loose and may use a range of methods such as marker, pencil, watercolour, collage and mixed media.

▲ Interior designers often need to represent multiple materials in a single interior. Application of tone and the use of cross-hatching can indicate where textures exist in an environment.

ISBN 9780170349994

13.3 METALLIC AND REFLECTIVE SURFACES

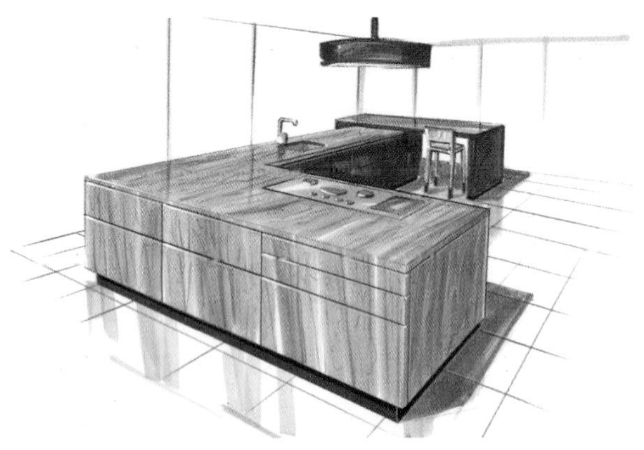

▲ When drawing high-shine or reflective surfaces, be sure to leave some areas white to emphasis their reflective qualities.

Materials that reflect light – such as glass and metal – can be challenging to draw. The textural characteristics of reflective surfaces are often smooth or slick.

Sallyanne Hunter

▲ To further enhance the appearance of reflective textures, you can identify the light source with a flare of light.

Metallic surfaces such as chrome have no colour of their own and only reflect the surrounding colours. To capture the appearance of metallic objects, colours should be crisp, intense and bright. The application of colour often depends on the shape of the object to be drawn.

Sallyanne Hunter

▲ A metal cylinder may use a series of bands of colour, which, along with a white highlight band, serve to reinforce the cylindrical and reflective nature of the object.

When drawing glass, illustrators and designers often use a series of horizontal or vertical lines to indicate an otherwise clear or transparent surface. This technique is ideal when working with pen or pencil.

Glass absorbs colour, so – when using other media such as markers – represent glass by drawing layers of subtle colour, such as cool greys and light blues. Remember that glass is transparent, so you may need to render what is behind the glass, as well as any reflections. Reflections on the glass appear as white, which can seem to be floating on the surface.

In glass objects, where the material is thickest, you will often see compressed reflections appearing as areas of blacks and whites; the base of a glass for example. Noticeable shadows are often cast from these more solid areas of glass objects.

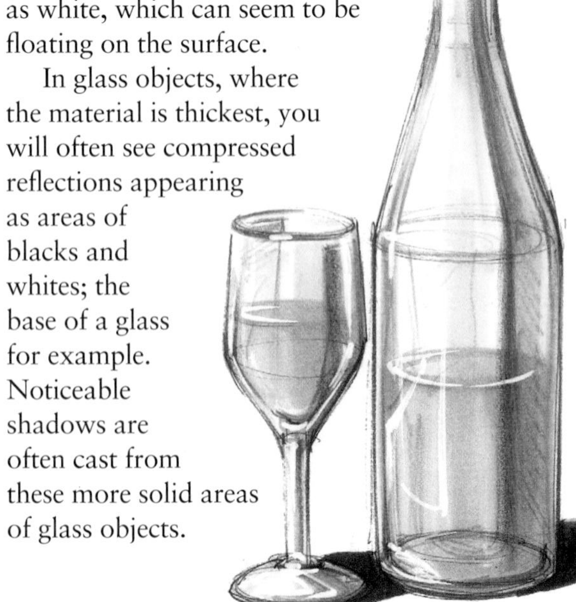

ISBN 9780170349994

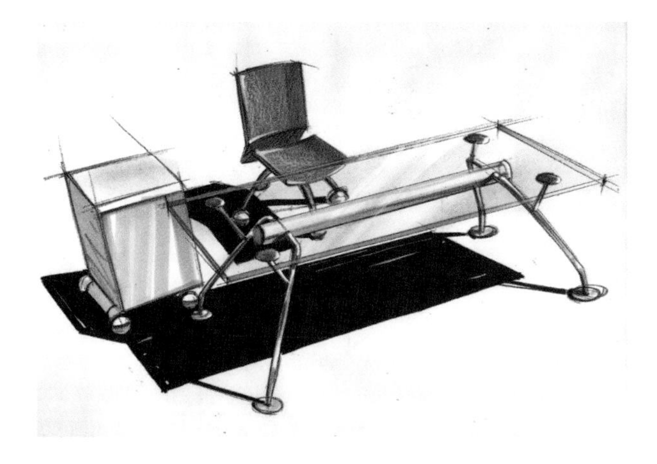

13.4 PLASTICS

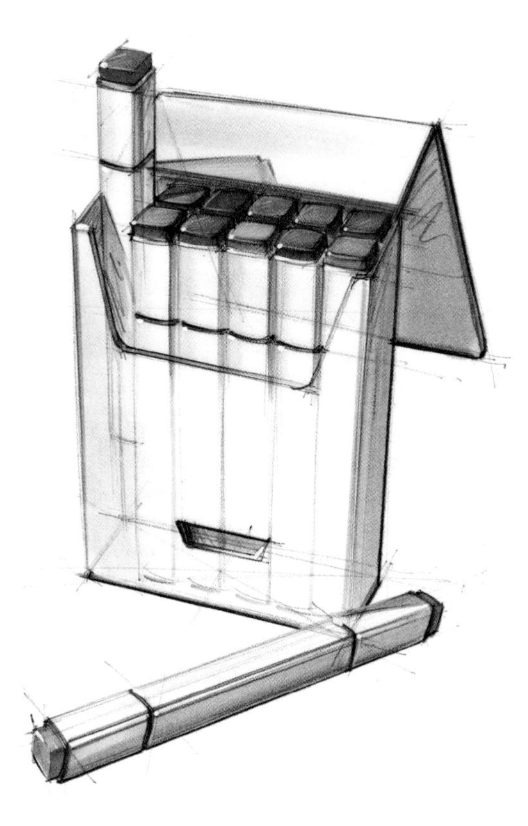

Acrylic materials and plastics often reflect light in the same way as other reflective surfaces. However, there are many matt plastics that show little or no reflection.

The properties of acrylics and plastics allow moulding and shaping into a wide range of shapes and forms.

Acrylic materials can be manufactured in a vast range of colours and textures. The colours of acrylic products often appear to be saturated and vivid. When rendering work from the lightest area to the darkest, build layers of colour to achieve a saturated appearance.

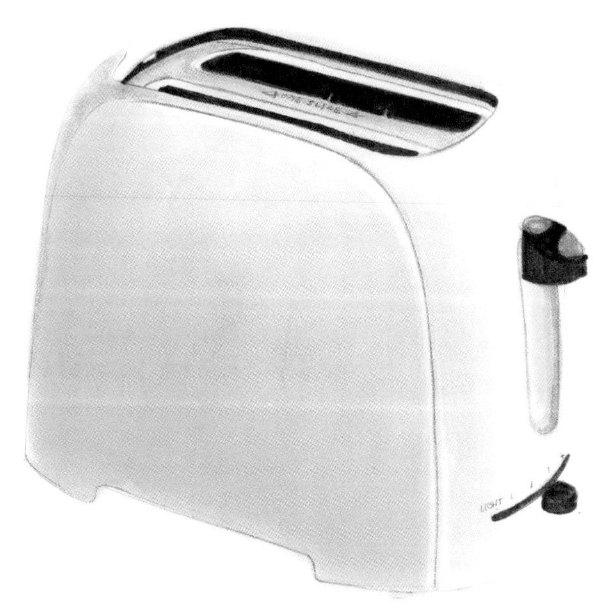

Sallyanne Hunter

ISBN 9780170349994

Leave some areas completely white to represent reflections and suggest form. Although plastics are not as highly reflective as metallic surfaces, there will always be highlight areas, and these should be indicated.

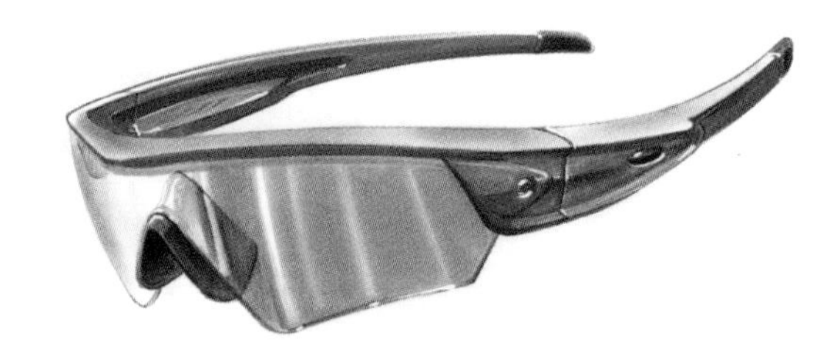

13.5 ECO MATERIALS

Eco materials include products that are made from recycled components or materials. They often feature colours and textures that are natural and that suggest an ecologically sustainable manufacture. Often, eco products are packaged in recycled substrates such as unbleached card or paper. Eco products tend to lack the glossy surfaces that traditional products might favour. In rendering eco products it is advisable to utilise a muted palette of browns and greens and focus on representing matt surfaces with little reflective qualities.

The rendering of eco designs generally focuses on the recycled nature of the materials, suggesting natural or fibrous textures rather than polished, shiny surfaces.

13.6 COMPOSITES

Composites are materials the combination of two or more materials, which together produce a new material. Composites are often created for their strength and durability. Examples of composite materials include concrete, fibreglass, carbon fibre and plywood.

▲ Wood composites include those created with plastics. Although they look like wood, they are usually more durable and weatherproof. In rendering composite wood products, it is feasible to feature some of the more reflective qualities of plastic as well as wood grain.

Although composites may appear to have the characteristics of a natural material, there may be slight visible differences. Invariably, as composite technologies develop, it becomes harder to tell the differences between natural products such as wood and wood-look products made with plastics.

Concrete can be polished or left in a natural state so rendering of concrete products will vary according to the design. A polished surface will have reflective qualities.

ISBN 9780170349994

Fibre products, such as carbon fibre, can present with both matte and reflective surfaces so their rendering needs to reflect the characteristics of the design product. The use of some highlight areas can achieve this.

13.7 CERAMICS

Ceramic products can be matt or gloss in texture. The surface qualities of ceramic designs are affected by the glaze and surface detail that is applied. Glazes include gloss, satin or matt but the object may also be left unglazed and in a natural state. Similar to plastics, gloss and satin glazed ceramics reflect some of their environment. Highlights assist in emphasising the areas that are reflective, while light source is key to emphasising the forms. Matt and unglazed ceramic products do not reflect and require thoughtful application of light and dark tones, along with textural details to appear three dimensional.

ISBN 9780170349994

EXTEND YOUR PRACTICE

RENDERING TECHNIQUES

Practice really does make perfect when it comes to rendering materials and textures. Use the following images to build your rendering techniques. Your teacher can download images for rendering practice from the NelsonNet teacher website. You may scan or photocopy this selection of line drawings.

For example: Use the colour version of this outdoor setting as a guide to adding colour and creating wood and metal effects to the template.

OUTDOOR SETTING

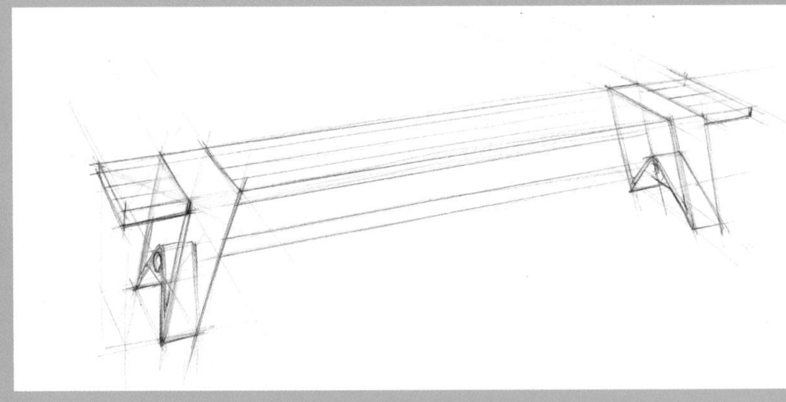

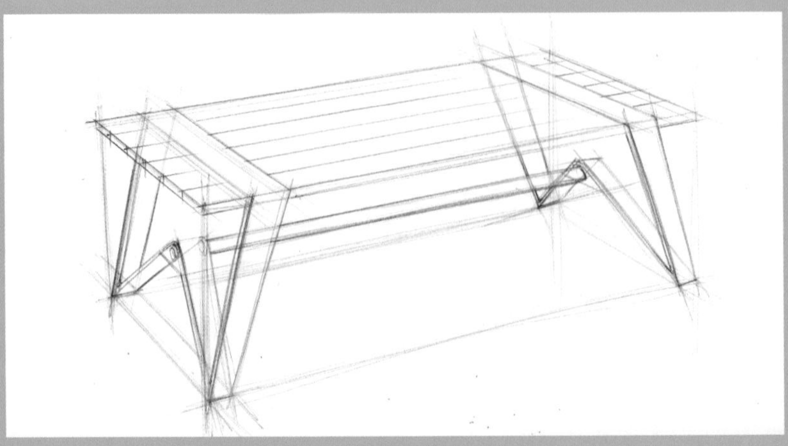

ISBN 9780170349994

DRINK BOTTLE
RUNNING SHOE
SCISSORS

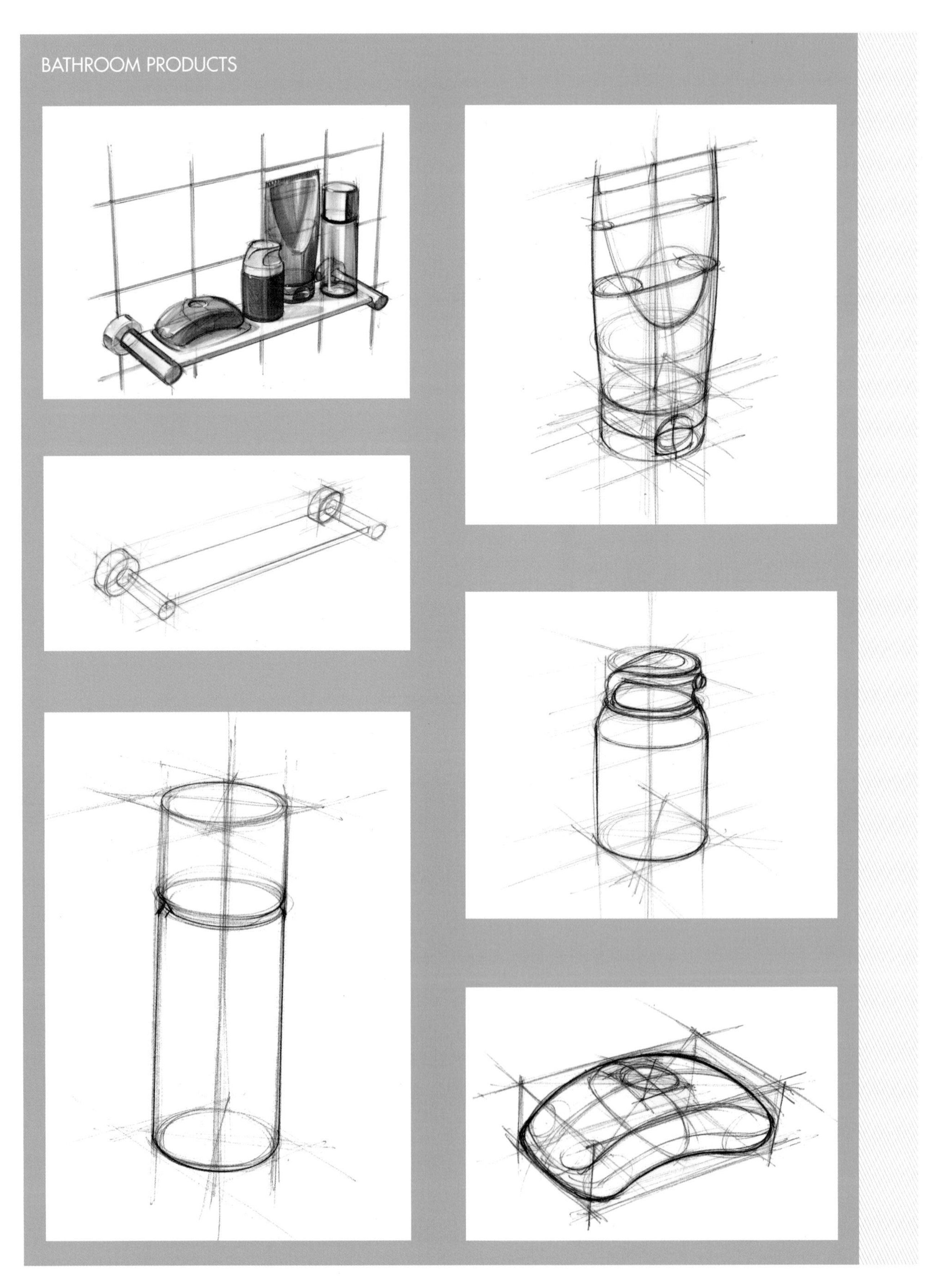

ISBN 9780170349994

PART D

GRAPHICAL REPRESENTATIONS

Graphical representation	Section D
Sketches	Chapter 14
Diagrams	
Maps	
Symbols and logos	
Cutaway views	
Exploded views	
Assembly drawings	
Animations and simulations	
Two-dimensional drawing including orthographic and development drawing, sections and details, architectural plans and elevations	Chapter 15
Three-dimensional drawing including axonometric and perspective drawing	Chapter 16

Graphical representations are the visual means used to convey ideas, messages, data and information. Each design problem requires the application of different graphical representations as part of the ultimate design solution. The selection of the most appropriate graphical representation usually depends on the requirements of the design brief. Choosing which graphical representation to apply is influenced by a number of factors, including:

+ Which graphical representation most effectively conveys the key visual information?
+ Which graphical representation best presents a solution to the design problem?
+ Which graphical representation provides the clearest visual means of communicating the idea or concept?
+ Which graphical representation is most relevant to the design task?
+ Many graphical representations feature rules and conventions in their production, serving to present information clearly and consistently. Others offer greater creative freedoms that allow the presentation of visual information to engage and stimulate the viewer. The appropriateness of each graphical representation is usually decided upon with understanding of the design area, function and target audience.

ISBN 9780170349994

CHAPTER 14
COMMON GRAPHICAL REPRESENTATIONS

'We should all draw, graphic designer or not, good or bad, because it's enjoyable and a really useful means of communication.'

Angus Hyland

14.1 SKETCHES

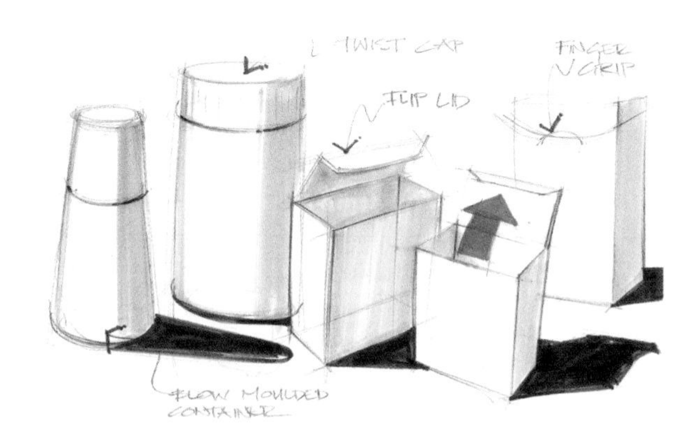

Design sketching is an important skill in all design areas. The ability to visualise ideas quickly can be extremely helpful, especially in the initial phases of the design process. Most professional designers sketch their first ideas on paper before moving onto digital media. The benefit of design sketching is its immediacy; ideas can flow quickly, be documented readily and without delay. Sketches can be the result of brainstorming, of team discussions or as the result of a client consultation. The ability to share ideas rapidly and immediately is beneficial to the progress of the design.

The qualities of sketching in design are as varied as the practitioners themselves. It is not essential to be a great illustrator to convey ideas and meaning through drawing. However, it is an important skill to develop and many tertiary design courses still require students to draw their early ideas.

ISBN 9780170349994

▲ In the creation of illustrations for this book, designer, Mark Wilken completed sketches during the briefing process. These sketches assisted Mark in generating ideas and ensured that he and the author had the same understanding of what was required for each illustration.

▲ Design sketch for a stylus, tablet, keyboard combo product

Design sketches are usually freely and loosely executed and do not represent a finished presentation drawing. Using loose but confidently applied line work and some rendering of tone and texture, design sketches can convey substantial visual information about form, textures and materials. Industrial design, in particular, sees the application of design sketching in the early ideation phase of the design process. Designers use sketching to explore concepts before undertaking more time-consuming and labour-intensive CADD representations.

There are techniques that can help you to build your own skills in sketching design ideas. The use of three-dimensional drawing methods is

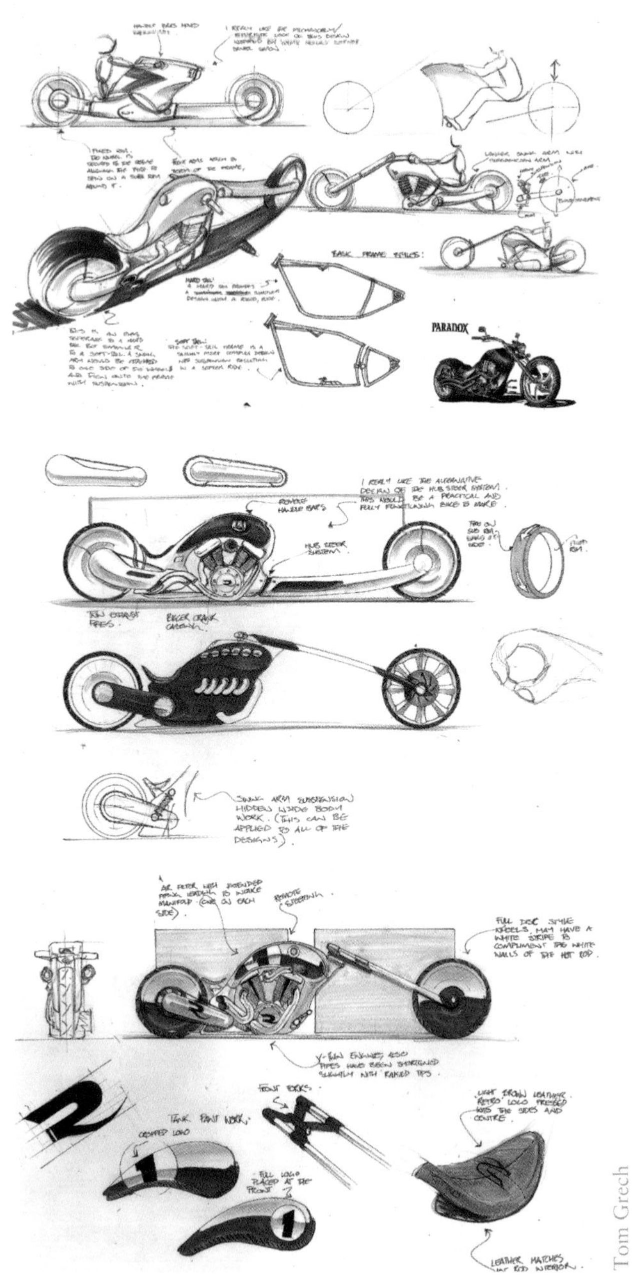

▲ This student used sketches to generate early ideas about the form and function of a custom motorbike.

still relevant in sketching as they assist in creating a high level of realism and realistic proportions. Selection of media is a first step in sketching. Using a basic collection of pencils and markers on bleed-proof paper can lead to good quality results. Popular with industrial designers, architects, interior designs and landscape architects, markers can produce instant results. They take some getting used to but once mastered, markers often become an essential tool for designers.

ISBN 9780170349994

DESIGN SKETCHES VS HAIRY SKETCHES

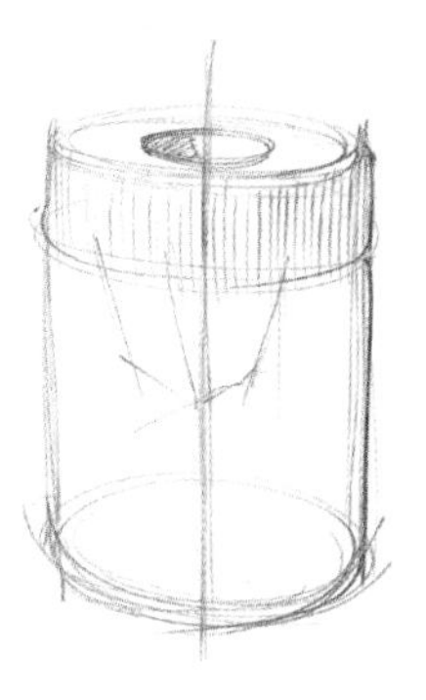

▲ Effective design sketches are drawn confidently and lightly. They have light construction lines and light line work. The line weights build gradually as the objects moves away from the light source.

▲ Sketches that are heavy handed appear to be 'hairy'. There is no obvious construction work, lines are consistently heavy and no consideration is given to the direction of the light source.

Product sketching, in particular, makes use of three-dimensional drawing methods such as perspective drawing and isometric drawing. The application of 'crating' techniques enables complex objects to be created from simple shapes and forms (see Chapter 16).

SHADOWING

Freehand shadows can be an effective means of 'grounding' a sketch and placing it within a given context. This provides a sense of realism and three-dimensional form.

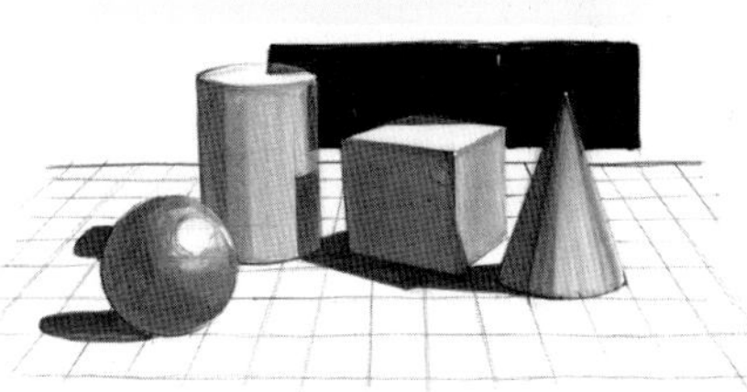

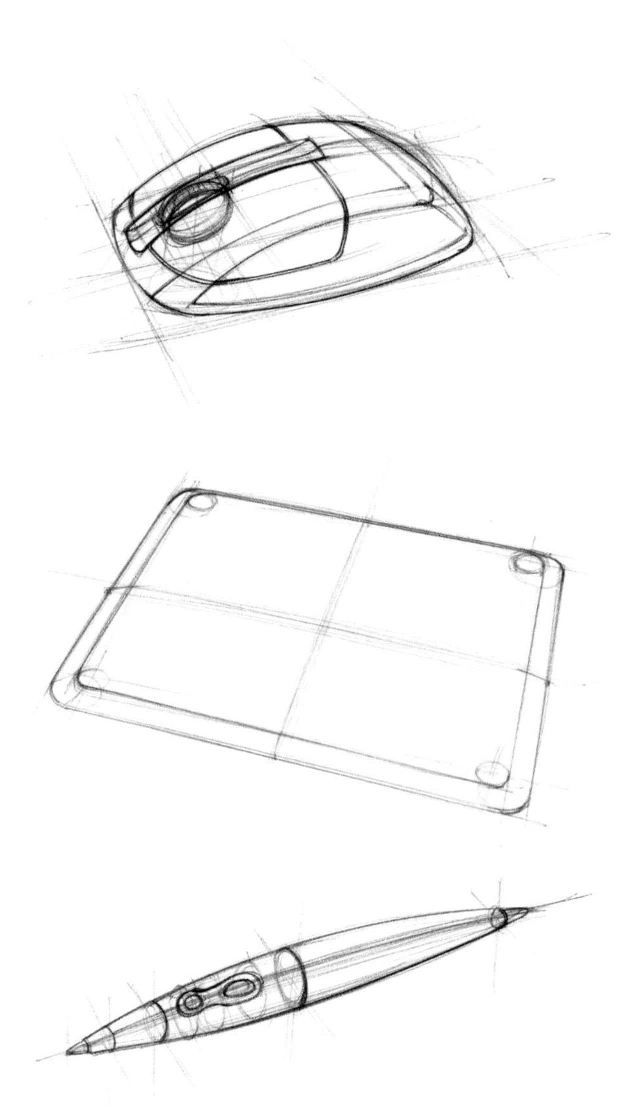

▲ Observe how simple forms have been used and combined to create these three objects. Applying perspective drawing techniques enables a simple hand sketch to appear realistic and proportional.

▲ Final images rendered and grouped together

ISBN 9780170349994

RENDERING SKETCHES

In sketching design ideas, the application of tone and texture can help to communicate important information about features and details. 'Rendering' is the term used to describe the application of tone and texture to create a three-dimensional appearance and/or to depict the surface details of an object.

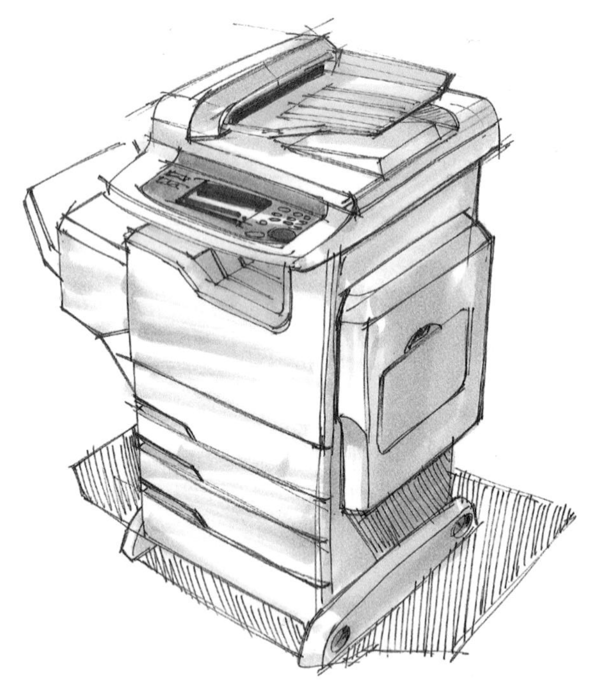

▲ Sketch of a multifunction printer rendered using fineliner and markers

Rendering is important when the communication of the form of an object is required. If a furniture designer, for instance, needed to illustrate a new line of chairs, it would be important to depict the characteristics of the fabrics and materials used in their production. The rendering of materials, using a range of media, can communicate a great deal of visual information about products and objects.

You may have seen illustrations of proposed designs labelled as an 'artist's impression'. Architects and interior designers often use these 'impressions' to help clients visualise what may otherwise seem to be a complex plan. The use of colour, line and tone – to demonstrate surfaces, texture and detail – helps to communicate ideas that might otherwise exist solely as instrumental drawings.

▲ Sketch of toy product with arrow indicating directional function

Effective rendering can be achieved through the application of an almost limitless range of media. Markers, ink pens, computer rendering, gouache and pencil are probably the most popular methods, with computer-generated images becoming increasingly common. It is also possible to achieve striking results with combinations of media, as well as pastel, collage, watercolour, cut paper and airbrush.

ESTABLISH A LIGHT SOURCE

Incorporating a light source into your sketches will add realism and create three-dimensional form. Natural or artificial light influences the appearance of objects, creating highlights and shadow areas. When light from the source hits an object it will often create a highlight, mid tones and dark tones, and cast shadows. Depending on the surface texture of the object, it may also reflect light.

ISBN 9780170349994

CROSSHATCHING

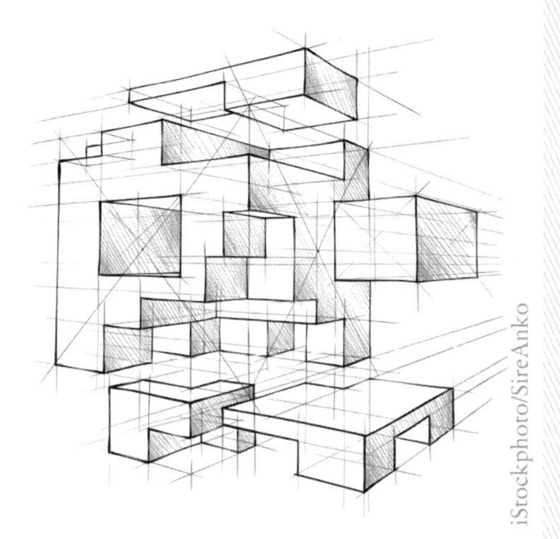

iStockphoto/SireAnko

Need to add tone or texture quickly? Crosshatching is a rapid method of applying tone to freehand sketches that can be very helpful in idea development and in communicating tonal information quickly. Vertical, horizontal and diagonal lines vary in proximity to one another, creating tonal and textural variations. This technique can also be used to suggest textures such as fabric, wicker and natural fibrous materials.

It is through the representation of light and dark areas that a three-dimensional form can be depicted. It is, therefore, important to identify the light source as the first step. In daylight, it is sometimes difficult to ascertain the primary source of light but invariably it will be a window or, if outside, the sun. In a dark space, a light globe or lamp will create a light source that will appear more clearly defined than the more diffused light of day, and will create sharp contrasts.

In rendering the form of objects, you may need to make an arbitrary decision about the primary light source, taking into account reflected light from other surfaces or secondary sources of light.

In the past, formal training in illustration involved learning 'rules' about light and shadow. In fact, in many classical paintings you can see how strictly these rules were followed, with very specific applications of light and cast shadow areas.

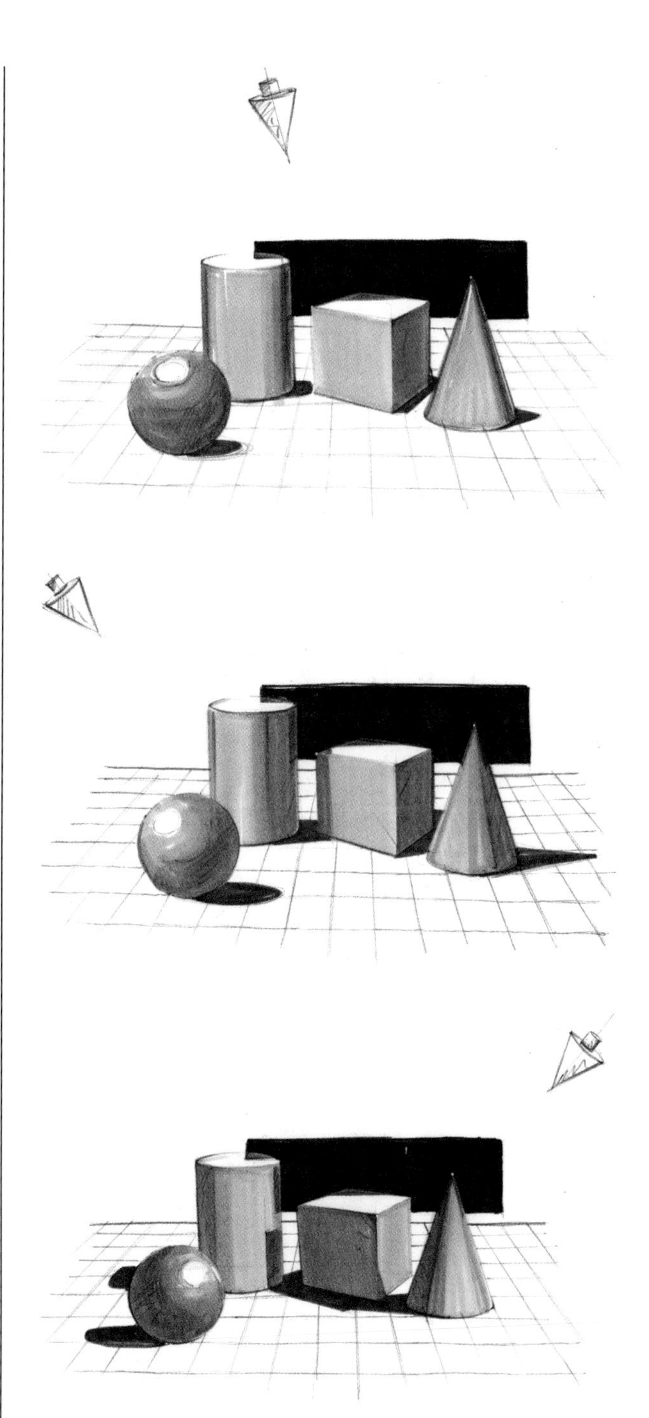

▲ The position of the light source will determine the application of tone and shadows.

Although it is still very important to understand the effect light has on an object, the application of tone is much more intuitive today, relying heavily on your observational skills and sensitivity to the subject matter.

ISBN 9780170349994

14.2 DIAGRAMS

Diagrams are often used when the visual representation of complex information is required. Diagrams present data in an accessible manner by using visual tools such as symbols, colours and explanatory drawings.

Diagrams are visual representations of information or data. They are designed to convey complex information in a visually clear and accessible manner. Diagrams often utilise symbols and design elements, such as colour, point and line, to convey detailed information visually.

The design of information graphics is a growing field of graphic design; the sophistication of design software and the reach of the Internet mean that complex information can often be explained through interactive diagrams. We live in a world that is filled with vast amounts of data and, for many people, data presented as a diagram offers a more easily understood graphical representation.

An increasingly common form of diagram is an infographic, which is usually found online and makes use of a range of highly illustrative design elements to convey information. From the term 'information graphics', infographics usually include more than one set of data or information and use simple but detailed illustrations. They may include graphs, maps, illustrations and symbols. Infographics can be applied to explain a process, timeline or event and are commonly used in textbooks, magazines, newspapers and online.

Many large infographics are published online and contain detailed information. The Internet is the ideal format for scrolling up and down through information and zooming into featured visual data. In fact, research has shown that infographics are some of the most popular content shared on social networking sites such as Facebook, Twitter and Reddit.

GRAPHS

The most common graphs serve to present detailed data in visual form and are usually divided into categories of line graphs, pie graphs and bar graphs. They use a scale and/or series of data sets that are indicated along two axes or, as in the case of circle diagrams, use divisions of a whole to communicate quantities of data. Graphs may use two-dimensional or three-dimensional means for visual impact.

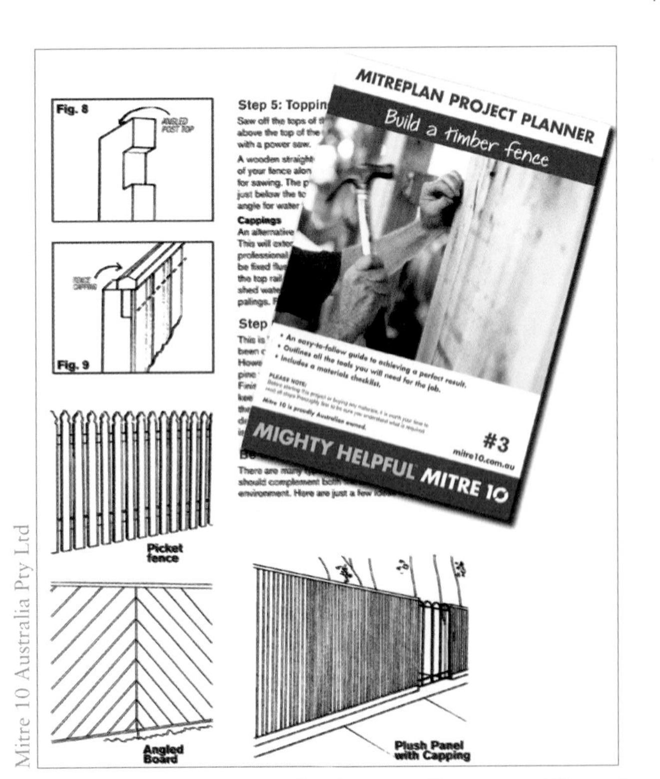

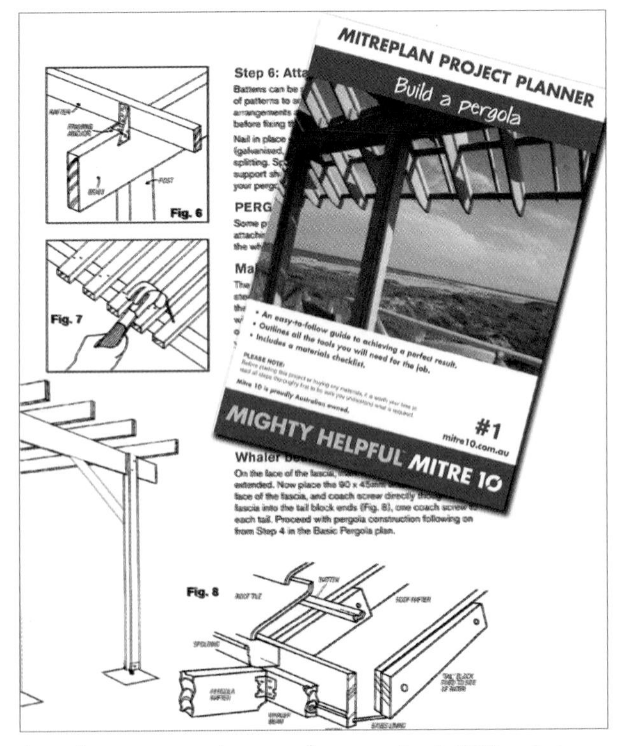

▲ These Mitre 10 project brochures use diagrams and illustrations to show the structure and means of construction in DIY projects. Generally, the target audience for these projects is not professional builders, so the visual detail is often simplified to guide the user.

ISBN 9780170349994

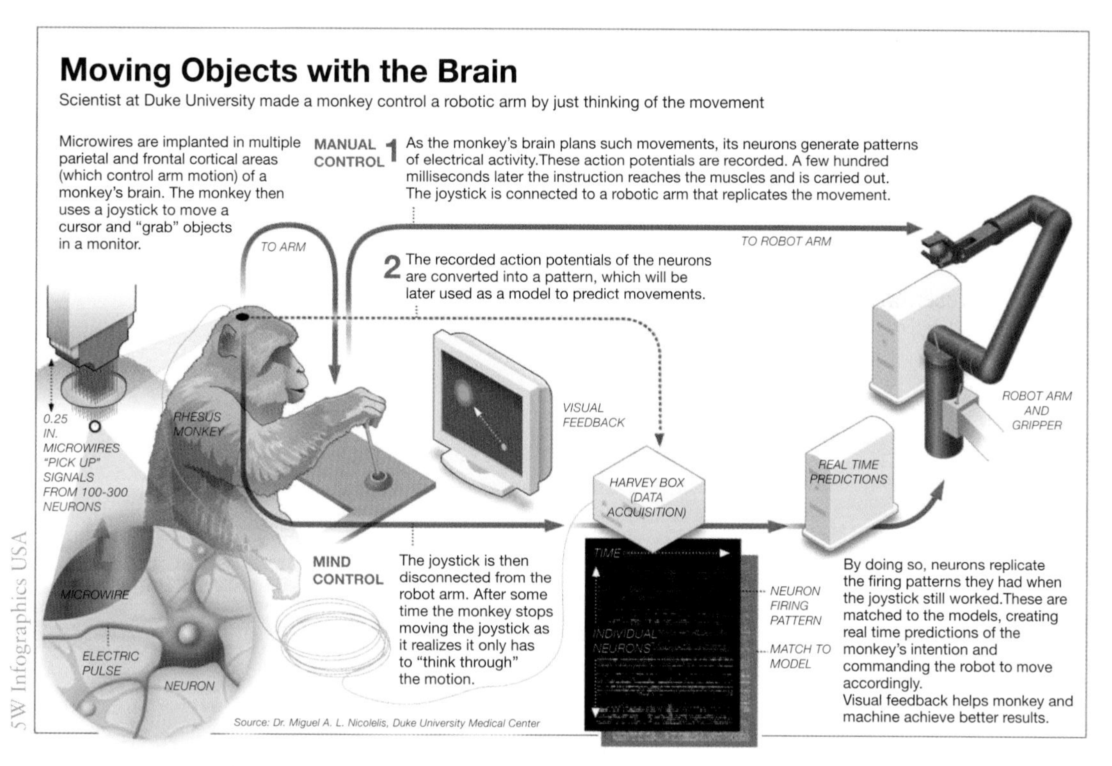

▲ Using a range of design elements and devices, such as pictograms, this online infographic displays a rich array of information about a serious issue.

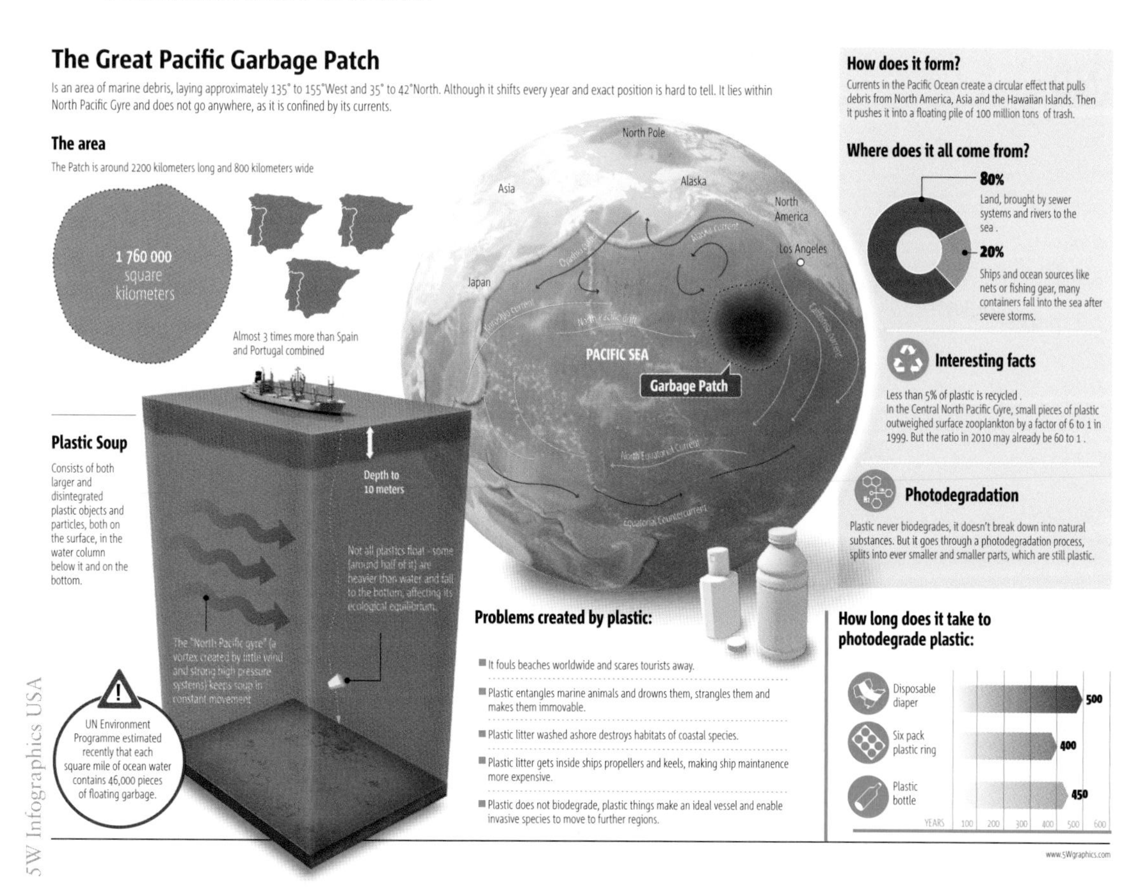

▲ Visual elements assist in making otherwise technical or complex information accessible and engaging for the audience.

ISBN 9780170349994

Shutterstock.com/TijanaM

Shutterstock.com/billdayone

◀ Pie graphs created using both two-dimensional and three-dimensional representations. It is important to decide which version is best suited to both the data and the context of the graph.

Pie graphs

Pie graphs (also known as circle graphs or circle diagrams), depict 'slices' of data, hence the name 'pie'. The full circle (360°) represents 100 per cent and the angle of each segment is found by multiplying their percentage value by 360°. For example, a value of 50 per cent is indicated by a segment of 180°. Pie graphs can be seen in both two-dimensional and three-dimensional forms.

Line graphs

Line graphs show changes over a period of time. They can be used to indicate trends and to track differences over a given period. When more than one line is featured, colour is usually used to distinguish each line.

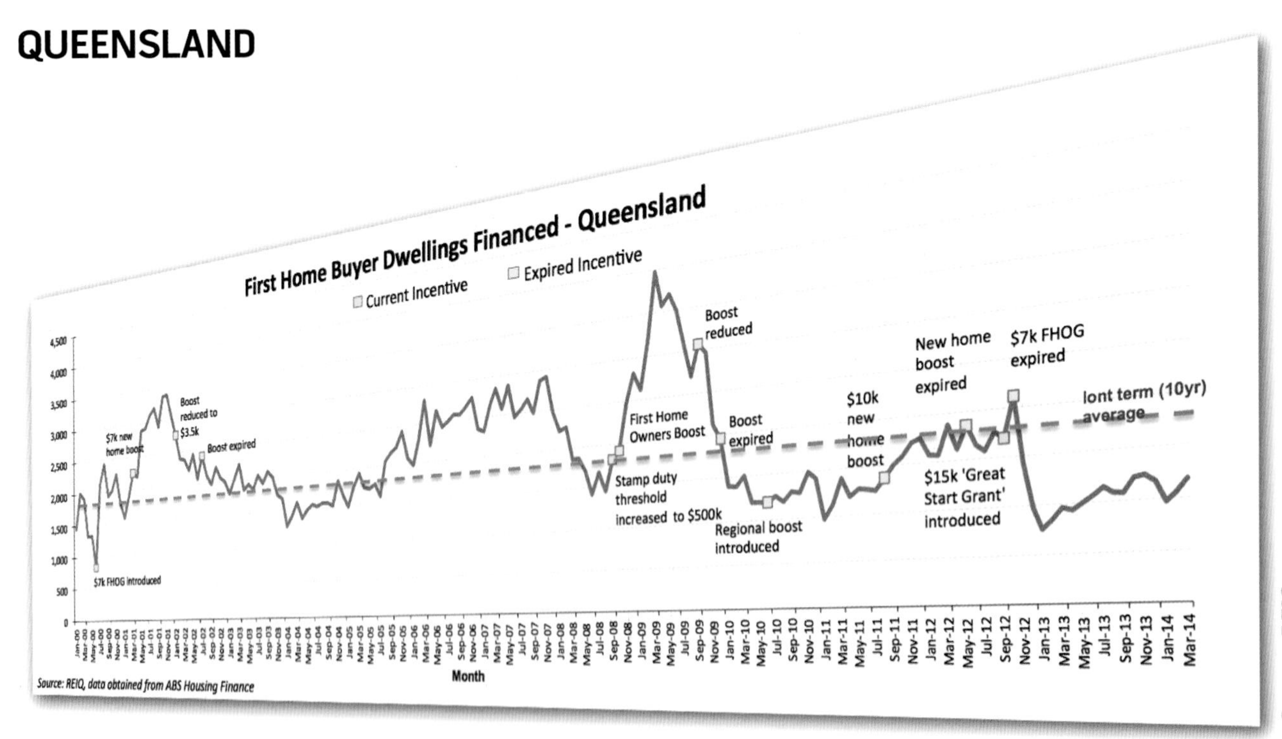

Courtesy REIQ

PropertyInvestor

REAL ESTATE TALK

▲ An REIQ line graph charts the rise and fall in numbers of Queensland first-home buyers over the past decade.

ISBN 9780170349994

Bar and column graphs

Bar and column graphs compare differences and similarities between data. Arranged either horizontally (bar) or vertically (column), these graphs enable visual comparisons so differences can be recognised quickly. More complex versions might use grouped bar or columns or divide each column further creating a 'stacked column graph'.

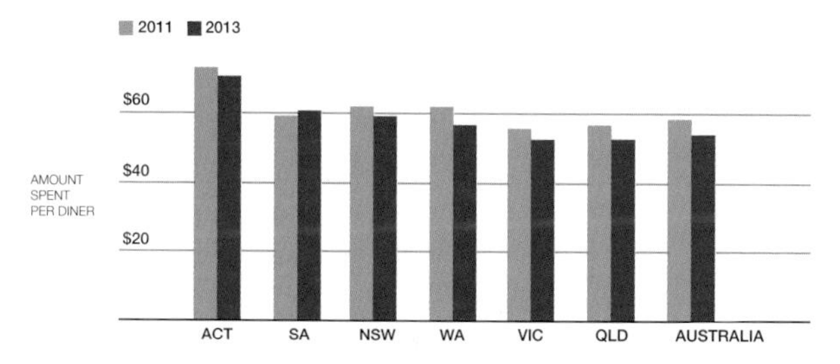

▲ Column graph illustrating the average amount restaurant diners in Australia (by state) spent on meals. The use of two coloured columns allows comparison between spending in 2011 compared with spending in 2013.

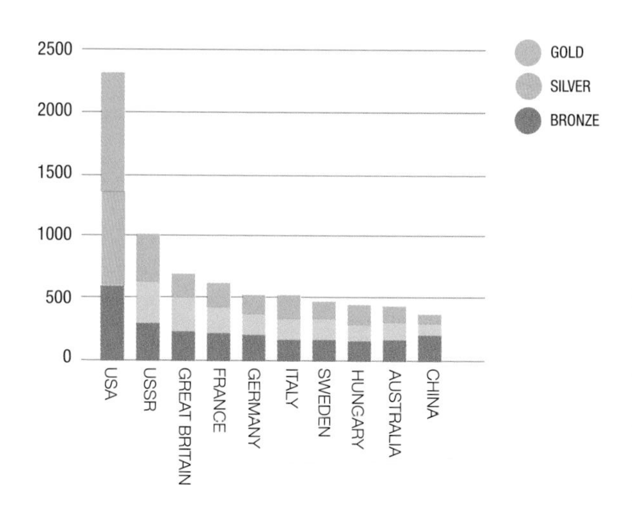

▲ Stacked column graph showing total medals won by country in the Olympic Games. The coloured division of each column allows the viewer to see the proportion of gold, silver and bronze medals within the total.

CHARTS

Flow charts

Flow charts are designed to visually explain the steps in a system or process. They use boxes of data (nodes) to identify the steps of a process and are designed to assist the viewer's understanding of 'what happens next'.

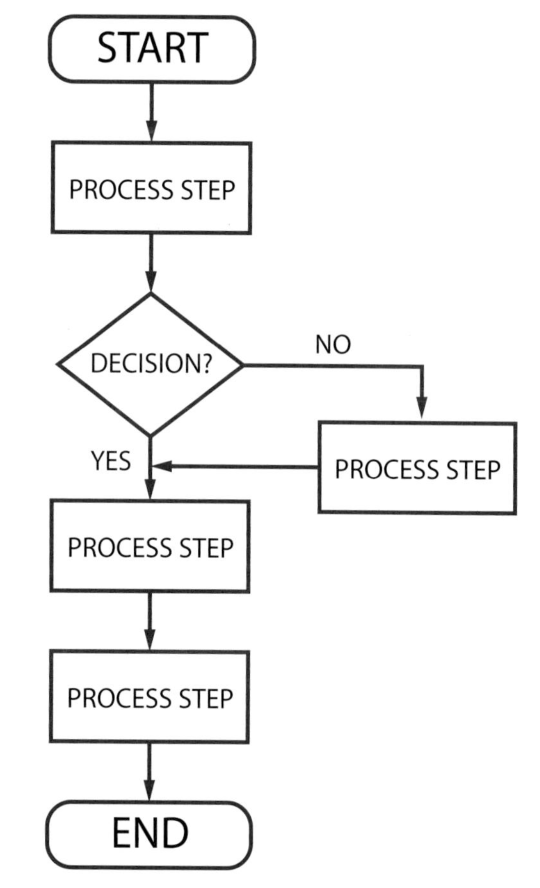

▲ Traditional flow charts use a series of shapes to indicate parts of the diagram. In this simple diagram, diamonds represent decisions made during the process.

Often used in the design of software systems, flow charts can be useful in explaining very complex technical systems. There are many types of flow charts, and some may use more visual means than others; however, the fundamental sequence-based appearance remains the same.

ISBN 9780170349994

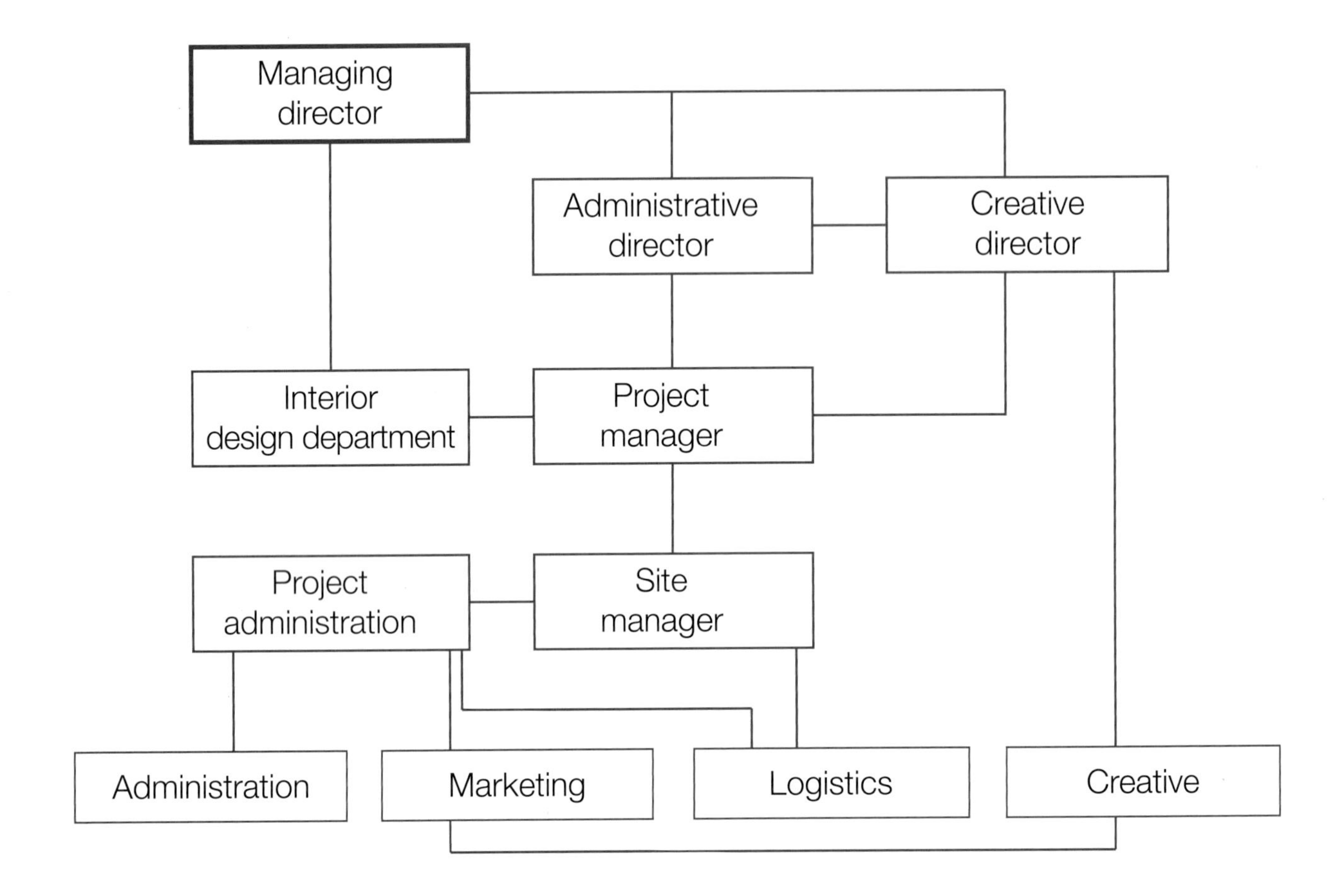

Organisational charts

Organisational charts are generally used to depict the relationships and hierarchy between roles within an organisation, institution or project team. They typically represent relationships between people but can be used more flexibly. A family tree is a good example of an organisational chart in action.

Timelines

Timelines are used to display events, images or data in chronological order. Often used to visually describe historical development over a period of time, they can feature illustrations and text in combination. Timelines can often be found in newspaper or magazine articles, history or science textbooks and online.

Gutenberg's 42 line Bible	Jenson Garamond *Aldus Italic* Aldus Roman	Baskerville Times	Didot Bodoni Franklin Gothic	Avant Garde Helvetica Gill Sans **Futura**	Calibri Myriad Georgia
1450 *Blackletter*	*1500–1700* *Classical* *Old Style* *Humanist*	*1700–1800* *Transitional*	*1800–1900* *Modern*	*1900–1960* *Modernist* *(Swiss Modern)*	*1960–* *Contemporary* *Digital*

▲ A simple timeline illustrating the chronological development of modern typographic styles

ISBN 9780170349994

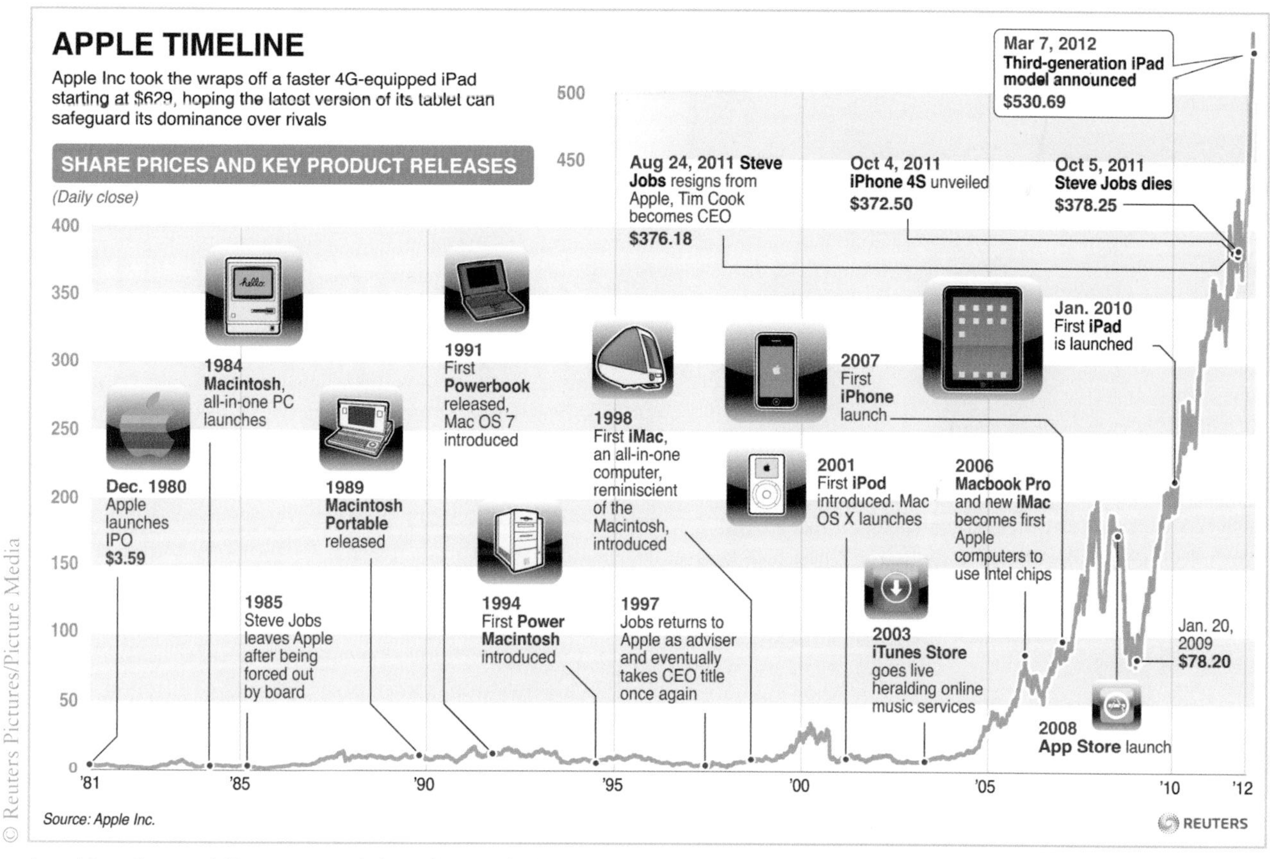

▲ Note the use of illustrations and the inclusion of a line graph in this diagram about Apple and the relationship between product launches and share price. The various elements of the design produce a diagram that is rich with information.

DON'T LOSE THE MESSAGE

Renowned information designer Edward Tufte warns against the overuse of visual 'decoration' in diagrams, which can distract from the key information and data that is being explained. Tufte is a world leader in information design and more information about his work and ideas can be found on his website.

Access all weblinks directly at http://nsg.nelsonnet.com.au.

14.3 SYMBOLS

Information designs often feature symbols. Also referred to as pictograms, they are often a simplified representation of a recognisable object or figure. Unlike a logo, where abstract imagery can also be applied, pictographic symbols are representative.

Commonly used in signage and diagrams, pictographic symbols help to convey the meaning of visual data in an unambiguous manner. Pictograms are universally recognised and, ideally, require no additional written explanation as to their meaning.

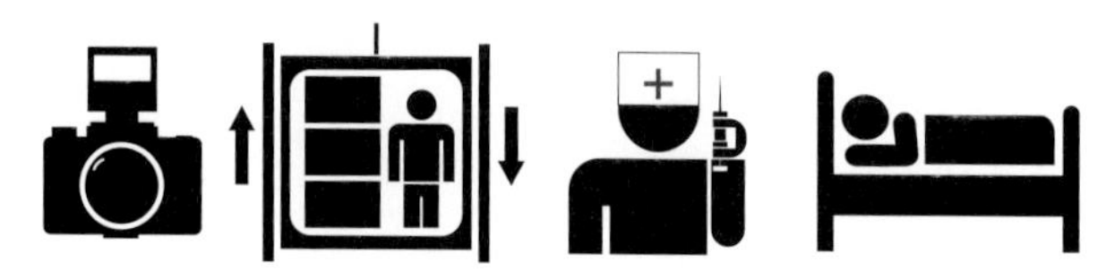

▲ Pictograms

Applied in information design, pictograms help the target audience understand information that would be more difficult or more time consuming to comprehend in written form.

▲ Airport directional signage using pictograms with minimal text

ISBN 9780170349994

INFOGRAPHICS

GOOD and Cool infographics are two of many websites that showcase infographics/diagrams. Each day new visual representations of information are posted on many varied topics.

Access all weblinks directly at http://nsg.nelsonnet.com.au.

Pictographic imagery is often used in web design and a system of images (also called icons) can help users to navigate a website. Although web icons are often used along with text, some icons are universally recognised, such as shopping cart and email icons.

▲ Web icons

Web icons are usually designed to fit the aesthetic appearance of a web page and may be two dimensional or three dimensional, feature gradients or effects, and display interactive changes to their appearance when clicked.

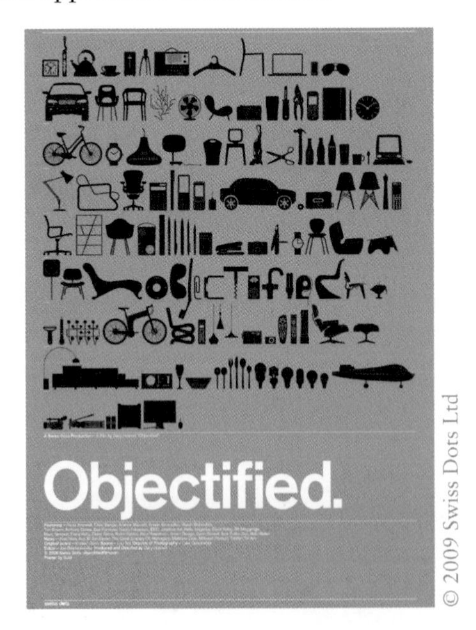

▲ Poster design for *Objectified*, a documentary film about product design by Gary Hustwit. The poster was designed by Michael C. Place of Build Design in London and includes pictographic symbols of iconic designs from the past 150 years.

Pictographic imagery is not exclusively designed for information communication. It is also applied in graphic design, fashion and publication design for visual effect and impact.

14.4 MAPS AND WAYFINDING

Maps and wayfinding devices, such as signage, are used together to create effective directional design. A map alone is just a map but when symbols and graphic indicators such as arrows are added, it can convey much more information. Maps are used for many purposes; to detail the features of a geographic area is its most recognised purpose but a map can also communicate information about a system (transport), points of interest (a three-dimensional graphic map) and direction (for example, from point A to B). Maps can be complex or extremely simple and the visual appearance is ultimately determined by the needs of the user.

Unlike a road map, transport maps do not feature geographical details. Instead, these simplified systems strip away unnecessary visual detail and provide a clear guide to destinations, directions and a visual scale of distance between stations. A colour key is used to distinguish the different transport lines.

ISBN 9780170349994

'Wayfinding' is an umbrella term for signage systems that help users navigate in an environment. Directional signage may assist users in finding the correct floor in a large building, or the exit point in a public park. The design of signage may feature text, pictograms, maps, arrows and other directional markers. In urban areas, wayfinding is implemented by city councils to help residents and visitors find their way around town. Effective wayfinding and signage systems are just that: systems. They use consistent type, icons and colours to ensure that users understand that individual signs are part of a larger communication system.

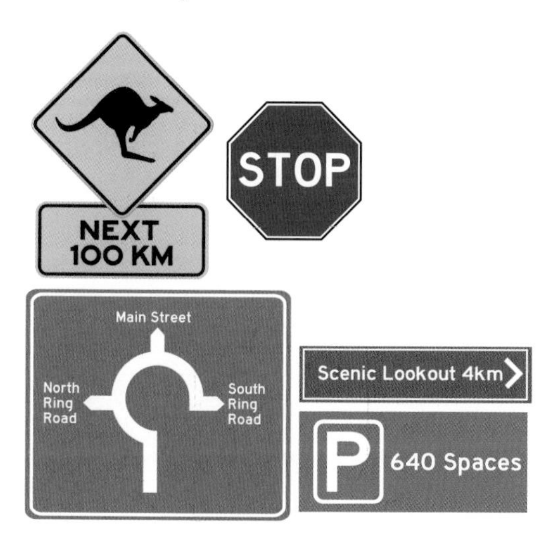

Road and highway signage systems are designed to assist road users by maintaining strict use of colour and text: yellow for caution, green for location and distance, brown for points of interest, blue for services and red for stop. Consistency is essential in roadway signage so that all road users grasp the same language.

DRIVING CHANGES VISUAL PERCEPTION

As driving speed increases, the peripheral vision of the driver decreases and foreground details fade in favour of looking further ahead. Perception of scale and speed decreases as the speed of a car increases, creating a greater dependence on signage for direction.

14.5 THREE-DIMENSIONAL DRAWING

CUTAWAY VIEWS

Cutaway views are used to describe the interior detail of objects or buildings. A cutaway view is often created in a paraline method, such as isometric, and exposes multiple levels of detail that could not ordinarily be seen.

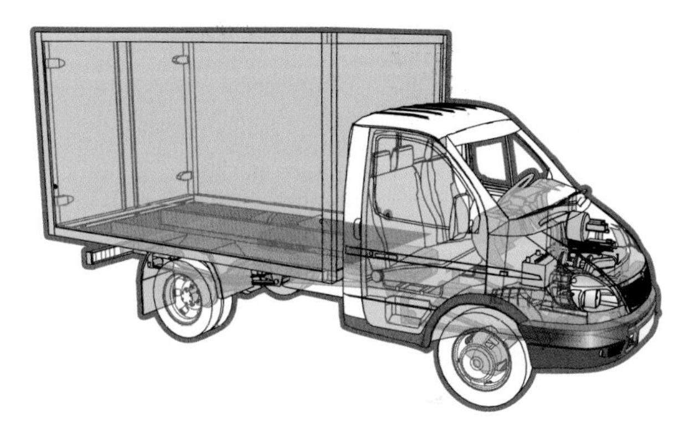

▲ Cutaway view of a truck

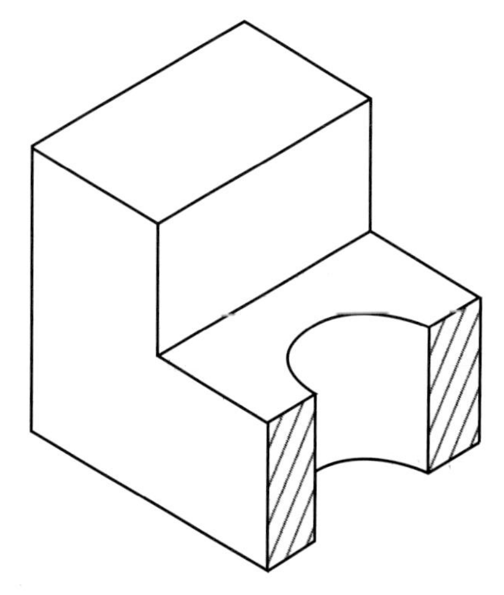

▲ When used in technical drawing, cutaway views use the same 45° crosshatching that is used in cross-section views.

In built environment design, a cutaway view enables a view of the interior appearance of a structure. Cutaway views are often used to explain the interior features, the structural details and the possible functions of a space.

EXPLODED VIEWS

An exploded view of an object is generally used to indicate the parts that make up a design product. An exploded view is most often drawn in a paraline method, such as isometric or perspective, and shows all parts in alignment with one another. The value of an exploded view is that it enables the viewer to clearly visualise the placement of parts of an object; it can represent aspects of an object that may be hidden from view when shown in an alternative drawing method.

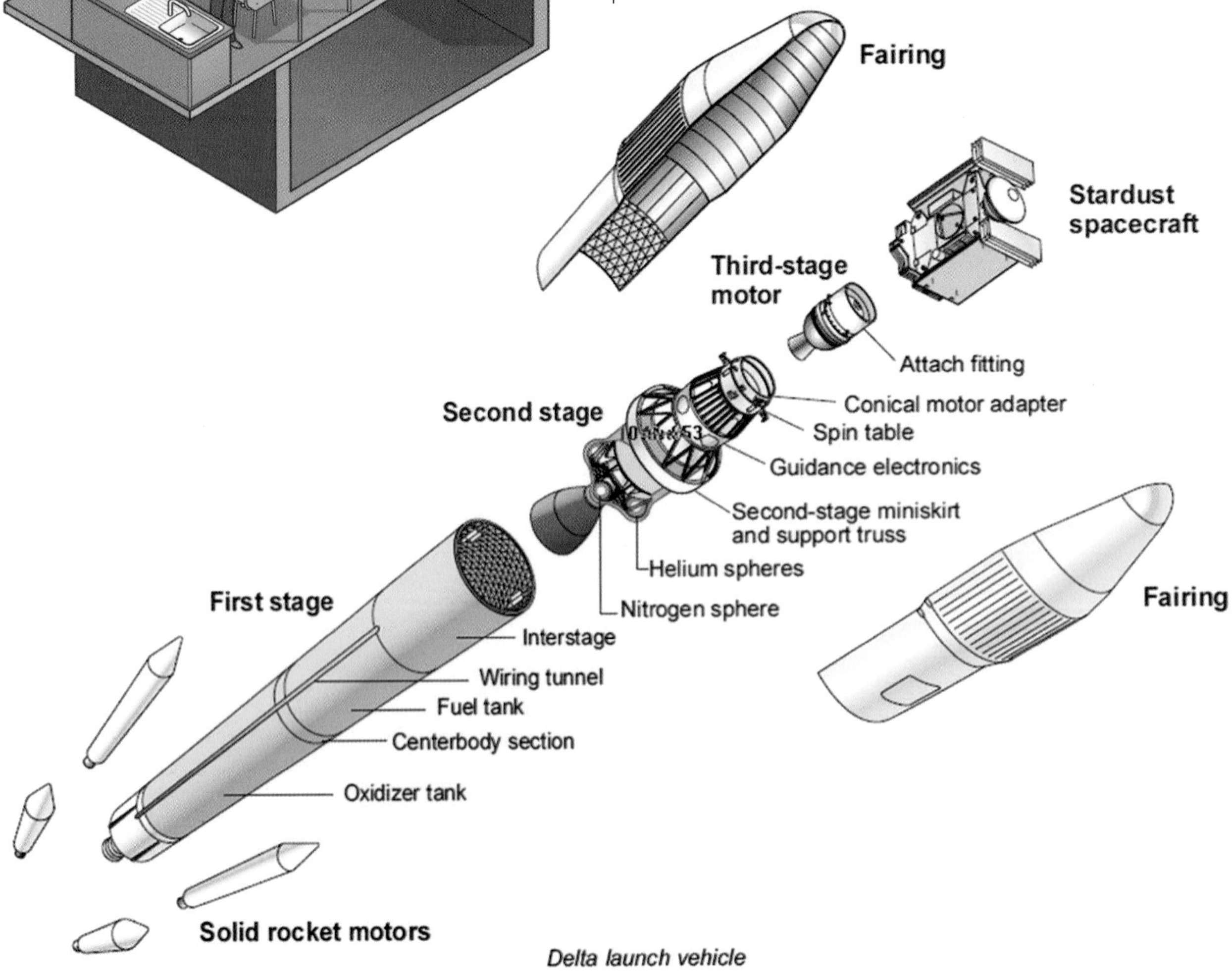

iStockphoto/vecstar

Delta launch vehicle

NASA JPL/Stardust

ISBN 9780170349994

▲ Isometric view of an alarm clock

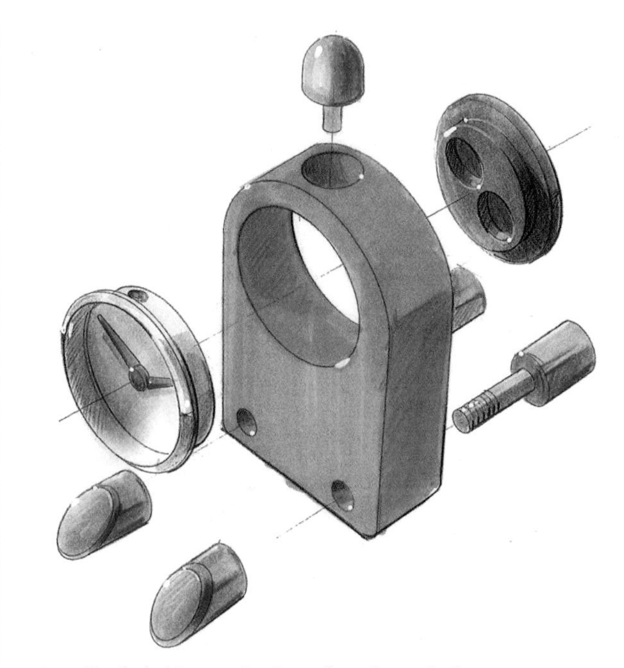

▲ Exploded isometric view of an alarm clock

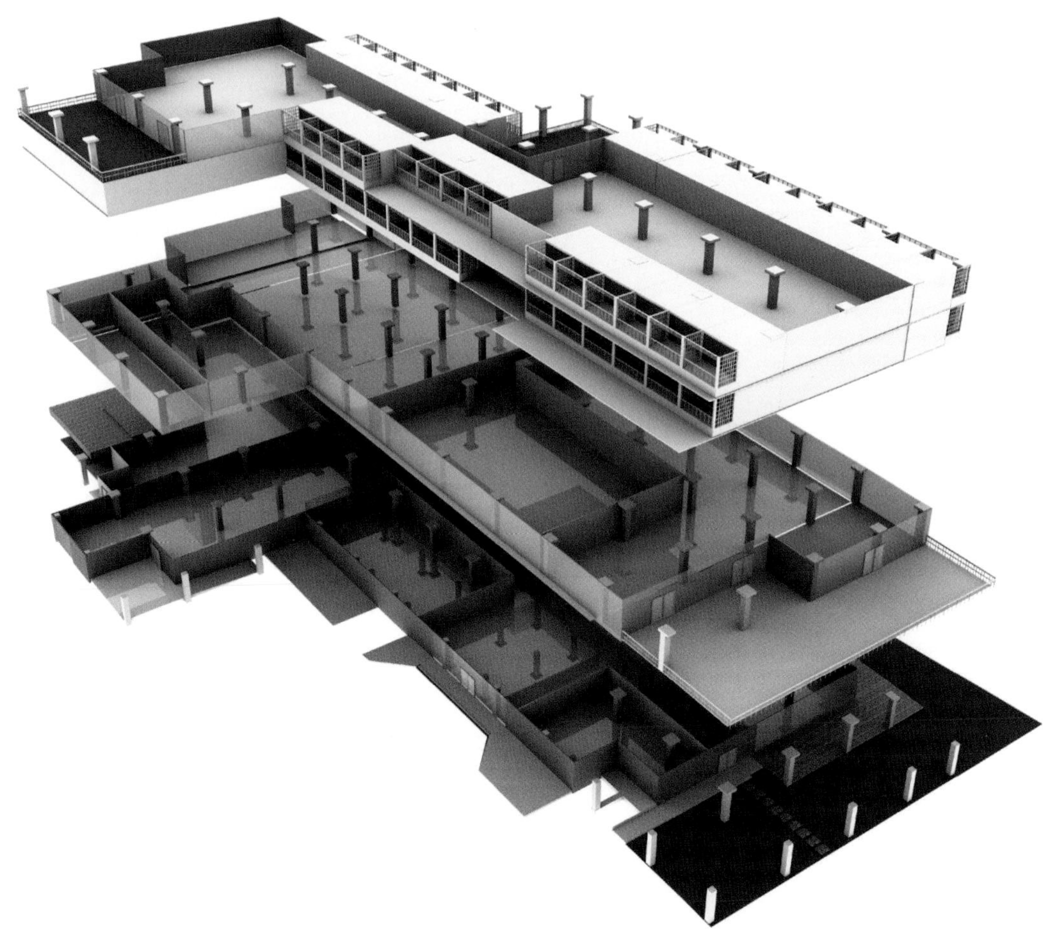

▲ Exploded view as applied in architectural design. The ability to see multiple levels of a building provides effective assessment of accessibility, aesthetics and functionality. Such a view enables non-design professionals to more easily visualise the three-dimensional form.

ISBN 9780170349994

ASSEMBLY DRAWINGS

Assembly drawings are drawings that illustrate how the individual components of a product fit together. Assembly drawings can be two dimensional or three dimensional, with three-dimensional versions commonly utilising an exploded view.

An assembly drawing features the illustration of the object to be assembled with numbered component labels for each part. These parts are then listed in a table of parts found on the drawing. An assembly drawing will provide all the necessary details, dimensions and joining techniques required to fully assemble all or part of the object.

Within an assembly drawing:

+ All the individual components have item numbers; for example, Item 1, Item 2 and so on.
+ The detail for these components is shown in the assembly drawing or in separate views, or on a separate detail drawing.
+ All the views necessary to gain a good understanding of the drawing are shown.
+ The dimensions to manufacture the object may also be included.

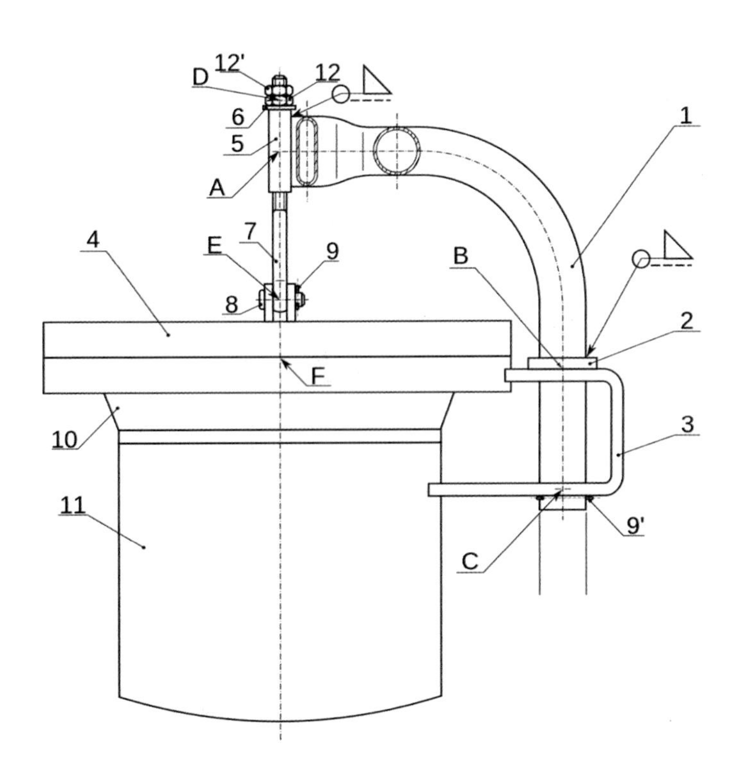

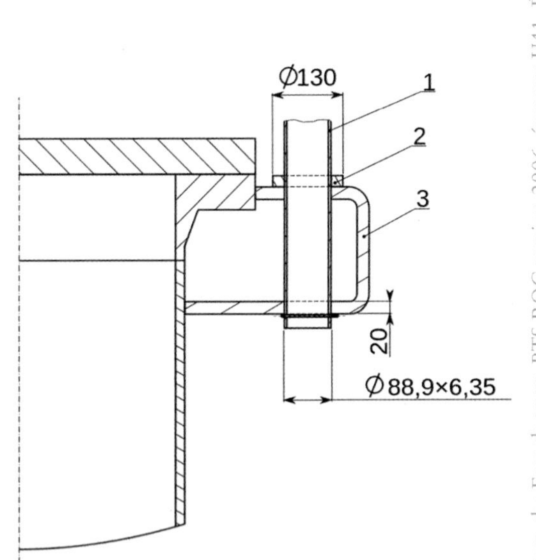

▲ Assembly drawing of support arm for a manhole cover

 ISBN 9780170349994

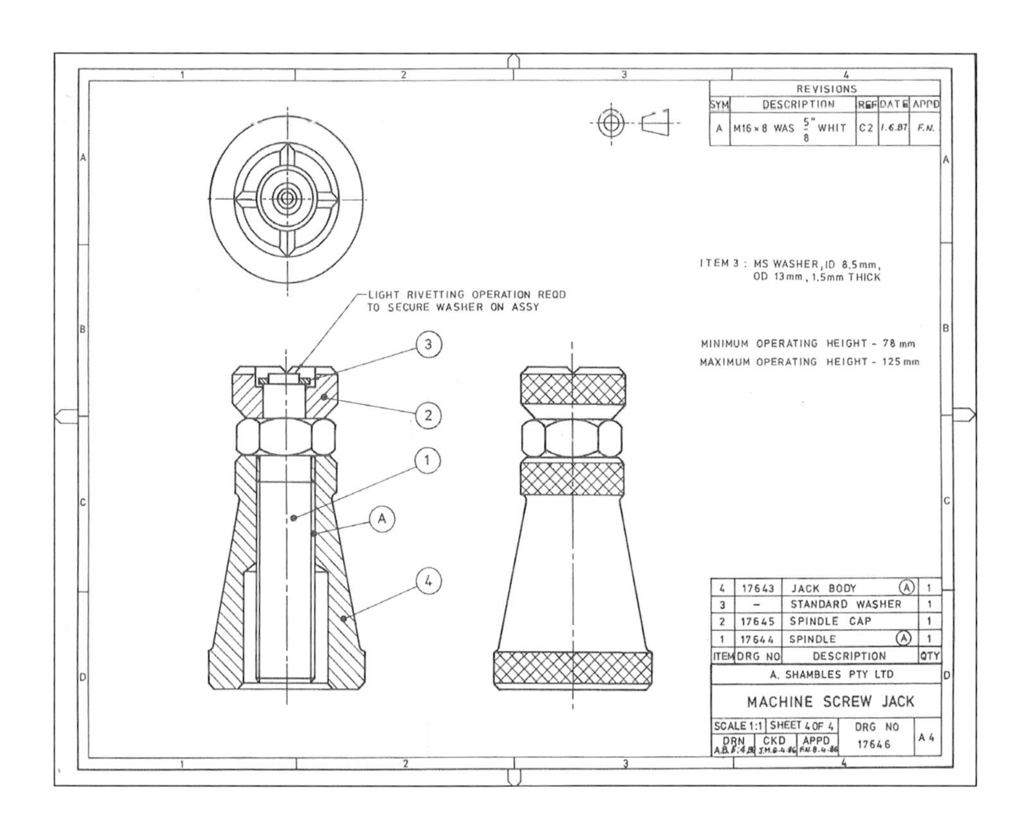

◀ An example of an assembly drawing. The parts list (bottom right corner) is present to identify each part of the object. Orthographic drawing conventions (see Chapter 15) are applied as per Australian Standards.

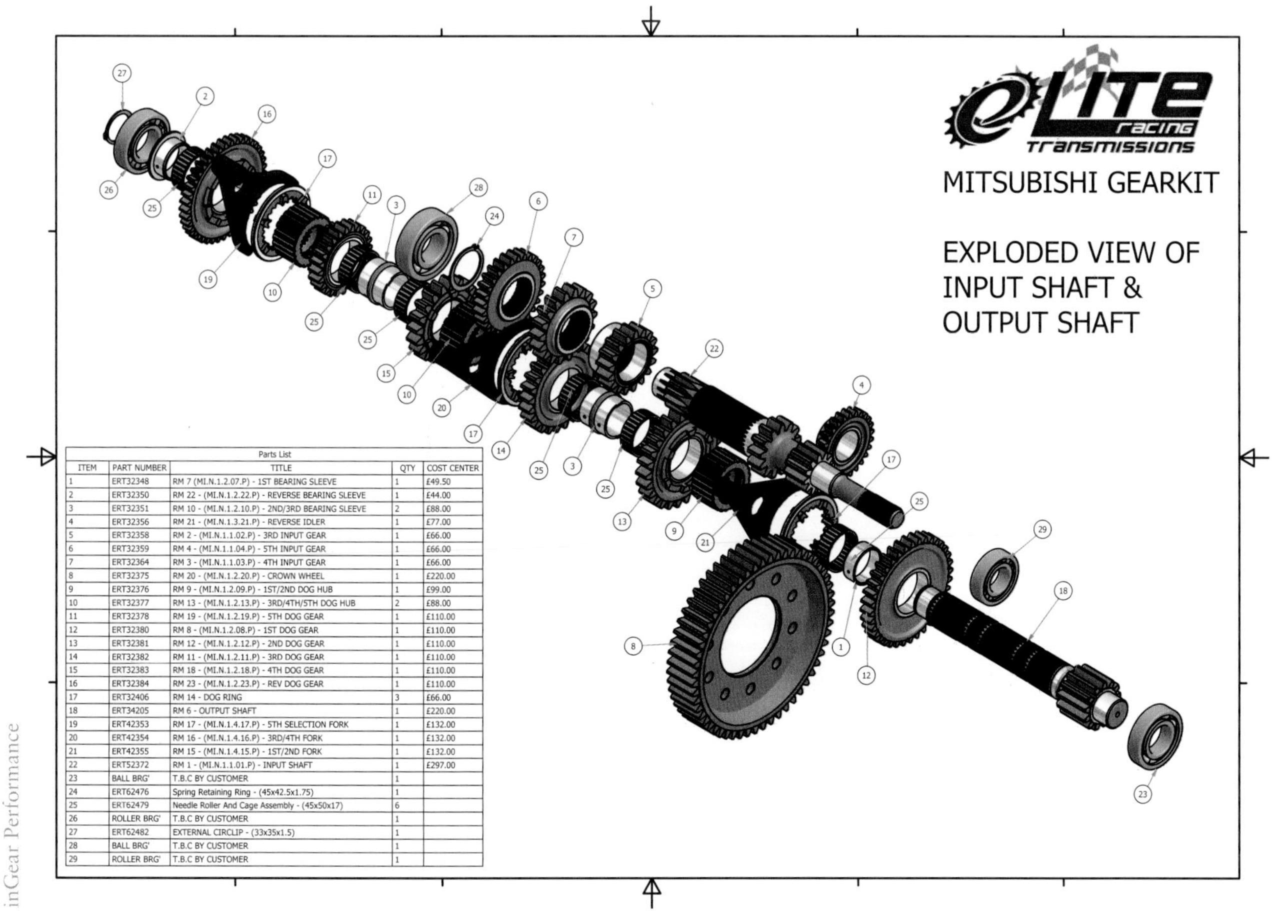

Parts List

ITEM	PART NUMBER	TITLE	QTY	COST CENTER
1	ERT32348	RM 7 (MI.N.1.2.07.P) - 1ST BEARING SLEEVE	1	£49.50
2	ERT32350	RM 22 - (MI.N.1.2.22.P) - REVERSE BEARING SLEEVE	1	£44.00
3	ERT32351	RM 10 - (MI.N.1.2.10.P) - 2ND/3RD BEARING SLEEVE	2	£88.00
4	ERT32356	RM 21 - (MI.N.1.3.21.P) - REVERSE IDLER	1	£77.00
5	ERT32358	RM 2 - (MI.N.1.1.02.P) - 3RD INPUT GEAR	1	£66.00
6	ERT32359	RM 4 - (MI.N.1.1.04.P) - 5TH INPUT GEAR	1	£66.00
7	ERT32364	RM 3 - (MI.N.1.1.03.P) - 4TH INPUT GEAR	1	£66.00
8	ERT32375	RM 20 - (MI.N.1.2.20.P) - CROWN WHEEL	1	£220.00
9	ERT32376	RM 9 - (MI.N.1.2.09.P) - 1ST/2ND DOG HUB	1	£99.00
10	ERT32377	RM 13 - (MI.N.1.2.13.P) - 3RD/4TH/5TH DOG HUB	2	£88.00
11	ERT32378	RM 19 - (MI.N.1.2.19.P) - 5TH DOG GEAR	1	£110.00
12	ERT32380	RM 8 - (MI.N.1.2.08.P) - 1ST DOG GEAR	1	£110.00
13	ERT32381	RM 12 - (MI.N.1.2.12.P) - 2ND DOG GEAR	1	£110.00
14	ERT32382	RM 11 - (MI.N.1.2.11.P) - 3RD DOG GEAR	1	£110.00
15	ERT32383	RM 18 - (MI.N.1.2.18.P) - 4TH DOG GEAR	1	£110.00
16	ERT32384	RM 23 - (MI.N.1.2.23.P) - REV DOG GEAR	1	£110.00
17	ERT32406	RM 14 - DOG RING	3	£66.00
18	ERT34205	RM 6 - OUTPUT SHAFT	1	£220.00
19	ERT42353	RM 17 - (MI.N.1.4.17.P) - 5TH SELECTION FORK	1	£132.00
20	ERT42354	RM 16 - (MI.N.1.4.16.P) - 3RD/4TH FORK	1	£132.00
21	ERT42355	RM 15 - (MI.N.1.4.15.P) - 1ST/2ND FORK	1	£132.00
22	ERT52372	RM 1 - (MI.N.1.1.01.P) - INPUT SHAFT	1	£297.00
23	BALL BRG'	T.B.C BY CUSTOMER	1	
24	ERT62476	Spring Retaining Ring - (45x42.5x1.75)	1	
25	ERT62479	Needle Roller And Cage Assembly - (45x50x17)	6	
26	ROLLER BRG'	T.B.C BY CUSTOMER	1	
27	ERT62482	EXTERNAL CIRCLIP - (33x35x1.5)	1	
28	BALL BRG'	T.B.C BY CUSTOMER	1	
29	ROLLER BRG'	T.B.C BY CUSTOMER	1	

▲ An exploded pictorial view of an object can be used as an assembly drawing. Each part is identified and its relationship between the parts around it is clearly illustrated in the three-dimensional view.

ISBN 9780170349994

14.6 ANIMATIONS AND SIMULATIONS

ANIMATION

Used in all design areas and sometimes referred to as 'motion graphics', animations combine series of images (either two or three dimensional) to create sequences that display motion. Animations may include sound and even offer some level of interactivity. Animations may show the workings of a product part, the method of utilisation of an object or instruct on the functionality of a design. Animation is also used in graphic design when print formats are not suitable for the communication of a message.

SIMULATION

Simulations are designed to provide the viewer with an *experience* of a design. Commonly applied in the design of spaces and some products, a simulation allows the user to engage with the design in a way that provides a 'virtual' experience of an environment. In product design, a simulation may offer insights into the 'actual' use of a product. The ability to observe computer-generated simulations of a product design can provide important testing information and identify concerns about safety or compatibility with the target user.

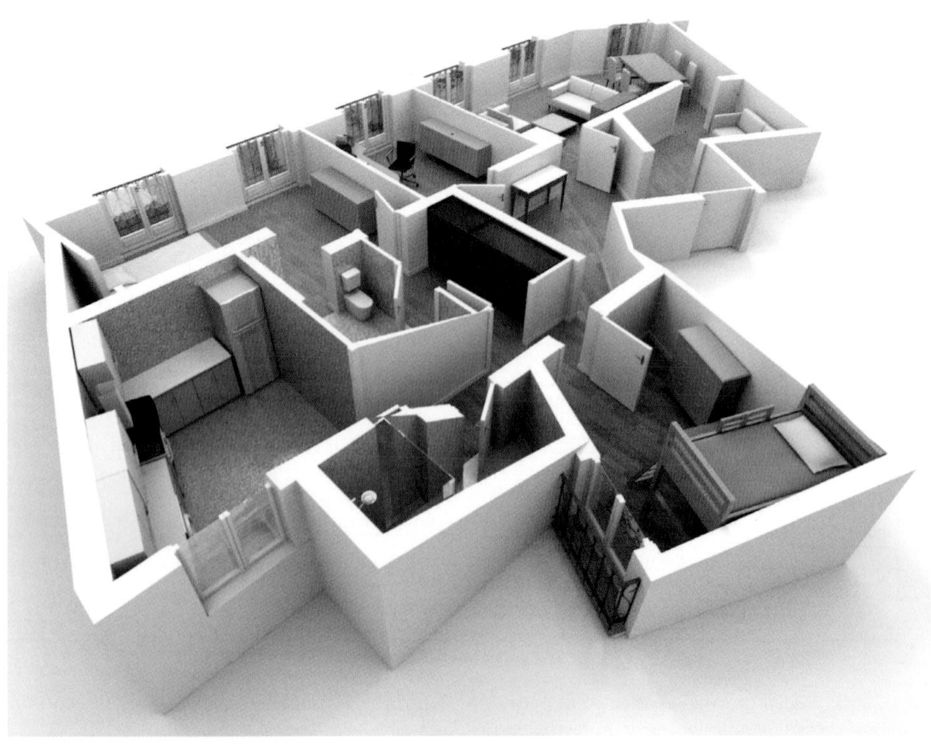

Shutterstock.com/Franck Boston

▶ Computer-aided design is often used to enable a client to visualise the finished three-dimensional space; fly-throughs and rendered three-dimensional models help with this process.

ISBN 9780170349994

CHAPTER 15
TWO-DIMENSIONAL DRAWING

'Design is the conscious effort to impose a meaningful order.'
Victor Papanek

Two-dimensional drawings provide a clear means of communicating information about the appearance, assembly, function or construction of an object. Two of the most commonly utilised methods of two-dimensional drawing are orthographic drawing and development drawing.

Imagine that your innovative new design for a bicycle is to be manufactured in a non-English-speaking country. You will need to ensure that the technical drawings you provide to the manufacturer are clear and contain all relevant details. Your drawings will need to convey information in a visual language that both you and the manufacturer can understand. Similarly, in the design of a building, the plans that are provided by an architect to a builder need to be clearly understood.

Two-dimensional drawing practice adheres to rules set out by a regulatory body, which create a consistent approach to the communication of technical information. These rules are known as 'standards' and are set in Australia by Standards Australia. The use of standards means that a design can be manufactured to precise specifications without misinterpretation or misunderstanding.

For students of Senior Graphics, the relevant standard is called *Technical Drawing for Students* (SAA/SNZHB1:1994) and is available for purchase online.

Other relevant Australian Standards include AS 1100.101-1992, AS1100.201-1992 and AS1100.301-2008.

ISBN 9780170349994

15.1 TWO-DIMENSIONAL DRAWING IN INDUSTRIAL DESIGN

An industrial designer needs to produce clear technical drawings so that an engineering firm can manufacture their product. It is also necessary for the designer and the engineer to speak the same technical language, so that the product can be manufactured successfully.

Over many years, as the information conveyed in technical drawings has become more complex, a universal technical language has evolved. Using recognised standards and conventions established by a regulatory body, it is possible for the visual information and ideas, concepts and finished designs to be clearly understood throughout the world.

Two-dimensional drawings provide a clear means of communicating information about the appearance, assembly, function or construction

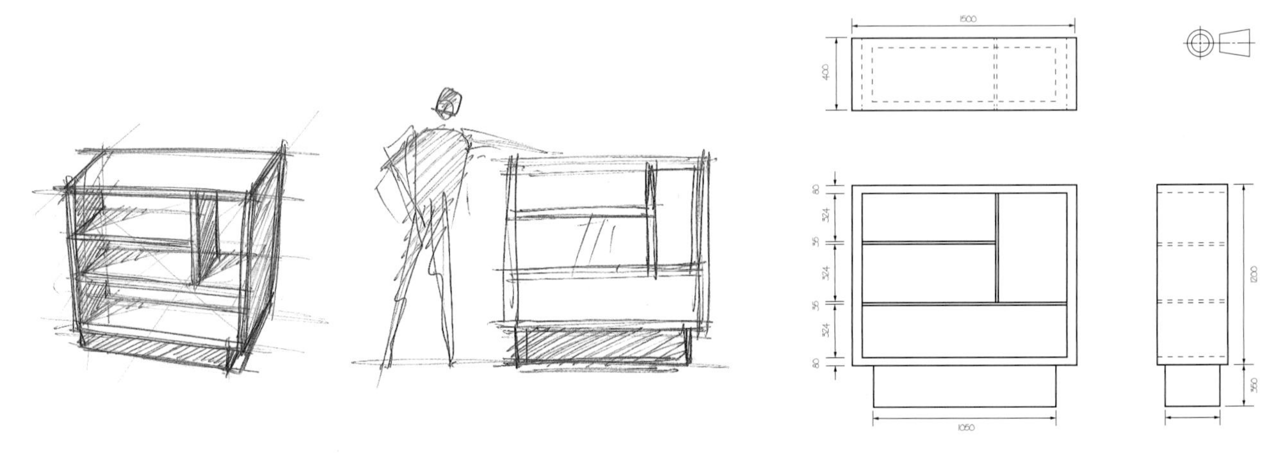

▲ Freehand sketch of shelving unit translated into an orthographic drawing. The orthographic drawing features visual and written information that assists in the manufacture of the product.

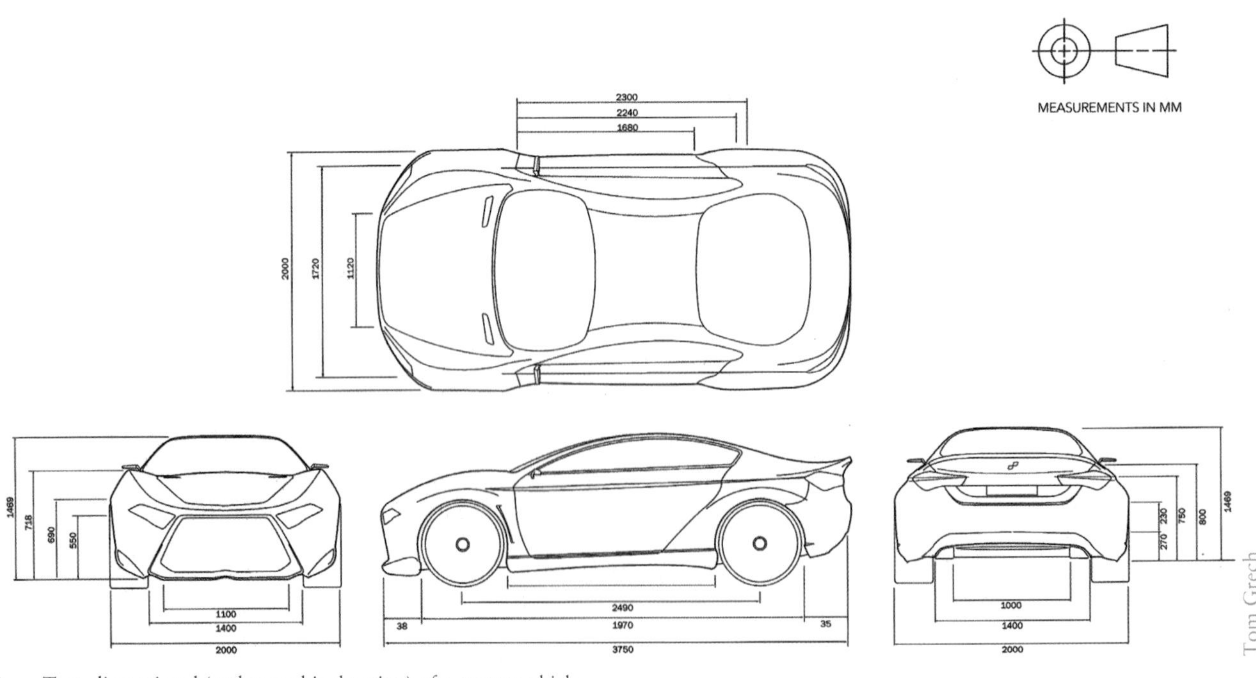

▲ Two-dimensional (orthographic drawing) of a motor vehicle

ISBN 9780170349994

of an object using multiple views of its form and structure. Two-dimensional drawing can be a test of your visual thinking skills as to draw an object that you perceive as having three dimensions (height, width and depth) in only two dimensions (height and width) can be challenging.

A two-dimensional drawing may be used at various stages of a design process; as freehand sketches of views of the object in order to explain the design concept, or as a finished technical drawing with dimensions and section view included.

ORTHOGRAPHIC DRAWING

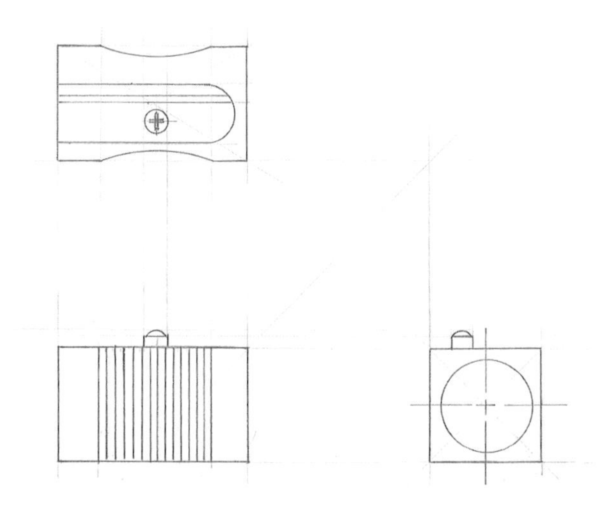

Orthographic drawing is sometimes referred to as multiview drawing. A series of drawings – known as 'views' – are drawn to show every part of the object clearly. Orthographic drawings are widely used by designers, engineers, builders, architects and manufacturers to specify the precise details of objects to be constructed or manufactured.

The application of orthographic drawing is strictly regulated by a set of formal standards. These guidelines, published by Standards Australia, ensure that everyone interprets the drawing in the same way. (Note that orthographic drawing is referred to as 'orthogonal drawing' in the Australian Standards.)

Orthographic drawings usually show the number of views needed to provide the maximum amount of information. The key is to ensure that there is enough visual information evident on the depicted views to avoid any confusion. Three views – the front, top and one of the sides – will usually provide enough information for the drawing to be read clearly and understood. Of course, there may be times when more than three views are necessary.

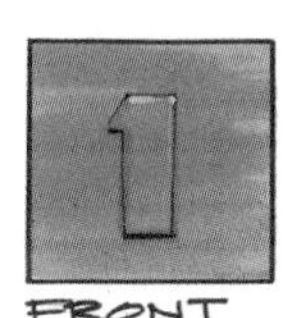

SEEING IN TWO DIMENSIONS

The best way to visualise an orthographic drawing is to imagine that the object is contained within a transparent box, and each part of the object can be seen on a different side of the box. This may help you to gauge how the object might look if the box is flattened out into a two-dimensional shape.

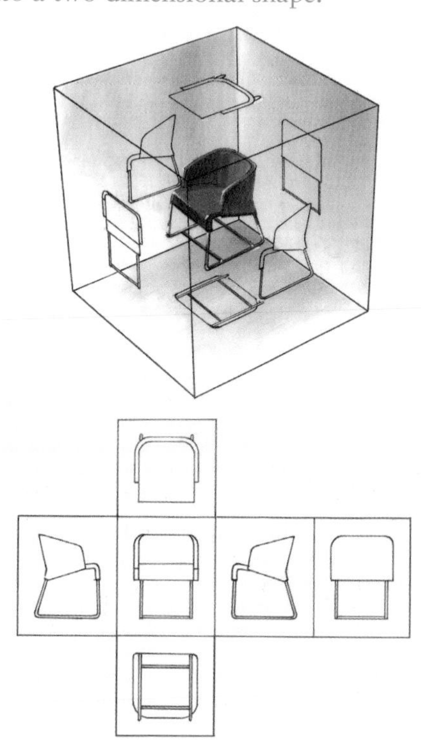

ISBN 9780170349994

The placement or arrangement of views in an orthographic drawing is extremely important. The common arrangement used in Australia is known as third-angle projection, which means that each view is positioned in the drawing so that it represents the side of the object in the view beside it. For example, the right-hand side view of the object is positioned on the right-hand side of the front view.

First-angle projection is used in some countries and represents objects quite differently. In a first-angle projection drawing, the right-hand side of an object would appear on the left-hand side of the front view. In the Australia, third-angle projection is the preferred method of orthographic representation, so *never* apply first-angle projection.

To indicate that third-angle projection has been used, a symbol appears on the drawing.

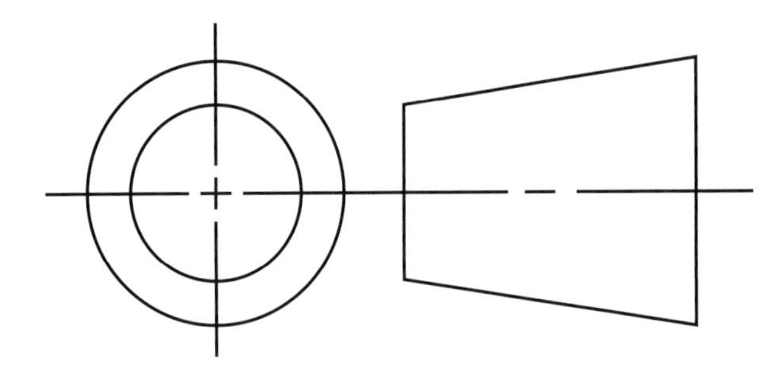

▲ The third-angle projection symbol. This should appear on all third-angle orthographic drawings.

KEY POINTS

When drawing in third-angle projection:

+ All views should be aligned
+ The top view is always situated above the front view
+ The right-hand side view appears on the right-hand side of the front, and the left-hand side appears on the left of the front view
+ You may be asked to appropriately label each view of an orthographic drawing; for example, FRONT VIEW, TOP VIEW, SIDE VIEW. If so, labels should be placed centrally under each view and written in upper-case type.
+ The third-angle projection symbol must always be included on your drawing.

FINDING THE FRONT VIEW

In many tasks, an arrow will indicate the front view of a three-dimensional object. This should be the view that shows the greatest amount of relevant detail about the length and height of the object. In an orthographic drawing of a car, for example, the front view would show one side of the car, rather than the actual front of the car (with headlights, windscreen, etc.).

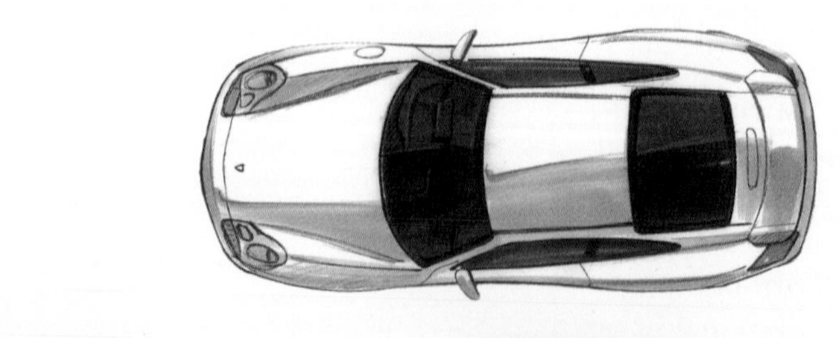

ISBN 9780170349994

A STEP-BY-STEP GUIDE TO ORTHOGRAPHIC DRAWING

A simple method of constructing an orthographic drawing of this three-dimensional form is outlined below. Remember to align your views correctly.

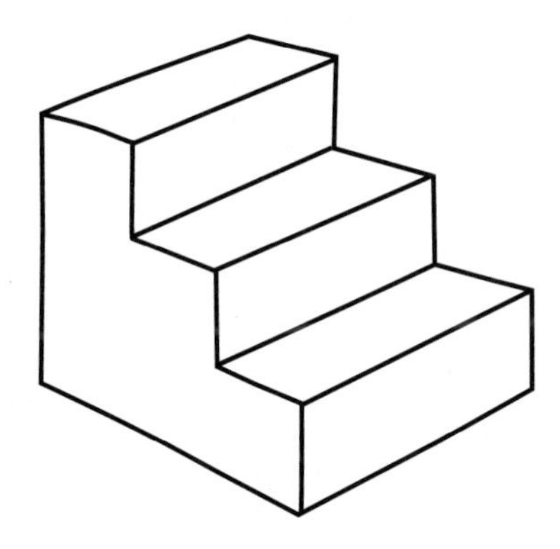

Step 1: Draw the FRONT view of the three-dimensional object.

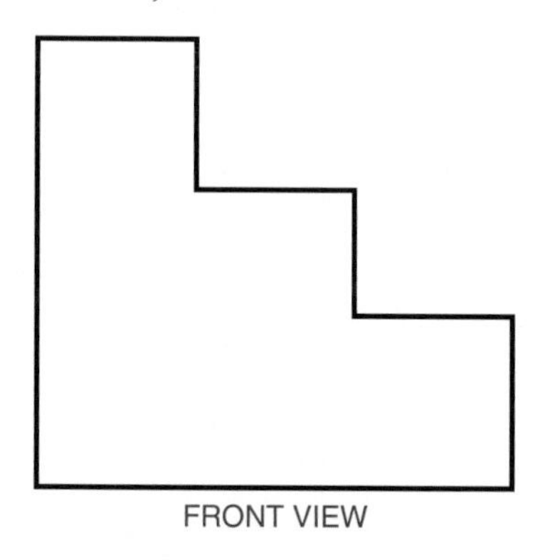

Step 2: Draw light projection lines to project all relevant detail. The TOP view is drawn approximately 40 mm above the FRONT view.

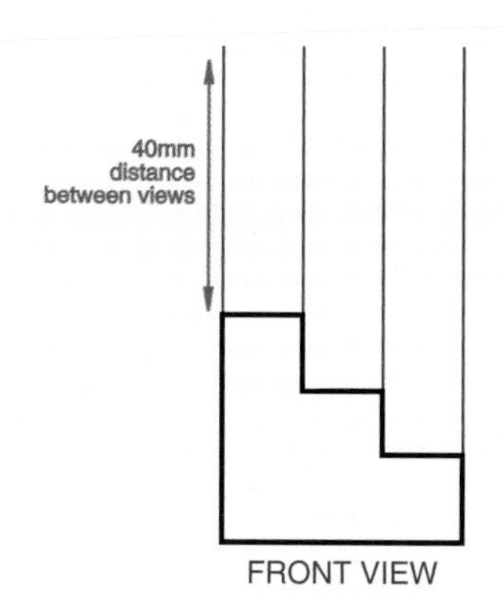

Step 3: Complete the TOP view using the projection lines as references.

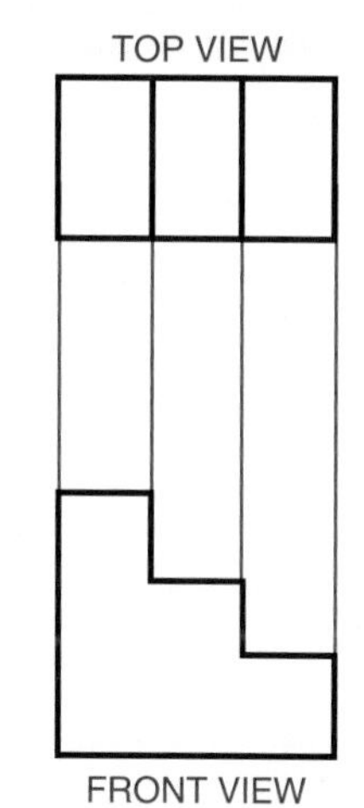

Step 4: Draw projection lines across from the FRONT view to form the SIDE view.

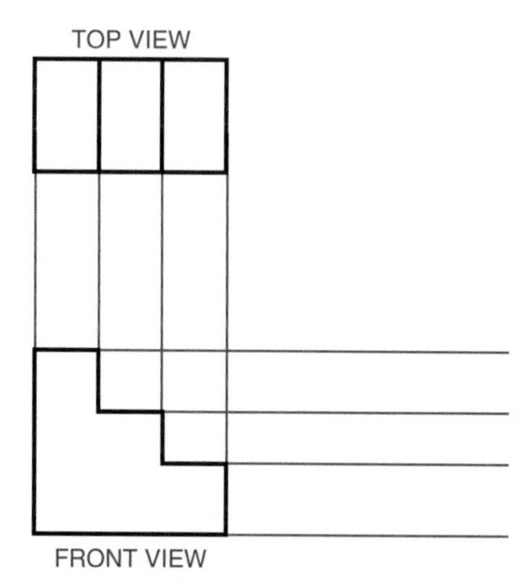

Step 5: Draw a 45° angle at the intersection of the last projection lines on the FRONT view.

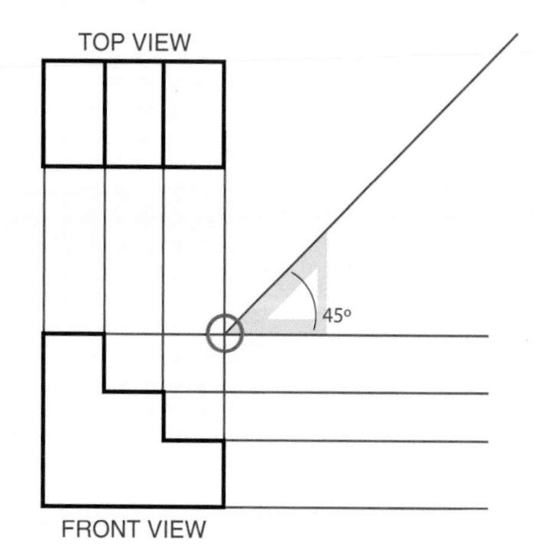

ISBN 9780170349994

Step 6: Project lines across from the TOP view. Where they intersect with the 45° line, draw vertical lines down to form the SIDE view.

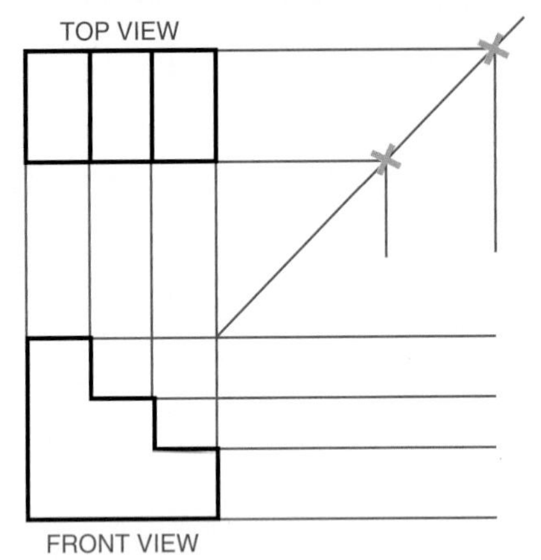

Step 7: The projection lines will make completing the SIDE view straightforward.

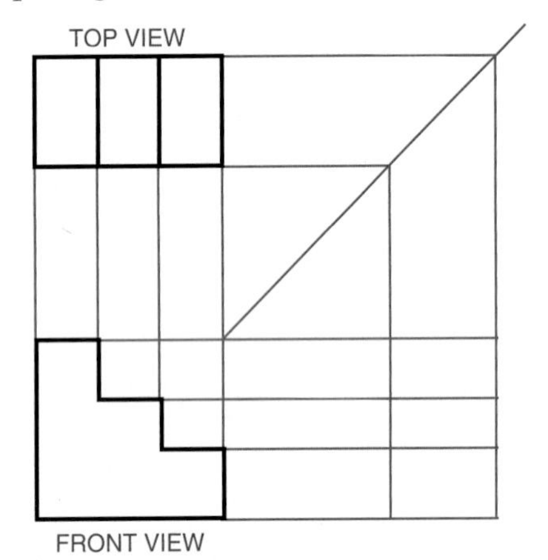

Step 8: Complete your drawing by outlining the SIDE view. Once your drawing is finished you can remove the projection lines or trace onto a clean sheet of paper for presentation.

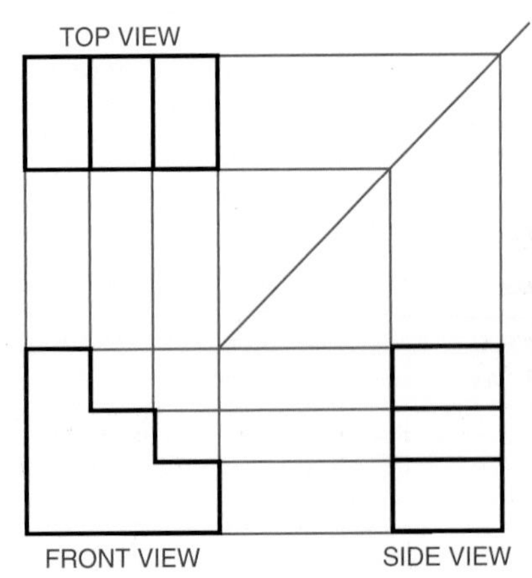

USE A GRID

If you are creating a hand drawn draft of your orthographic drawing before moving onto a CADD program, try using grid paper for your initial drawings, as this can help you to align views correctly and keep your line work accurate. Once you have the drawing exactly right on the grid paper, you scan it to use on the computer.

Line conventions

As in plan development drawing, the use of line in orthographic drawing is significant. The width of lines indicates essential information such as whether details are visible or hidden. Generally, only two line thicknesses – thick and thin – are used within one drawing. When providing extremely detailed information, architects and drafting professionals sometimes use medium lines as well.

Orthographic drawing	Line type	Application
	A thick continuous line	Used to show visible outlines
	A thin continuous line	Used to draw: • dimension lines • projection lines • hatching • short centre lines • general symbols • fictitious outlines.
	A thin broken line	Used to show hidden details
	A thin chain line	Used to show centre lines
	A thin broken line, thick at ends	Used to show a cutting plane
	A thin freehand continuous line	Used to show a break in a view, especially when drawing large objects

▲ Line conventions and applications in orthographic drawing

ISBN 9780170349994

Scale

Every orthographic drawing must be drawn in proportion to the original three-dimensional object. The scale must be applied consistently throughout the drawing and indicated on the page.

Australian Standards provides recommended scale ratios for a consistent approach in two-dimensional drawing. The first numeral in the ratio indicates a measurement on the drawing, while the second numeral indicates the equivalent measurement on the actual object. For example, a scale of 1:5 indicates that one unit of measurement on the drawing represents 5 units on the actual object, so 1 mm on the drawing is equal to 5 mm on the object itself.

An orthographic drawing that shows actual size should indicate a scale of 1:1. Drawings that depict objects at smaller than actual size should be drawn using one of these scale ratios:

1:2	1:20	1:200	1:2000
1:5	1:50	1:500	1:5000
1:10	1:100	1:1000	1:10 000

For drawings that depict an enlargement of the original object, the following scales apply.

2:1 5:1 10:1 20:1 50:1

When creating an orthographic drawing, you will need to establish which scale is most appropriate and enables you to depict the maximum amount of detail.

Hidden detail

Sometimes an object features details that are internal and cannot be seen by viewing external surfaces. In these circumstances, orthographic drawing is vital in depicting otherwise hidden details.

Hidden details are indicated by a thin broken line. It is essential that this line convention be followed so that the detail is clearly understood and not mistaken for an outline.

Hidden details might appear on any of the featured views.

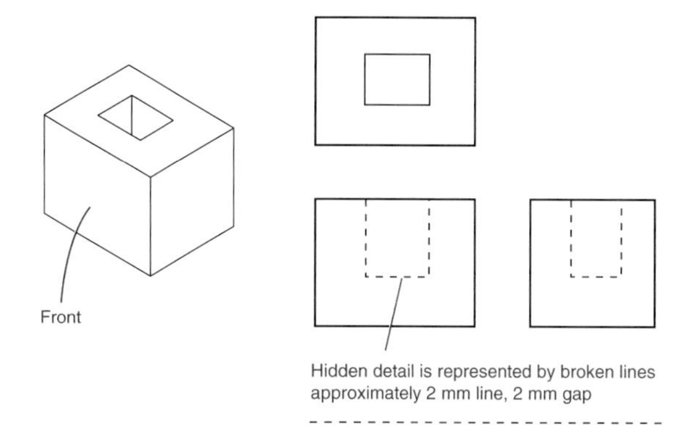

Cross-sections

The cross-sectioning of an object allows for the internal details to be clearly shown. A cross-section is literally a 'cut-through' view of an object. An imaginary cut is made at an appropriate point to display the relevant information.

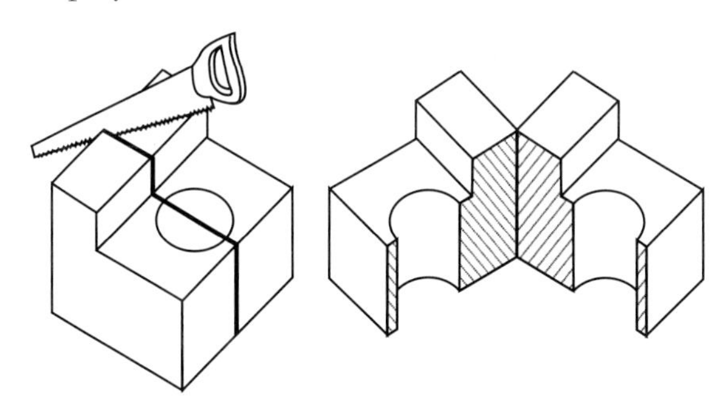

The cutting plane

The path of the imaginary cut is indicated on the orthographic drawing by a cutting plane. The cutting plane is shown as a broken line with thick lines at each end. Narrow arrowheads touch the end of the cutting plane, indicating the direction of the cut and the subsequent direction of view.

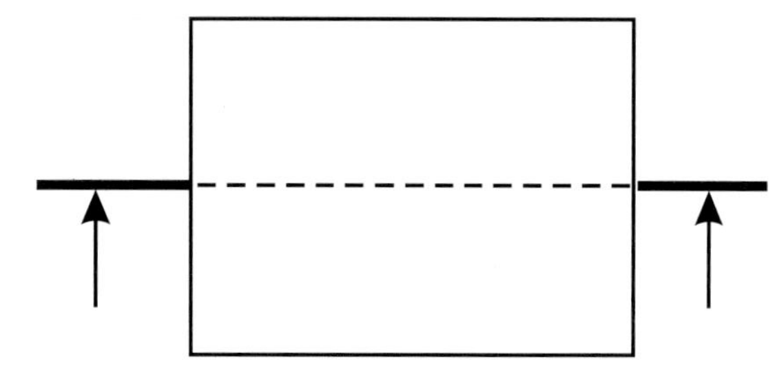

▲ A broken line indicates the cutting plane. The start and end of the cut are indicated by a short bold line.

The cutting plane is always labelled clearly. The first cross-section is labelled Section A-A, and subsequent sections are B-B, C-C and so on.

ISBN 9780170349994

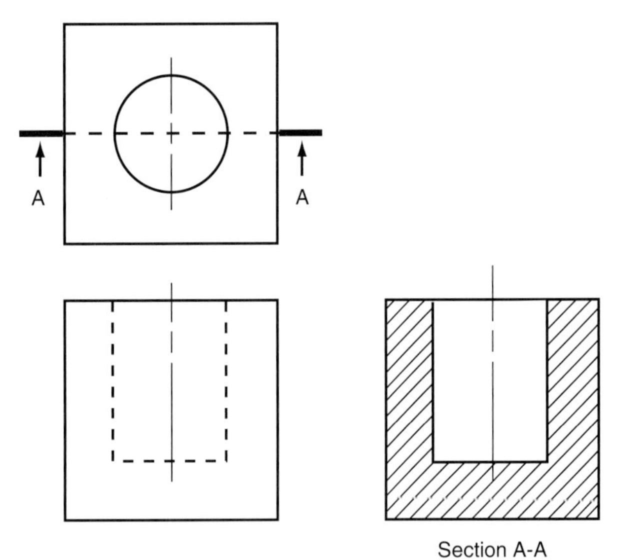

▲ The sections must be clearly labelled as Section A-A, Section B-B and so on, as required.

The section view is defined by the direction of the arrows on the cutting plane and appears on an orthographic drawing, at the side of the regular views.

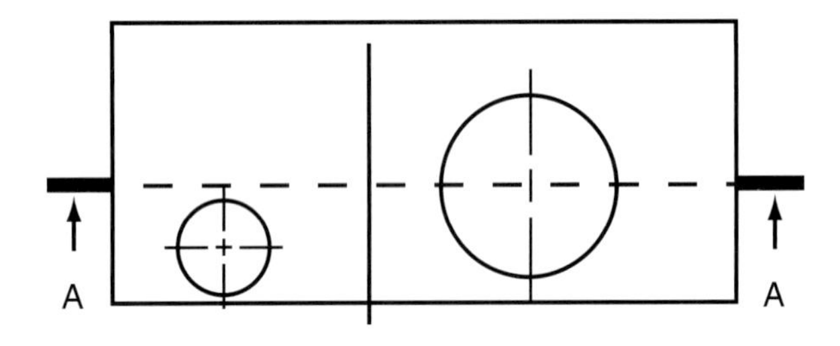

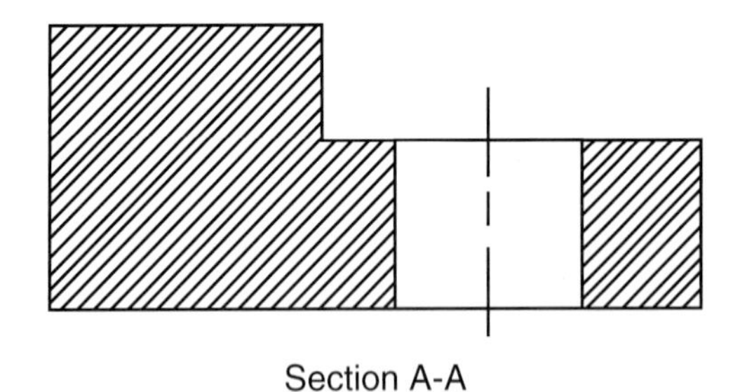

▲ Note that the direction of the arrows indicates the view that should be shown on the cross-section.

Crosshatching

In order to show the solid areas of the object that have been 'cut', hatching is used. The hatching should be drawn at 45° and be suitably spaced relative to the area covered. It is recommended that wide spacing be used as long as the clarity of the technical information is not compromised. On smaller drawings, narrower spacing may be necessary. Hatching lines should be fine in order to distinguish the section from the outline.

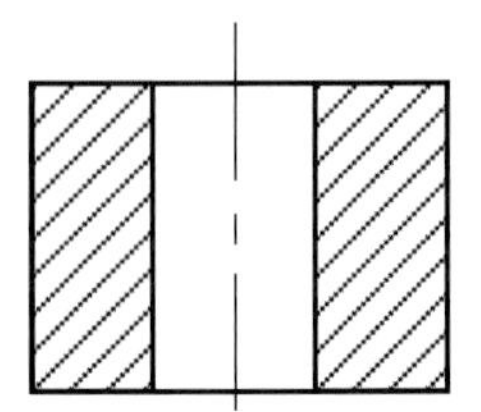

▲ The distance between hatching lines should be consistent and drawn at 45°.

Crosshatching on adjacent solid planes

Where adjacent parts of an object meet, the direction of hatching should be reversed.

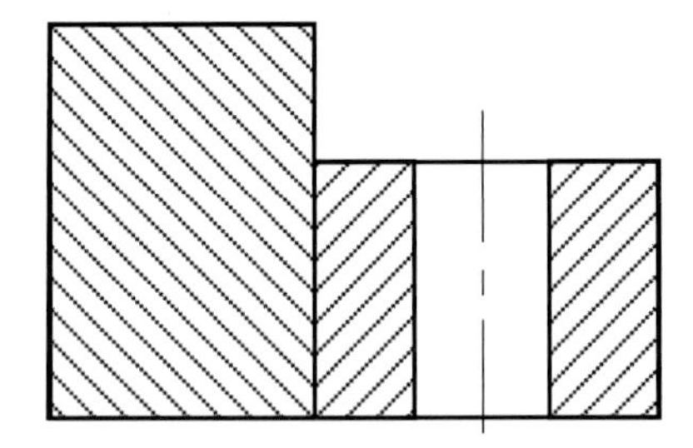

Indicating large sections

Where very large sections are concerned, hatching can be limited to the areas around the outline to indicate a cut plane.

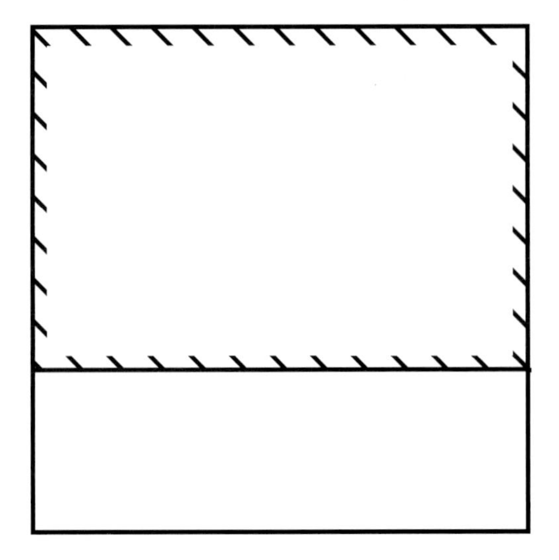

ISBN 9780170349994

Cylindrical objects

Cylindrical objects can show internal detail by using a half section rather than a full section.

DIMENSIONING ORTHOGRAPHIC DRAWINGS

Dimensioning is the placement of measurements on an orthographic drawing. Like other aspects of this drawing method, there are strict conventions to be followed when dimensioning.

The key to successful dimensioning is to make sure you include every relevant measurement that is crucial to the manufacture or construction of the object. However, it is equally important not to 'over-dimension', and each dimension should appear only once.

When you add dimensions to your drawing, make certain that you only use those dimensions necessary to clearly define the object.

Sometimes very complex objects are drawn in third-angle orthographic drawing so it is important that the details of the drawing are very clear. Carefully plan your dimensioning so that the viewer can interpret the drawing quickly and without ambiguity.

The elements of dimensioning

+ *Projection lines*: These thin lines are placed outside the outline of the object and define the area being dimensioned. Projection lines do not touch the outline but sit approximately 2 mm from the outline edge.
+ *Dimension lines*: These are also thin lines and have thin, closed arrowheads at either end. The head of each arrow touches the projection line, defining the dimensioned area. Dimension lines are drawn 10 mm from the outline of the object.

Where there are multiple dimensions in the same area, dimension lines should remain 10 mm apart. The smaller dimensions are drawn closest to the outline.

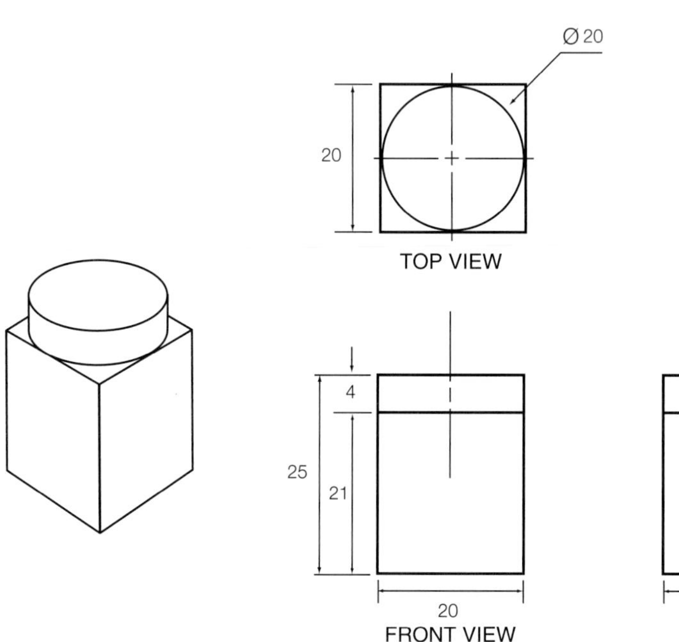

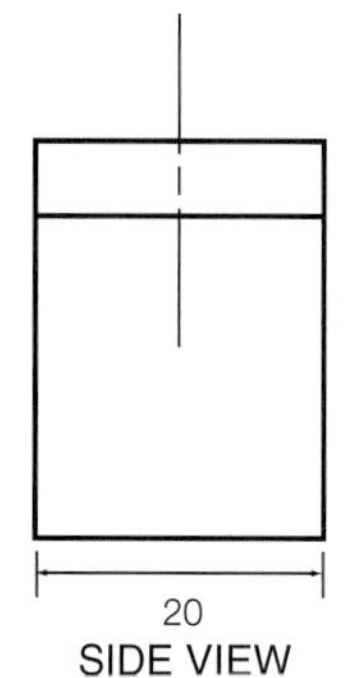

MEASUREMENTS IN MM
SCALE 1:5

▲ Dimensioned third-angle orthographic drawing

ISBN 9780170349994

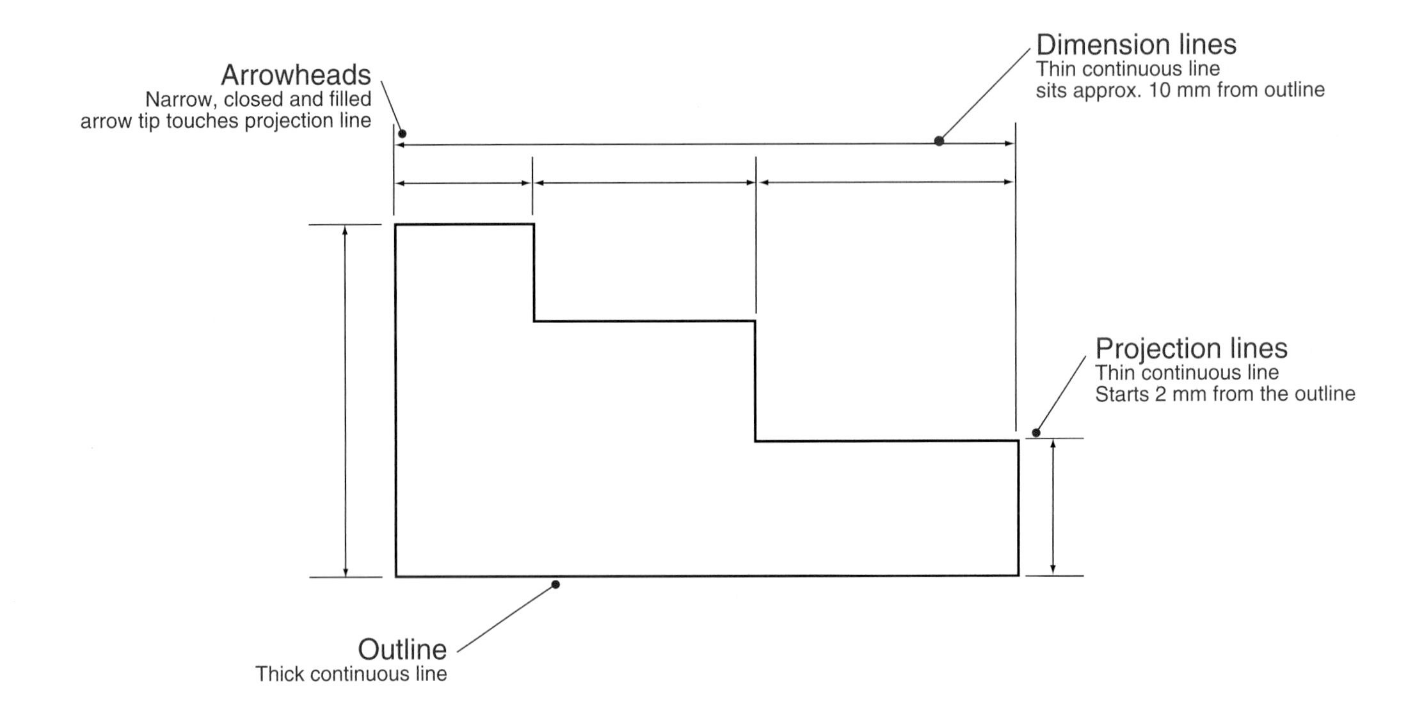

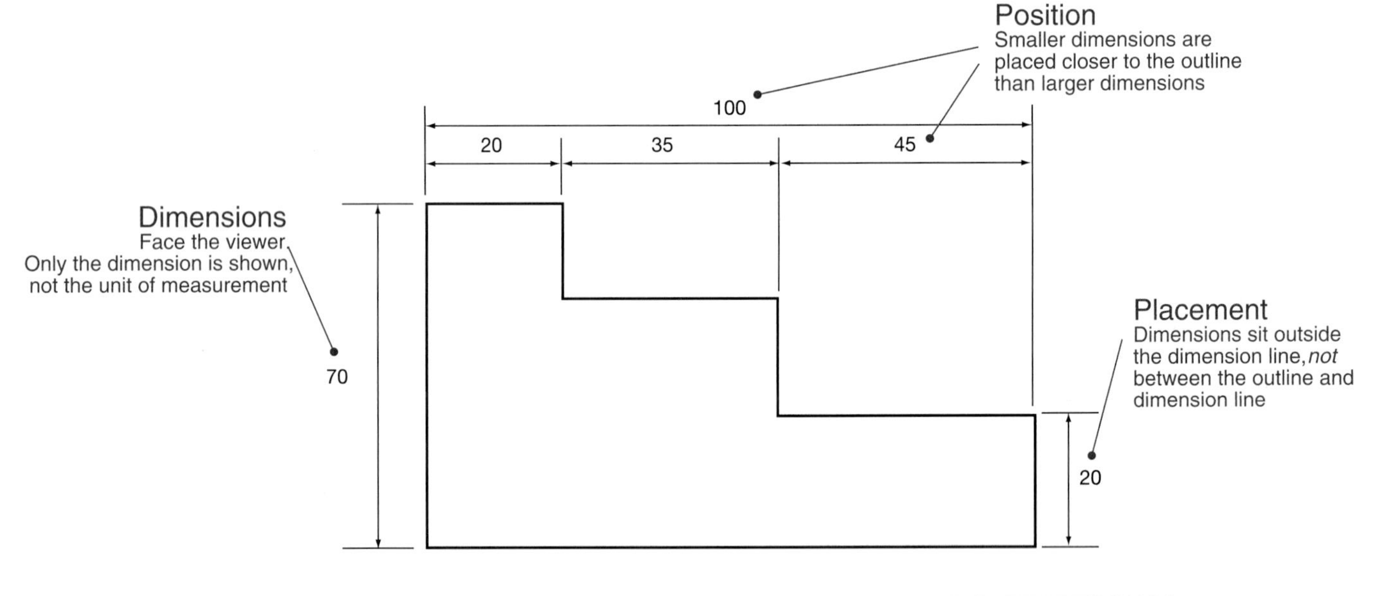

MEASUREMENTS IN MM

ISBN 9780170349994

▼ Arrowheads

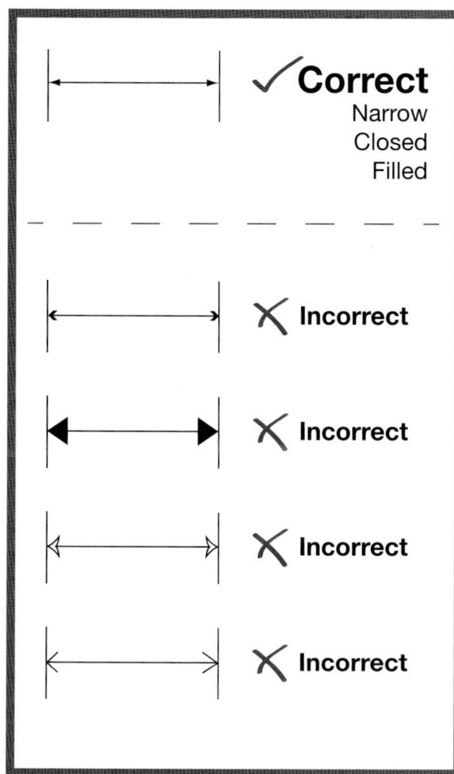

▼ Dimensions

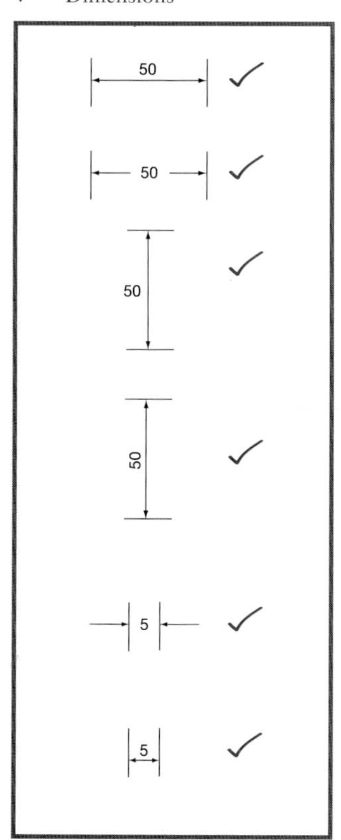

The dimension of the object (not the drawing) is placed on or inside the dimension line. The dimension is never placed between the outline and the dimension line. The method you use when placing your dimensions is up to you (or your teacher) but clarity must be your priority. Preferred methods of representing dimensions are illustrated.

All measurements are in millimetres, and this should be indicated on your drawing by the words 'MEASUREMENTS IN MM' in UPPER-CASE letters. Don't add 'mm' to each dimension. Remember that the key is to keep the drawing as clear and uncluttered as possible.

DIMENSIONING CIRCULAR DETAILS

Centre line

A circular or other symmetrical feature on an object is indicated by the application of a centre line. The centre line is a thin chain line that is placed through the centre of the symmetrical feature.

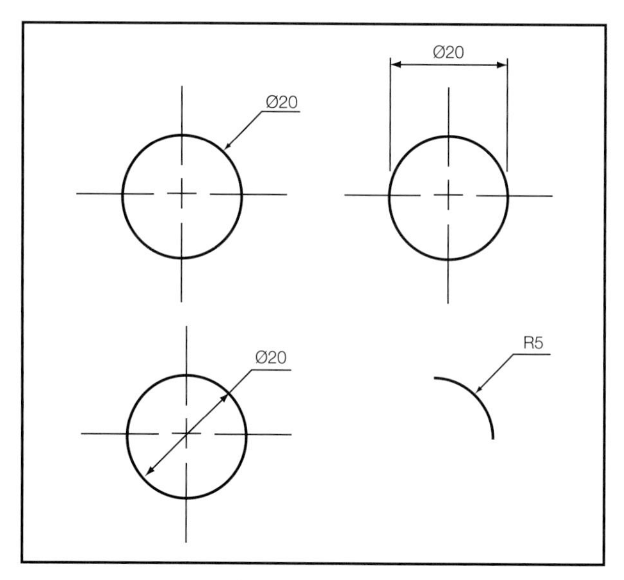

▲ Dimensioning circles

+ When dimensioning a full circle, use the symbol Ø for diameter.
+ When dimensioning part of a circle (an arc), use R for radius.
+ The dimension lines with a single arrowhead used in the dimensioning of circular details are known as leaders.

ISBN 9780170349994

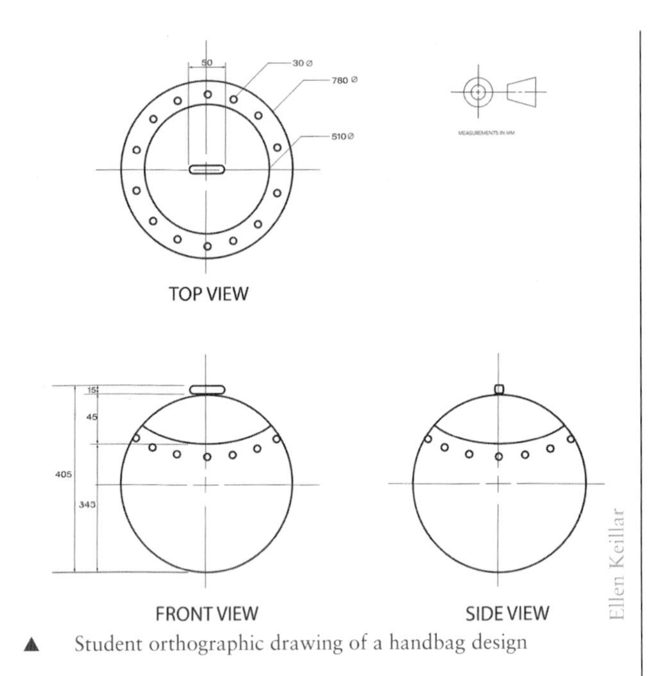

▲ Student orthographic drawing of a handbag design

Ellen Keillar

MULTIPLE DIMENSIONS

When you are working with complex objects, follow these guidelines.

+ The smallest dimensions are placed closest to the outline.
+ Larger dimensions sit at least 10 mm away from the smaller dimension.

At all times, remember to keep your orthographic drawing as uncluttered and clear as possible.

DIMENSIONING SECTIONS

Dimensions specific to the section can be shown on the section view but the standard rule for dimensioning applies: Don't over dimension. Dimensions that appear on the three regular views do not need to be shown on the section view.

Leaders – the continuous thin lines with an arrowhead – are used to indicate dimensions of any details that may otherwise be awkward to represent. Leaders are also used for notes within an orthographic drawing.

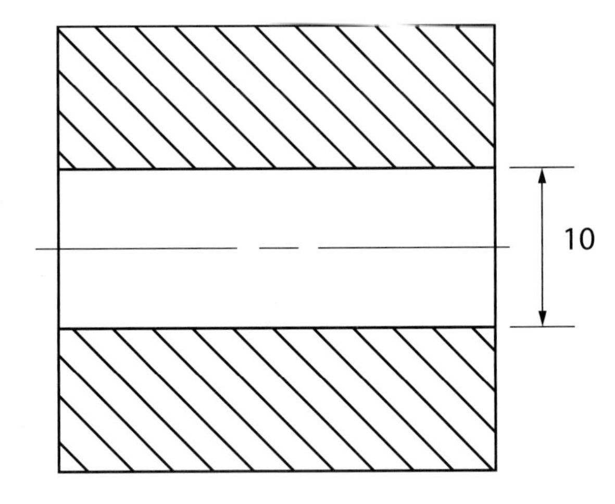

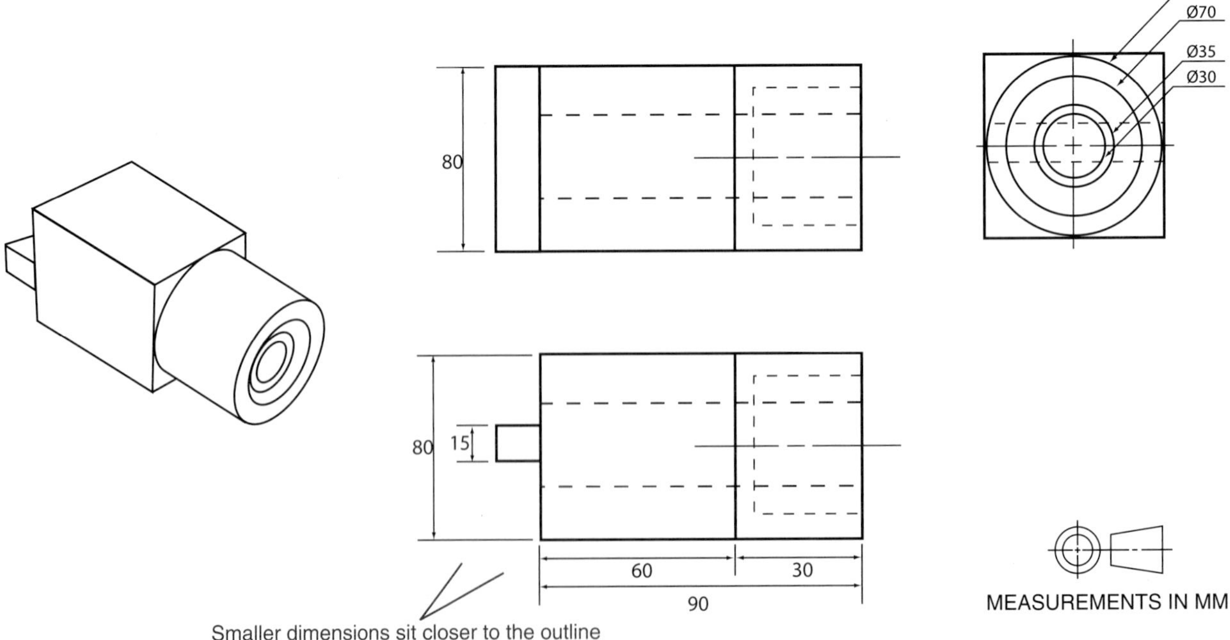

▲ Dimensioning of a complex shape

ISBN 9780170349994

USING LEADERS

- A leader has a narrow arrowhead, similar to a dimension line.
- The tip of the arrowhead touches the outline.
- The leader is drawn at 45°.
- Leaders should be short and never overlap. (Dimensions may be repeated to avoid this.)

DIMENSIONING DETAILS

Angular features

When indicating the angle of a feature, dimensions should be expressed in degrees and decimal parts, such as 33° or 33.5°. If the angular measurement is less than one degree, include a zero before the decimal point; for example, 0.3°.

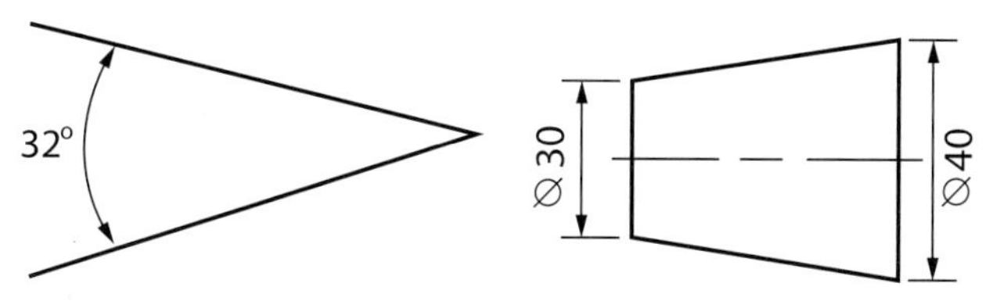

▲ Note that the dimensions are set vertically. This is often applied to save space on a drawing. Keep in mind that the placement of dimensions should be consistent across a drawing.

Tapered features

The following symbol indicates a tapered feature. The symbol indicates the direction of the taper and the ratio of taper to the diameter, and should be used whenever a tapered detail is shown in an orthographic drawing.

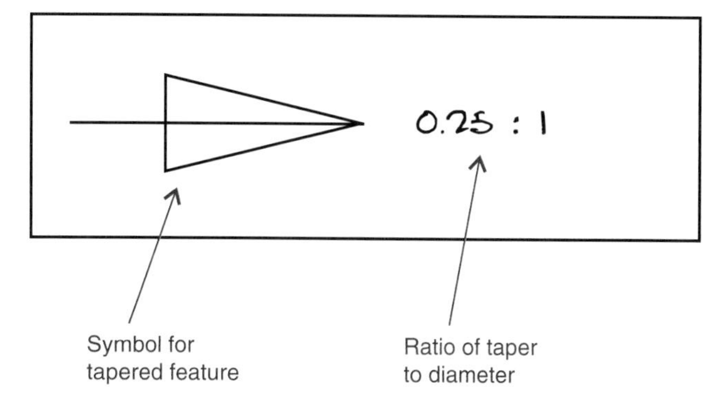

Dos and don'ts of dimensioning

Do	Don't
Ensure that your projection lines and dimension lines are thinner than the outline.	Write 'mm' next to individual dimensions. Write the sentence 'MEASUREMENTS IN MM' elsewhere on your drawing.
Keep your line widths consistent.	Cross dimension lines or projection lines with other lines unless it is absolutely unavoidable.
Dimension on the view that shows a detail most clearly.	Use a centre line as a dimension line.

ISBN 9780170349994

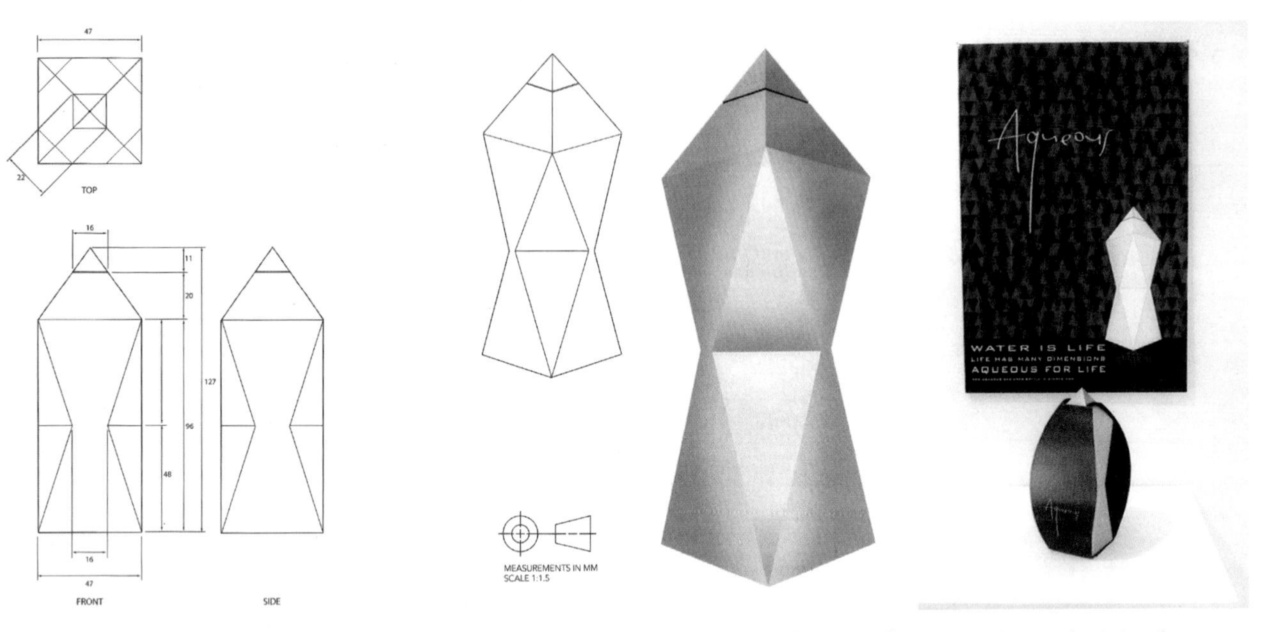

Sarah Mason

▲ In this student's design for a drink bottle, she created two-dimensional and three-dimensional representations and a 1:1 scale model with packaging.

15.2 DEVELOPMENT DRAWINGS

Development drawings are a method of two-dimensional drawing used when an object is to be manufactured from a single piece of material. Also known as a net or packaging net, these drawings provide information about the form of an object to be created from material such as cardboard or sheet metal. Examples include cardboard boxes used for packaging and a point-of-sale display.

Within development drawings, the representation of lines has great significance. Each line has a different meaning and it is essential that the person viewing the drawing can understand the meaning of each line. Where to cut? Where to fold? What to discard and what to keep?

+ Broken lines indicate the folds of an object.
+ A solid line indicates the cutting edge.
+ If the object is to be glued, a row of black dots indicates the glue area.
+ Areas where adhesive is required, or where folded areas interlock to create the form of the object, are called tabs.

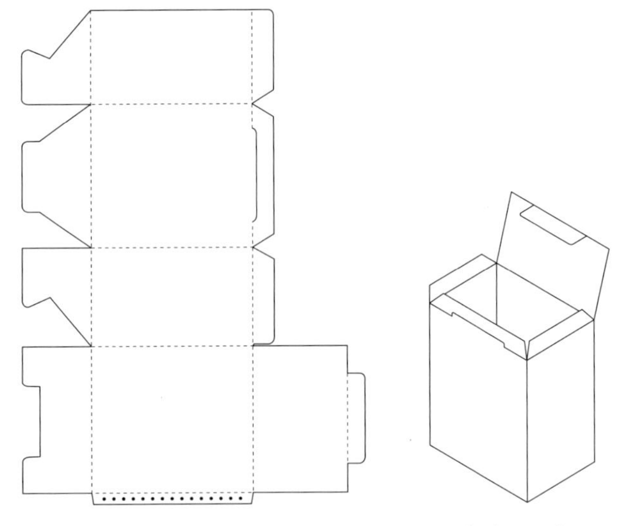

▲ Development drawing of packaging. Note the line and symbol conventions; black dots indicate glue lines, broken lines indicate folds and solid lines indicate cuts.

DIELINE

The Dieline is a site dedicated to the design of packaging. View award-winning and innovative packaging designs from around the world.

Access all weblinks directly at http://nsg.nelsonnet.com.au.

ISBN 9780170349994

TWO-DIMENSIONAL DRAWING IN BUILT ENVIRONMENT DESIGN

Technical drawings of floor plans and elevations are used in environmental design, which, along with computer-generated three-dimensional images, create representations of spaces and structures that do not yet exist.

Drawing for the environment – in architectural design, interior design and landscape design – involves the application of a range of conventions. These are standard approaches that enable the viewer to understand the meaning of a drawing. The depiction of a door on an architectural plan, for example, has a standard appearance that helps us to understand what it represents.

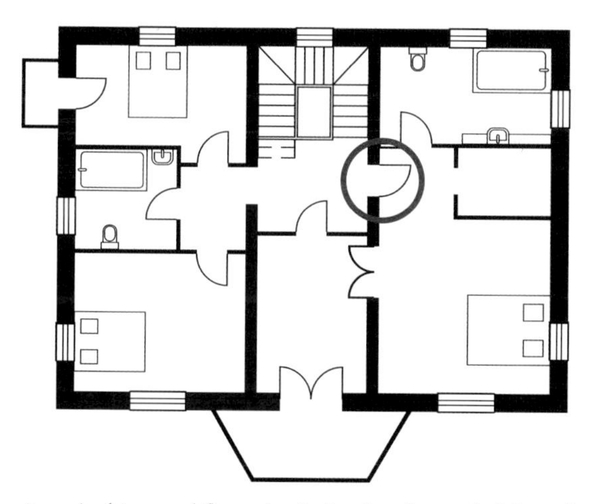

▲ Architectural floor plan indicating the symbol for a door

Two-dimensional methods are applied at various stages of the design process and may be used to visualise ideas in the early stages, as well as assist construction with refined technical drawings in the latter stages.

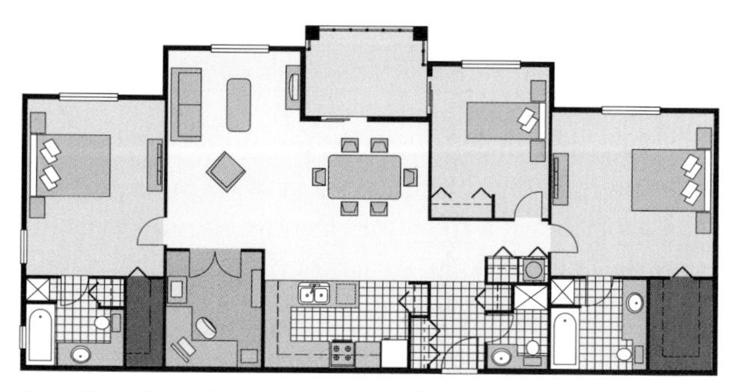

▲ Two-dimensional drawing of a floor plan

PLANS AND ELEVATIONS

In environmental design, plans and elevations are typically used to convey visual information about a three-dimensional design. Plans are the equivalent of the top view in orthographic drawing, and elevations show the front and side views.

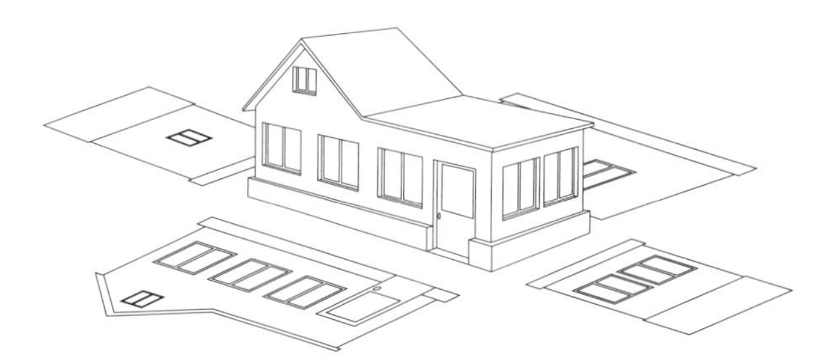

Plans and elevations designed for the purpose of construction are usually line drawings whereas plans and elevations designed for presentation can be much more detailed and may include, colour, shadows, textures and backgrounds. Plans are often used in real estate advertising to enable potential purchasers or tenants to see the floor plan of a property.

▲ Real estate advertising board showing floor plans

ISBN 9780170349994

Drawing the plan view

There are a number of conventions used in representing architectural details in two dimensions. One of most common conventions is the use of symbols to describe features. Although their visual appearance may differ slightly if the drawing is completed by hand or via digital means, the meaning remains the same.

Symbol conventions

Symbols should be drawn to the same scale as the plan and, where possible, be used without a text label or abbreviation. Many details of a plan, such as domestic appliances (dishwasher, refrigerator, wall oven), are indicated by a rectangle with a diagonal line and may require an abbreviation for clarity.

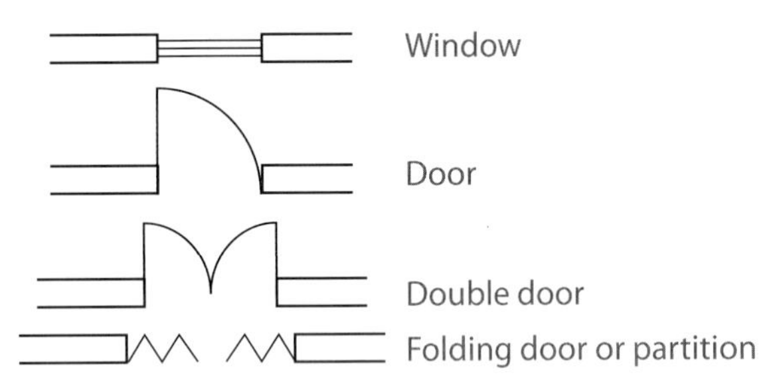

▲ Door and window symbols

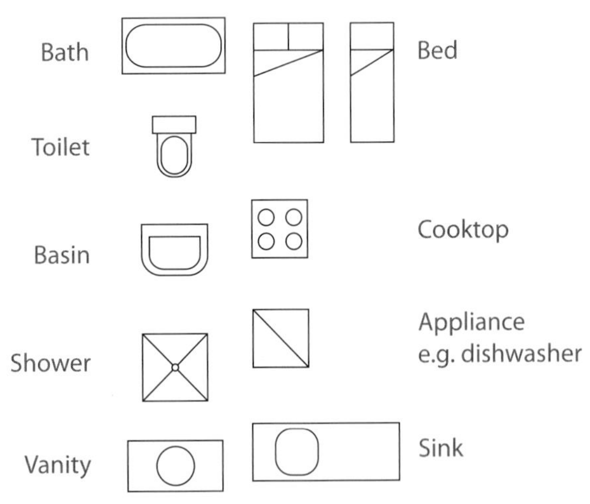

▲ General symbols. Note that the symbol used for general appliances, such as dishwashers and fridges, requires an abbreviation (see next table) to identify its purpose.

CADD programs offer many examples of architectural symbols and these can assist in adding meaning and function to the drawing of a space. However enticing it may be to fill a plan view with details, always remember to err on the side of clarity.

Abbreviations

Abbreviations are used to label the features and fixtures used in a design.

▼ Examples of abbreviations used in plan and elevation drawing

Word	Abbreviation
Aluminium	AL
Bookcase	BC
Brick veneer	BV
Brickwork	BWK
Cement render	CR
Ceramic tile	CT
Clothes drier	CD
Washing machine (clothes washer)	CW
Cooker	C
Corrugated	CORR
Cupboard	CPD
Dishwasher	DW
Door	D
Down pipe	DP
Floor waste (bathroom/laundry)	FW
Heater	HTR
Hot water unit	HW
Linoleum	LINO
Open fireplace	OFP
Refrigerator	R
Roller shutter	RS
Shower	SHR
Stainless steel	SS
Terracotta	TC
Toilet	WC
Urinal	U
Vinyl	V
Wardrobe	WR
Weatherboard	WB
Window	W

Note that these abbreviations do not include all materials that are likely to be used in the construction of a dwelling. Details about fittings that do not affect the structure of a building, such as floor coverings, wall coverings and interior decorations, are usually listed on a separate document.

ISBN 9780170349994

Line conventions

Line conventions are important in environmental drawing. As in orthographic drawing, the width of lines communicates different information. Although it is a combination of bold and fine lines that are generally applied in plan and elevation drawings, medium lines are sometimes utilised where a detail needs to be differentiated.

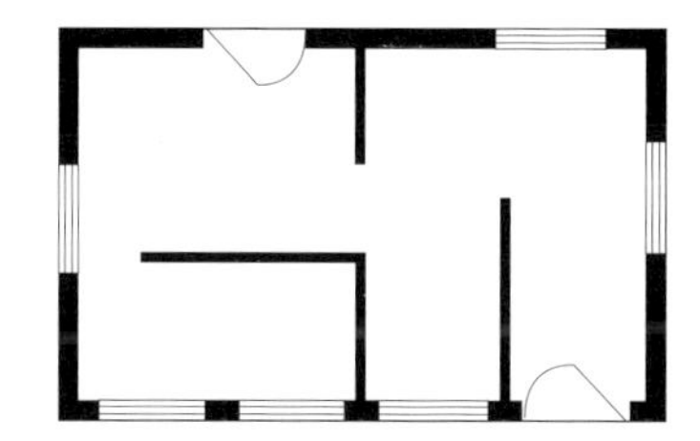

Bold lines (3 mm) indicate the outlines of structural walls and thin lines (1 mm) indicate interior walls, windows and doors. A black, filled shape, or thick outline, is used to identify wall thickness.

Scale

The same scales that are used in industrial design are also applied to architectural drawing:

Reduction scales

1:2	1:20	1:200	1:2000
1:5	1:50	1:500	1:5000
1:10	1:100	1:1000	1:10 000

Enlargement scales

2:1	5:1	10:1	20:1	50:1

The most common scale applied in architectural drawing is 1:100.

Drawing elevations

Unlike orthographic drawings, which are identified as 'Front', 'Top' and 'Side' views, elevations are usually named for the direction they face; for example, 'North', 'South', 'East' and 'West'.

To create elevations, it is necessary to have a completed floor plan drawn to scale. The floor plan is used for the projection of lines to create the elevation views. The elevations are usually drawn to the same scale as the floor plan.

Additionally, you will need to establish the roof height of your structure prior to drawing the elevations.

A STEP-BY-STEP GUIDE TO DRAWING EXTERIOR ELEVATIONS

Step 1: Establish the ground level, window height, door height and roof height of your structure and indicate these with horizontal lines.

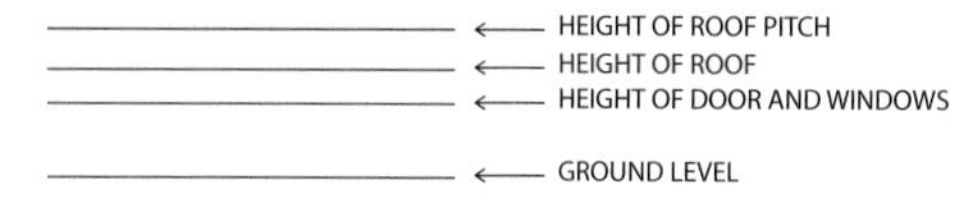

Step 2: Place your floor plan above the horizontal lines.

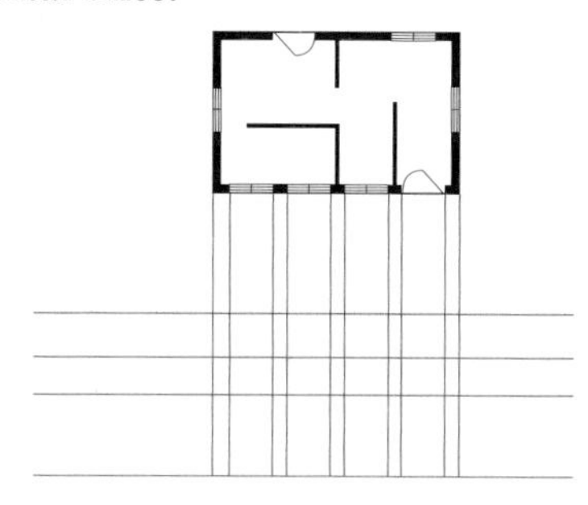

Step 3: Project vertical lines down from the plan view to indicate features such as doors, windows and external walls. Using the vertical and horizontal lines as a guide, add the details of the elevation view.

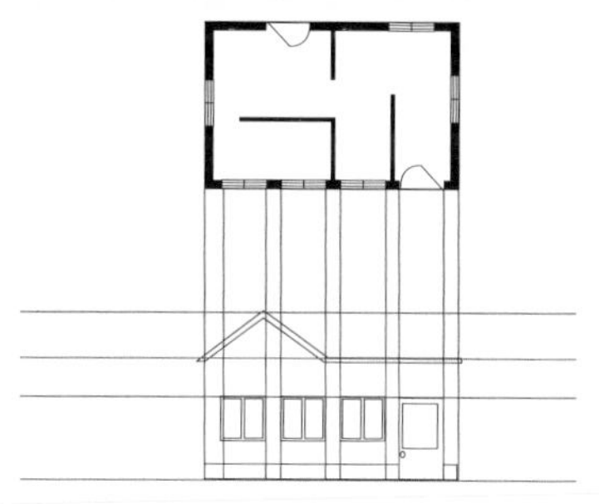

Step 4: Rotate the plan view and project lines to create the next elevation.

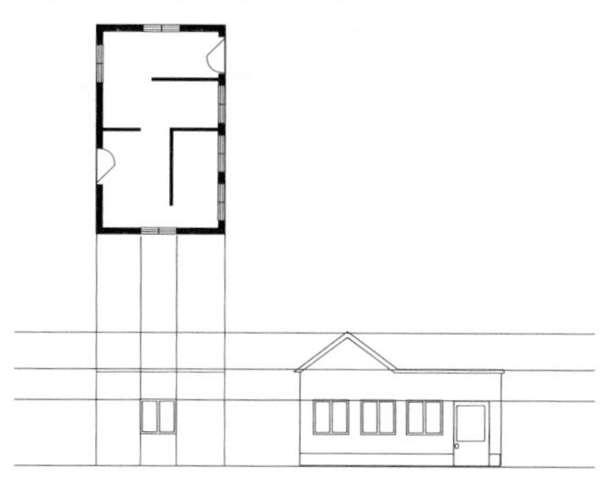

ISBN 9780170349994

Step 5: Continue the process until all elevations are complete.

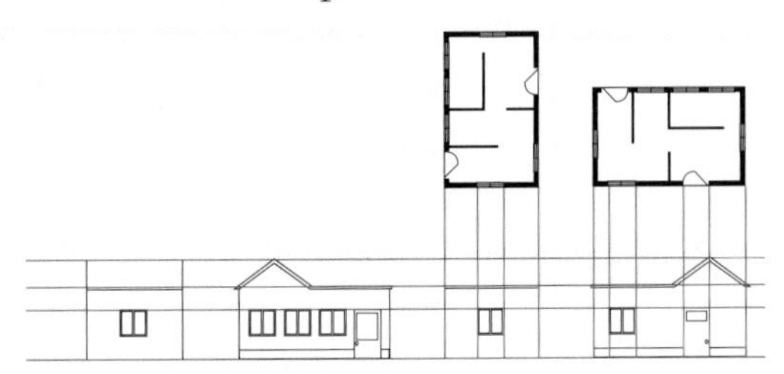

Step 6: Label each elevation in upper-case letters.

A STEP-BY-STEP GUIDE TO DRAWING INTERIOR ELEVATIONS

Interior elevations can be constructed in a similar way to external elevations. The main difference is that the ceiling, rather than the roof, height needs to be indicated.

Step 1: Indicate (horizontal) height and ground lines then project (vertical) lines from the relevant section of the floor plan.

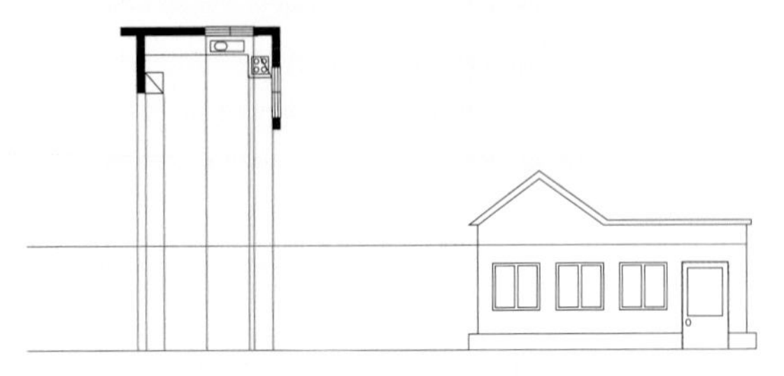

Step 2: Complete the interior elevation.

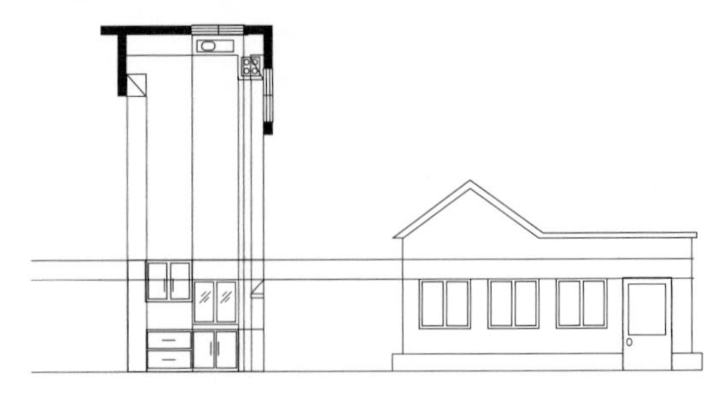

Step 3: Title the elevation in upper case. Usually the t itle explains the room purpose, e.g. KITCHEN, LAUNDRY. When there are multiple elevations of the same room, the title may also indicate direction; for example, KITCHEN WEST ELEVATION.

Dimensioning plans and elevations

There are usually many more dimensions on an architectural drawing than on an orthographic drawing so the method of dimensioning differs.

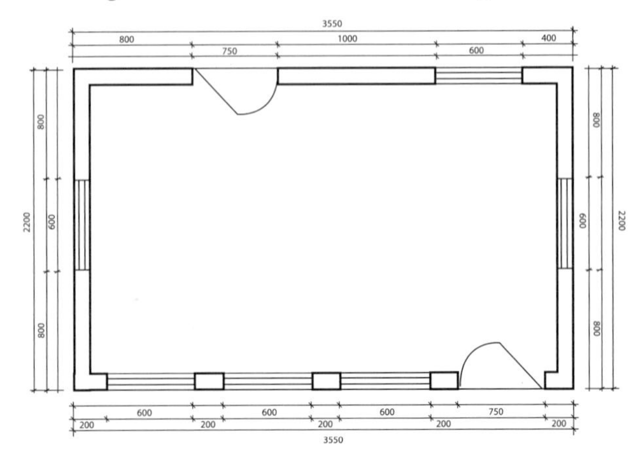

ISBN 9780170349994

Projection lines extend from the outline by 2 mm. Dimension lines are required for all features of the structure. Smaller dimensions sit closest to the outline; for example, doors and window details.

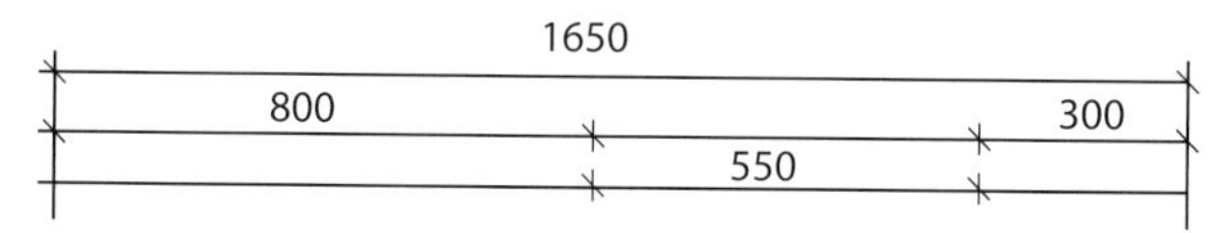

Instead of arrowheads, 45° strokes (2 mm long) are used to indicate the beginning and end of a dimensioned area.

Section views

As in orthographic drawing, section views are used to show detail that cannot be seen on a regular plan or elevation view. Section views can expose the structural features of a building and depict the internal configuration of spaces. Typically, section views use a thick outline to indicate the cut surfaces and progressively lighter line weights to show interior details.

Rendering plans and elevations

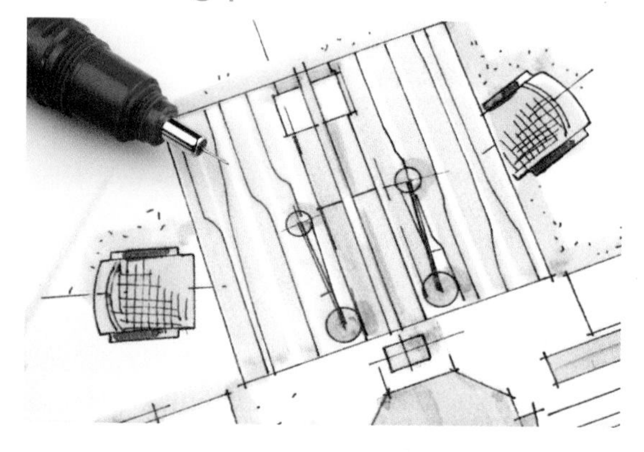

In environmental drawing, many exterior and interior elevations feature representations of materials. Architects and interior architects often render elevations to convey visual information about the materials and textures that will appear in the design.

CADD programs also offer rendering options and options for depicting textures such as wood, metal, stone, glass and fabric.

From simple dot and line rendering to depict textural detail to fully rendered representations of each material used, the use of rendering is often determined by the purpose of the final drawing.

For more information about representing materials, see Chapter 13.

Landscape design

In landscape design, plan views are an integral part of the design process. A landscape plan view provides an overview of a site and may indicate areas for landscaping, construction and planting. It may be a hand-drawn or computer-generated line drawing or fully rendered presentation.

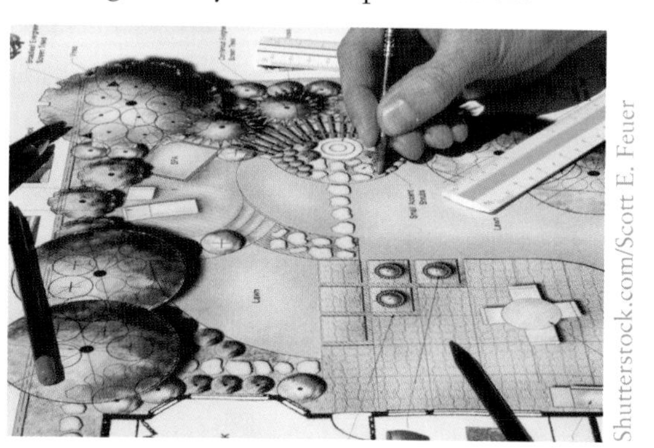

Shutterstock.com/Scott E. Feuer

A landscape plan often includes annotations that indicate the nature of a planting scheme or the specifics of the materials to be used in the construction of a feature. Elevations are also used to illustrate the appearance of views within the landscape design. Like architectural drawings, landscape designs are drawn to scale.

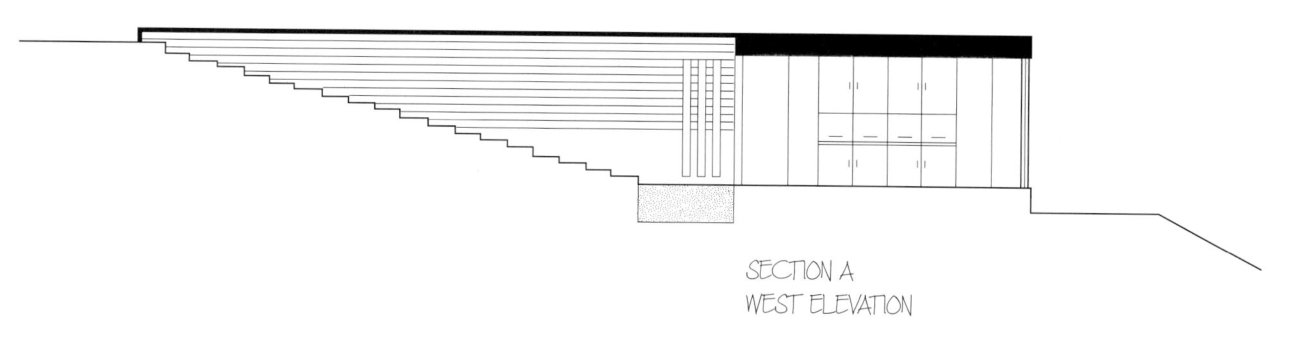

ISBN 9780170349994

CHAPTER 16
THREE-DIMENSIONAL DRAWING

'To design is to communicate clearly by whatever means you can control or master.'

Milton Glaser

Three-dimensional drawing represents how we see objects. We are accustomed to observing the length, width and depth of objects. Also known as pictorial drawing, axonometric drawing and perspective drawing are the most common methods applied in the Senior Graphics syllabus.

16.1 AXONOMETRIC DRAWING

Axonometric drawings are commonly used by industrial designers. They are used in the design of products, in engineering and mechanical drawings. Axonometric drawings include isometric, dimetric, trimetric and planometric drawings. Axonometric drawings are constructed of lines that remain parallel and do not converge at any given point. For this reason they are sometimes referred to as paraline drawings. These drawing methods are an effective visual means of representing the form and features of a three-dimensional object.

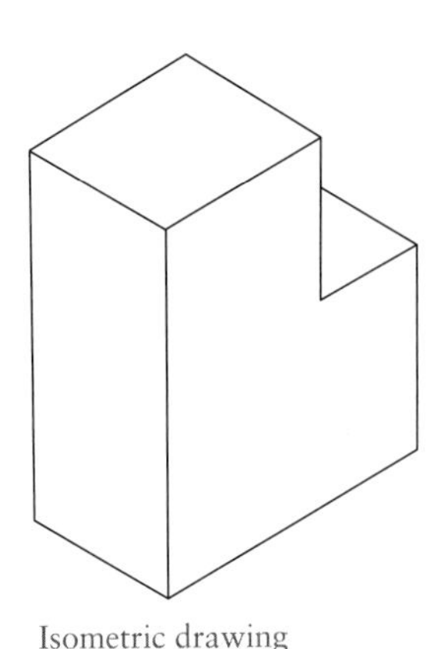
Isometric drawing

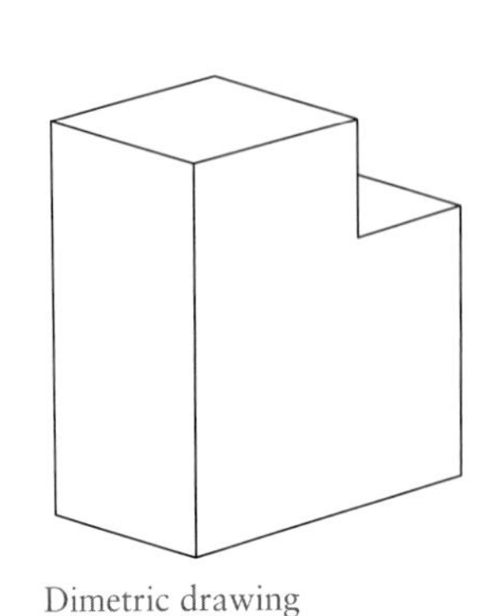
Dimetric drawing

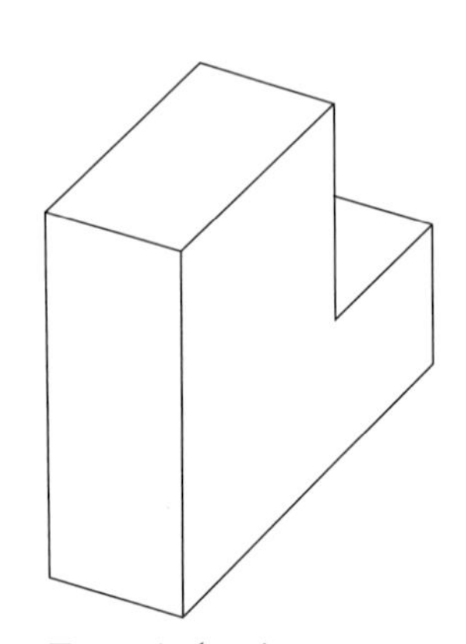
Trimetric drawing

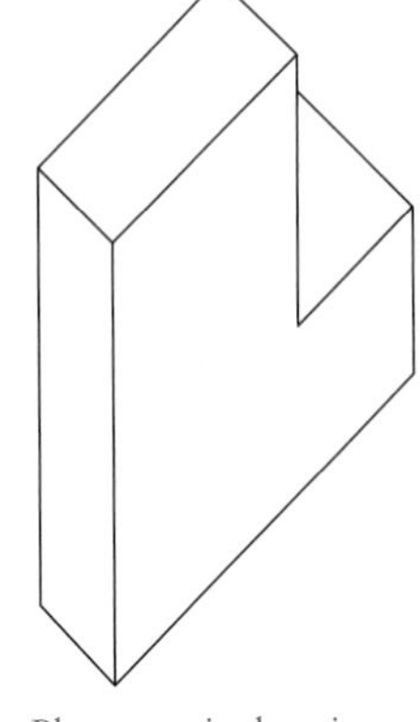
Planometric drawing

ISBN 9780170349994

ISOMETRIC DRAWING

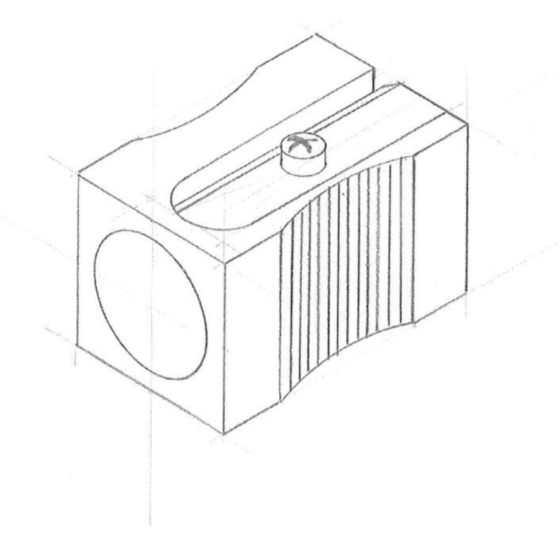

In industrial design, isometric drawings are arguably the most commonly used pictorial drawings because of their relatively simple construction. Isometric drawings are quicker to draw than two-point perspective, yet look somewhat similar and may provide similar information. In an isometric drawing, the height (or corner) of the object faces the viewer, and the width and depth of the object recede (remaining parallel) at 30°.

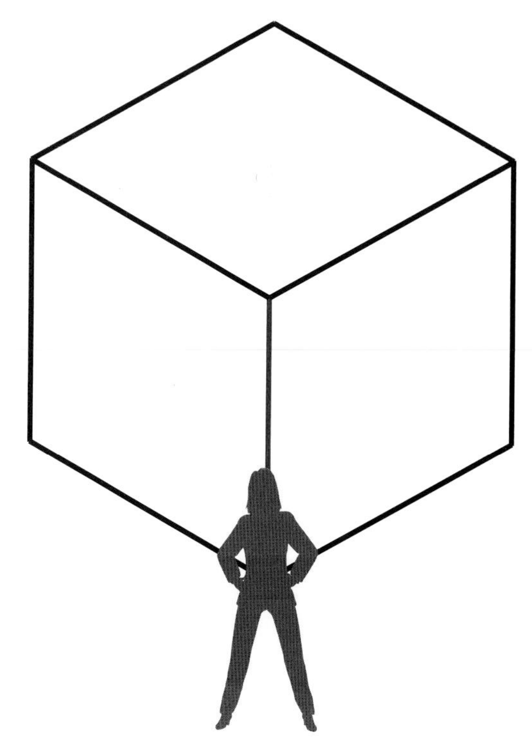

▲ Viewer position: isometric drawing

Isometric drawings are frequently used in industrial design where it is important to show as much detail as possible about the object or product using a three-dimensional representation. Isometric drawing allows a designer to depict the form and details of an object in a manner that conveys true proportions.

Isometric drawings are commonly applied in technical drawings in addition to orthogonal representations. Drawings of engineering components and the like sometimes include a three-dimensional version of the object in isometric as well as two-dimensional technical drawings to provide detail about the appearance of a completed product or part.

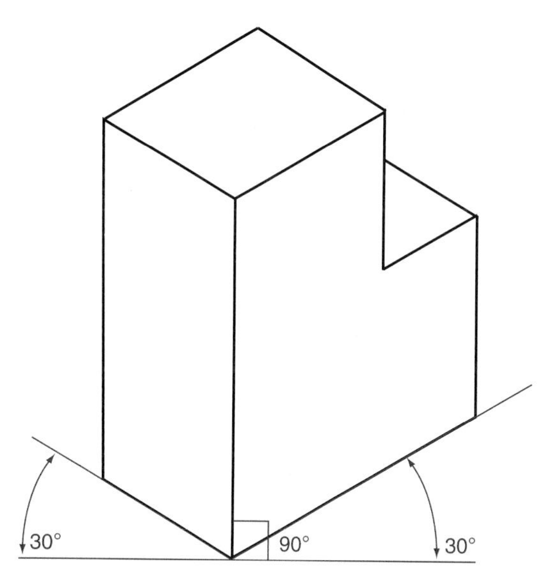

▲ Isometric drawing

ISOMETRIC GRID PAPER

If you are drawing by hand, you might find it helpful to draw your isometric drawings on isometric grid paper, which is predrawn with 30° and 90° lines in non-reproducible (i.e. will not photocopy) light blue.

ISBN 9780170349994

A STEP-BY-STEP GUIDE TO ISOMETRIC DRAWING

Step 1: Begin by drawing the height of the object facing you. This will be at 90°.

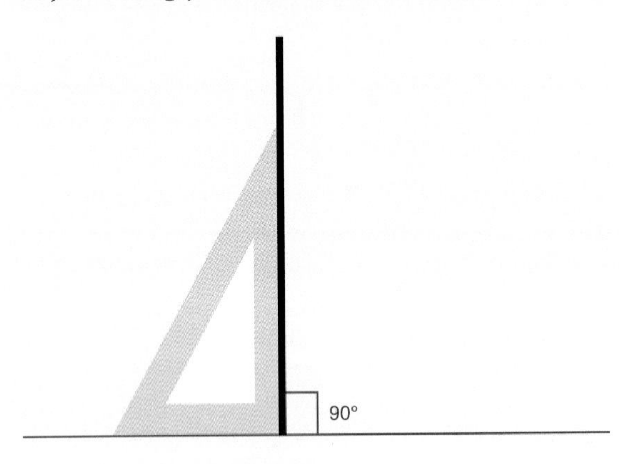

Step 2: Using a 30° set square, draw two 30° lines at the base of the vertical height line.

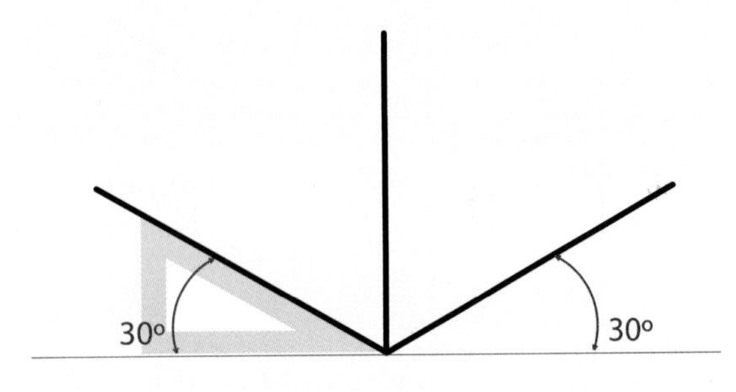

Step 3: Measure and indicate the length of the object on the 30° lines and the height of the object on the vertical line.

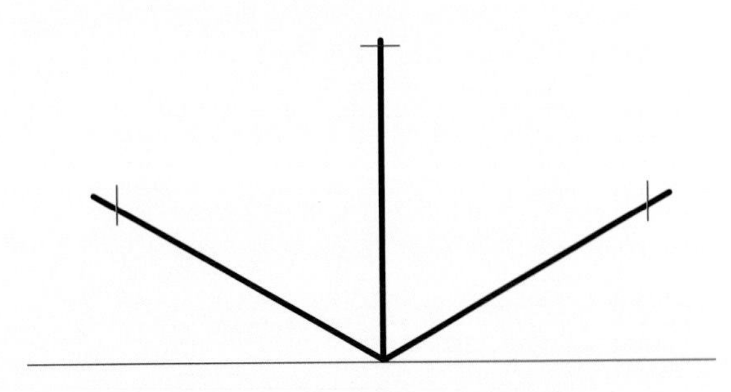

Step 4: Draw vertical lines from the 30° lines and measure and indicate the height.

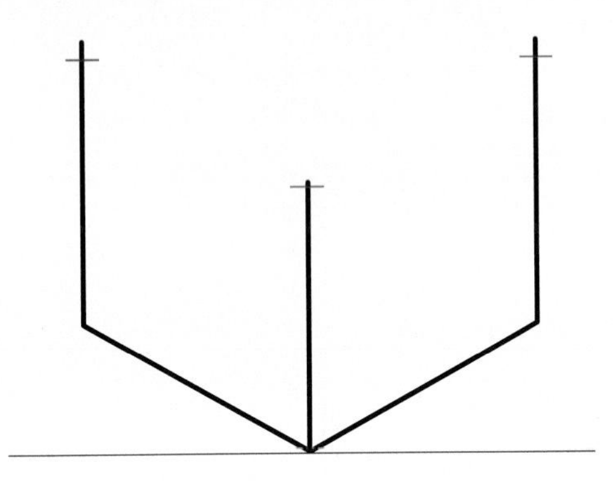

Step 5: Draw 30° lines from the top of the vertical line.

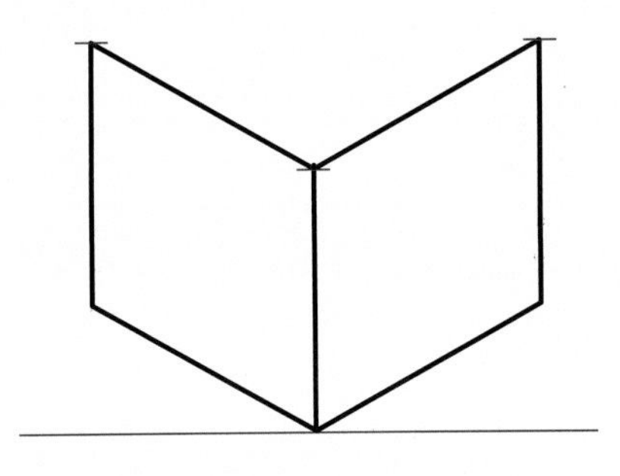

Step 6: Draw 30° lines parallel to the base of the object to complete the top. Erase any excess lines to tidy the drawing.

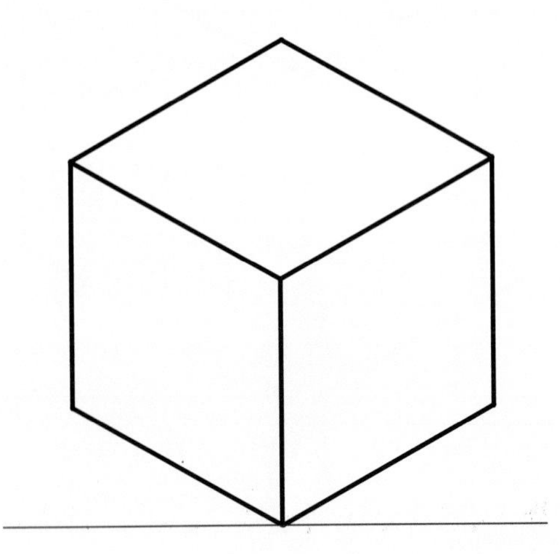

ISBN 9780170349994

DIMETRIC DRAWING

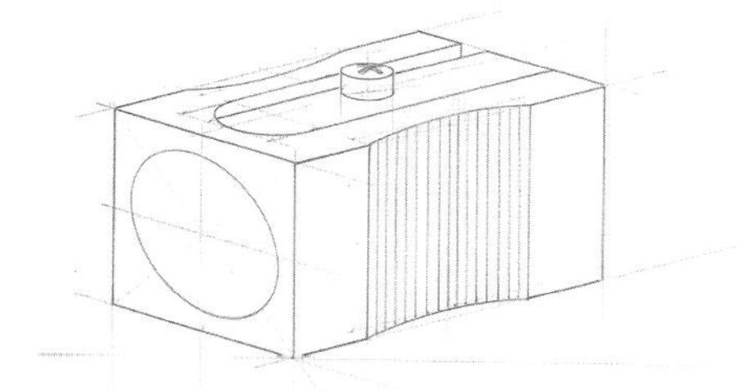

Using the same basic principles as isometric drawing, dimetric drawings depict lines that remain parallel to one another.

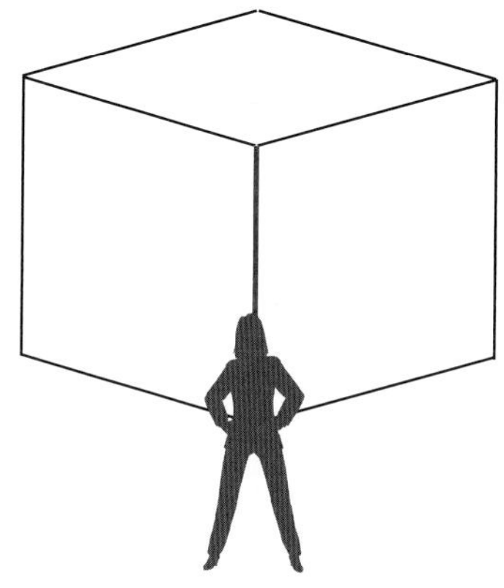

▲ Viewer position: dimetric drawing

However, in dimetric drawing the height (or corner) of the object faces the viewer, and the width and depth of the object recede (remaining parallel) at 15°. This angle means that dimetric drawings display a foreshortened view of an object, which can appear slightly more realistic than the 30° rotation of an isometric view.

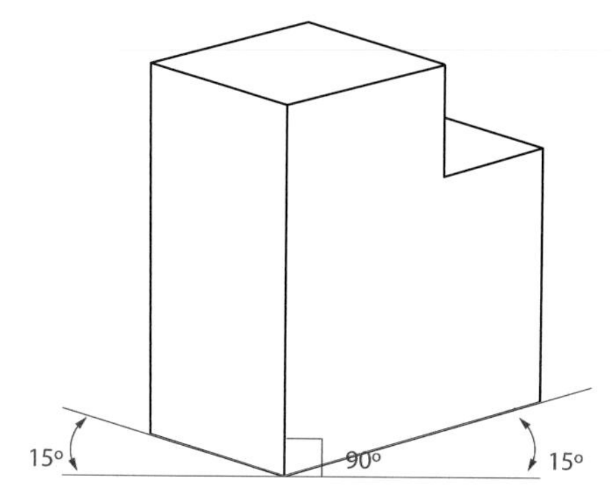

▲ Dimetric drawing

A STEP-BY-STEP GUIDE TO DIMETRIC DRAWING

Step 1: Begin by drawing the height of the object facing you. This will be at 90°.

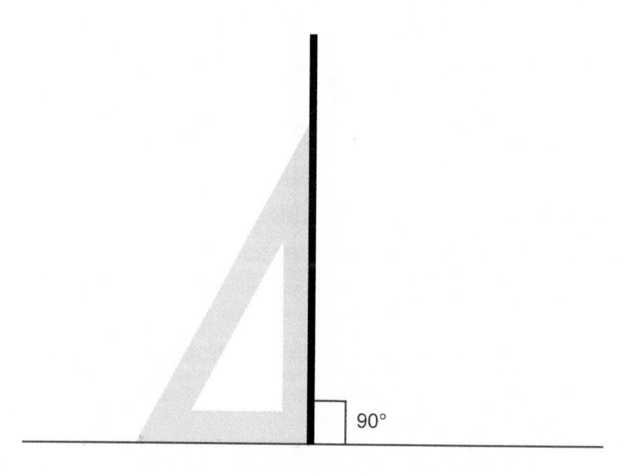

Step 2: Using a protractor, identify 15° and draw two 15° lines at the base of the vertical height line.

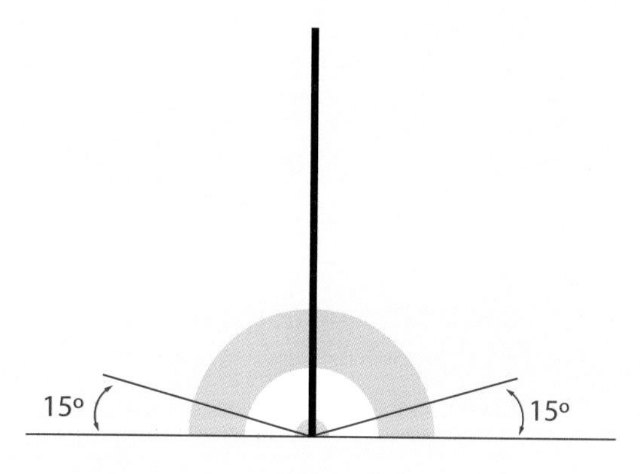

Step 3: Measure and indicate the length of the object on the 15° lines and the height of the object on the vertical line.

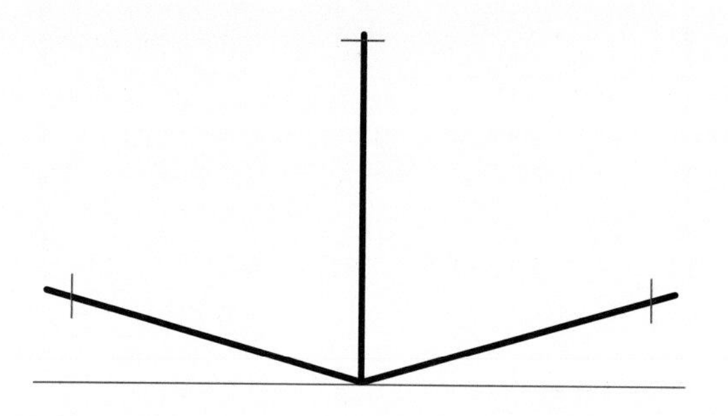

ISBN 9780170349994

Step 4: Draw vertical lines from the 15° lines and measure and indicate the height.

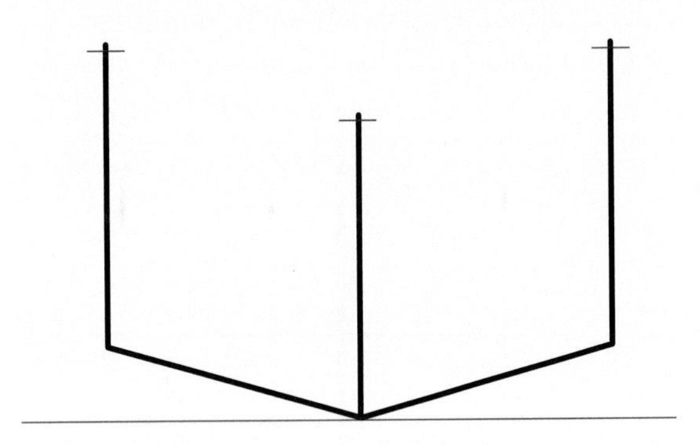

Step 5: Draw 15° lines from the top of the vertical line.

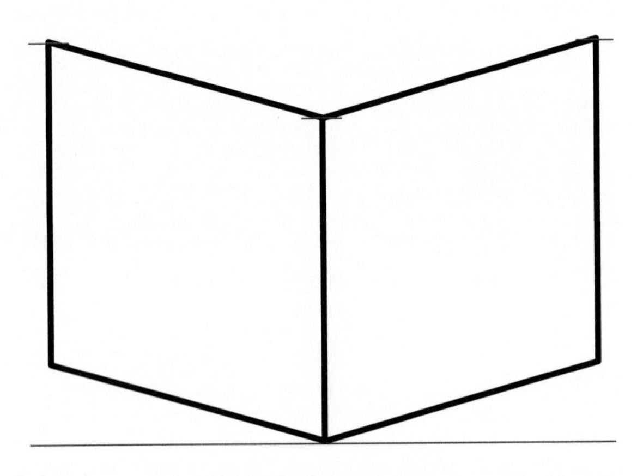

Step 6: Draw 15° lines parallel to the base of the object to complete the top. Erase any excess lines to tidy the drawing.

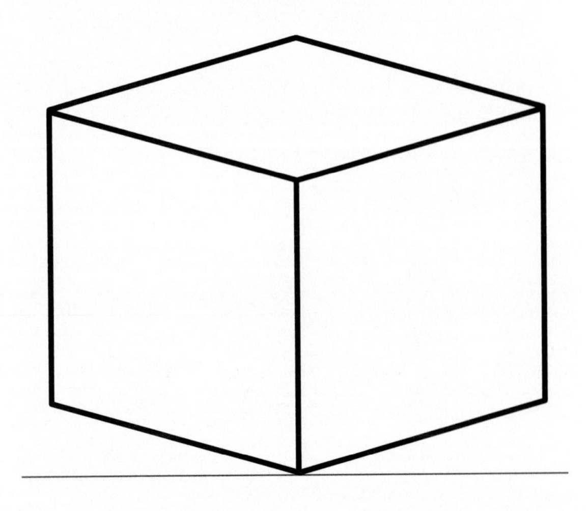

TRIMETRIC DRAWING

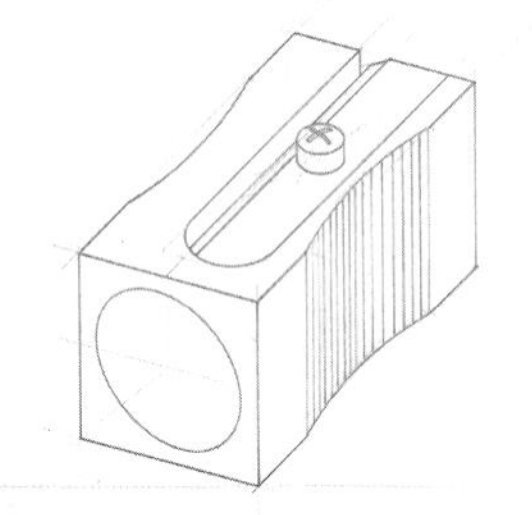

Trimetric drawing is the least commonly applied axonometric method. Unlike isometric and dimetric projections, trimetric images are not equally foreshortened and are drawn with a combination of 15° and 45° angles.

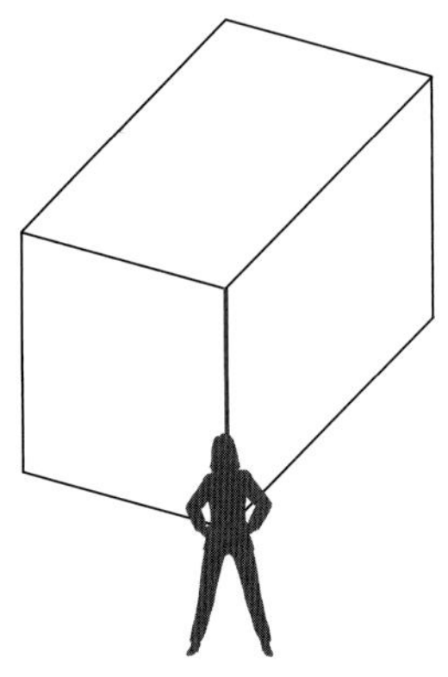

▲ Viewer position: trimetric drawing

Due to the combination of angles and unequal foreshortening, trimetric drawings can seem to present a distorted and unrealistic image. Trimetric images are rarely seen although they have been used in some game design including the SimCity series.

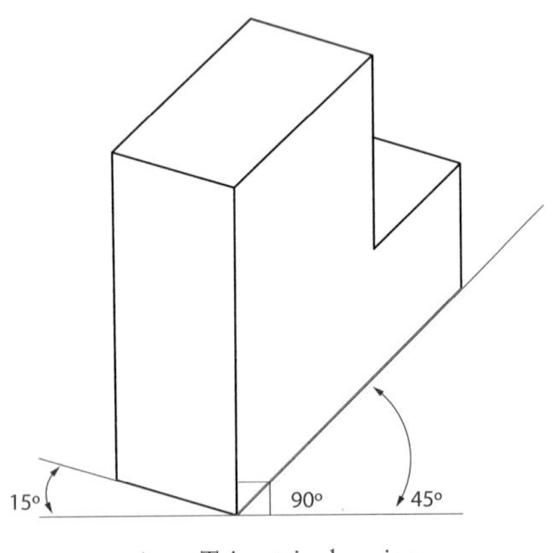

▲ Trimetric drawing

ISBN 9780170349994

A STEP-BY-STEP GUIDE TO TRIMETRIC DRAWING

Step 1: Begin by drawing the height of the object facing you. This will be at 90°.

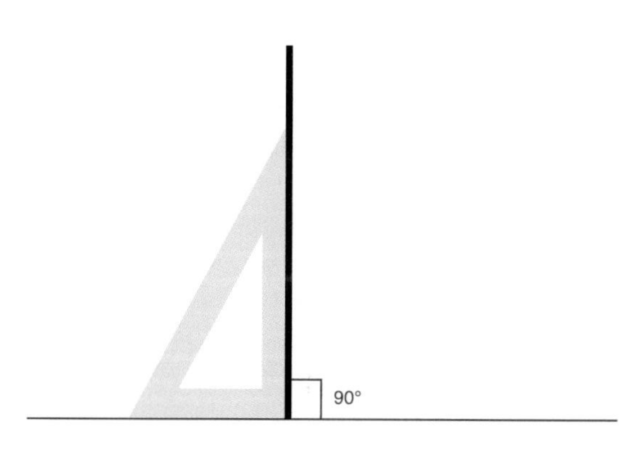

Step 2: Using a protractor, identify and draw 15° and 45° lines at the base of the vertical height line.

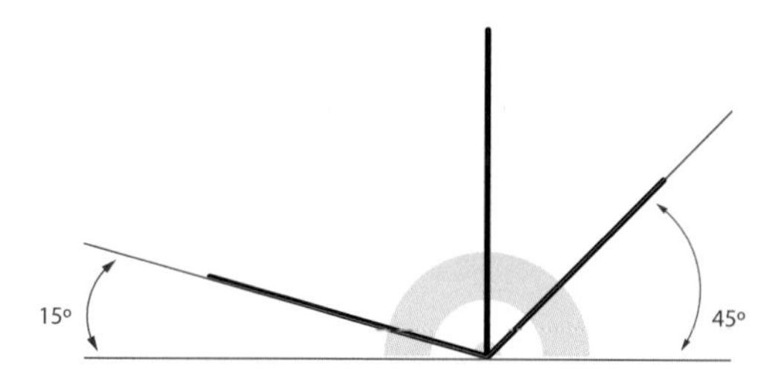

Step 3: Measure and indicate the length of the object on both the 15° line and the 45° line, and the height of the object on the vertical line.

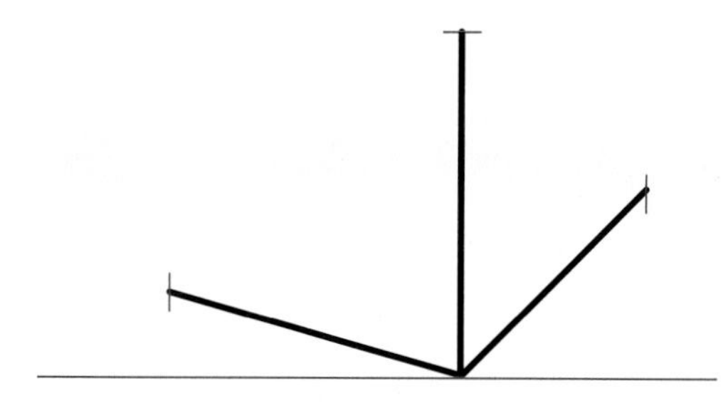

Step 4: Draw vertical lines from the 15° line and the 45° line and measure and indicate the height.

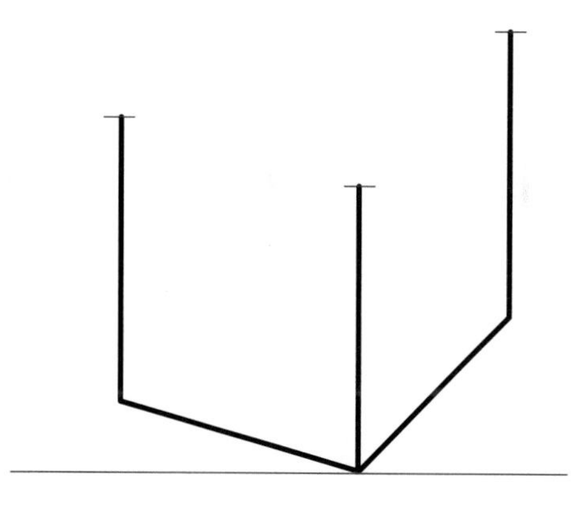

Step 5: Draw a 15° line and a 45° line from the top of the vertical line. These should be parallel to the lines you have already drawn at the base of the object.

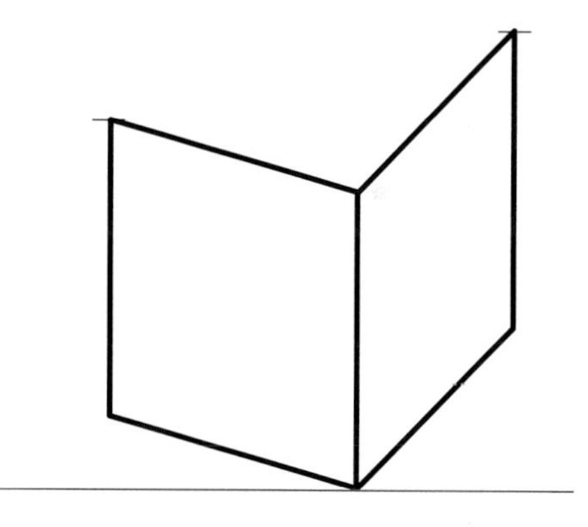

Step 6: Draw a 15° line and a 45° line parallel to the base of the object to complete the top. Erase any excess lines to tidy the drawing.

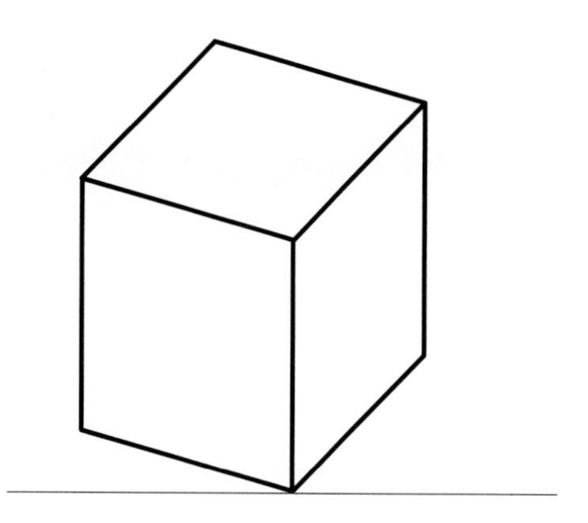

ISBN 9780170349994

PLANOMETRIC DRAWING

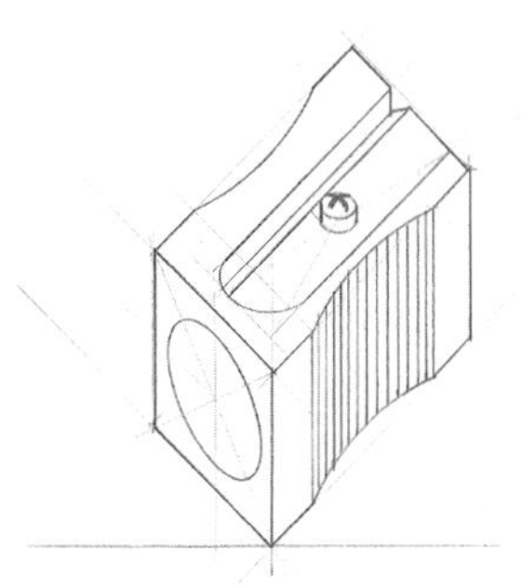

Planometric drawings are similar in their construction to isometric drawings, but the length and depth of the object is drawn at 45°. The height of the planometric object faces the viewer and all sides recede at 45°. The elevation of 45° gives the viewer a 'bird's-eye' view of features while retaining three-dimensional qualities.

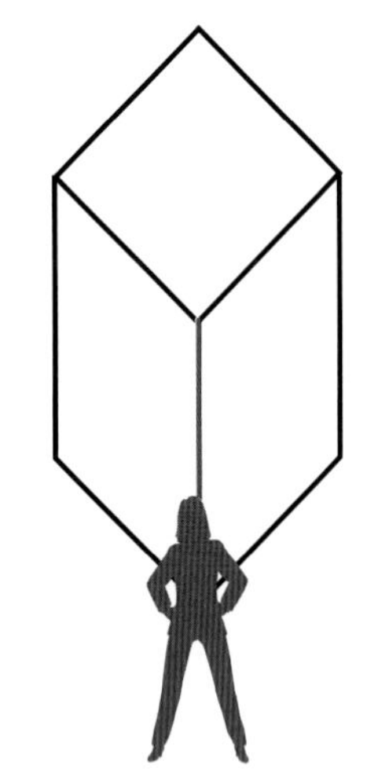

▲ Viewer position: planometric drawing

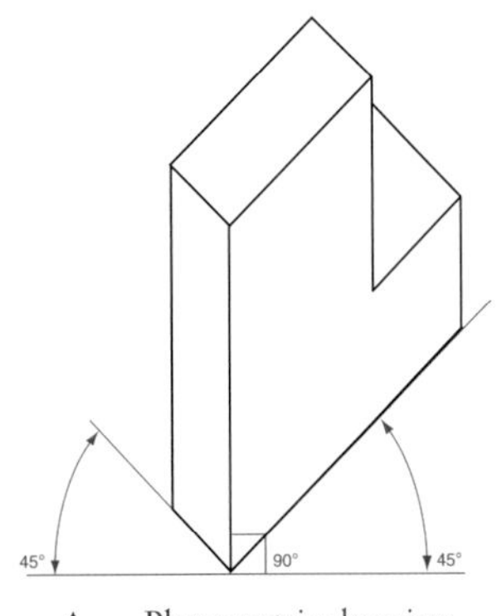

▲ Planometric drawing

A STEP-BY-STEP GUIDE TO PLANOMETRIC DRAWING

Step 1: Begin by drawing the height of the object facing you.

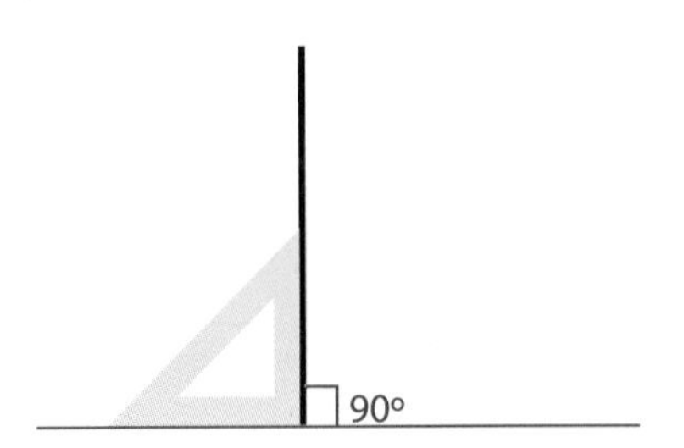

Step 2: Using a 45° set square, draw two 45° lines at the base of the vertical height line.

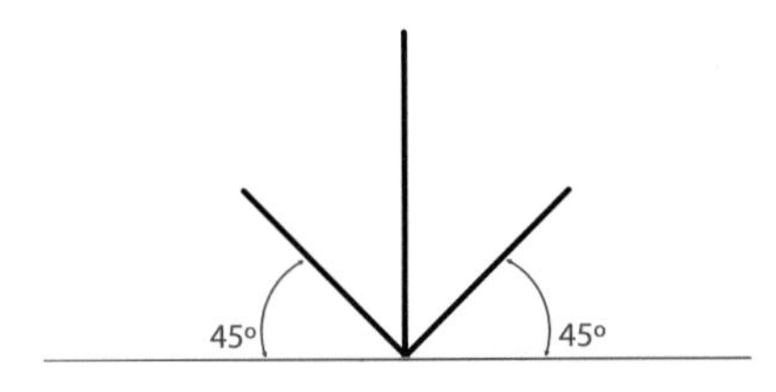

Step 3: Measure and indicate the length of the object on the 45° lines and the height of the object on the vertical line.

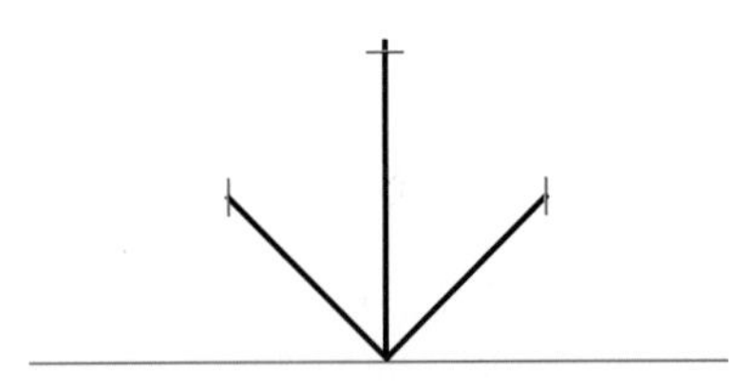

Step 4: Draw vertical lines from the 45° lines and measure and indicate the height.

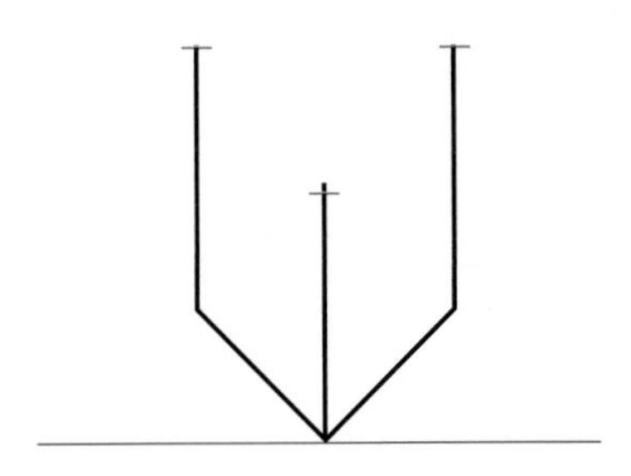

ISBN 9780170349994

Step 5: Draw 45° lines from the top of the vertical line.

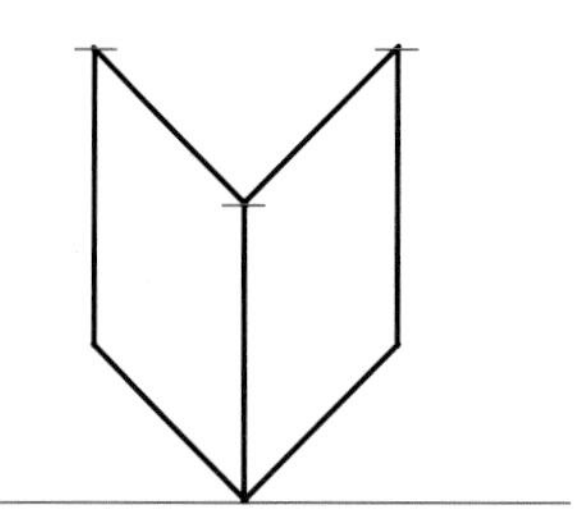

Step 6: Draw 45° lines parallel to the base of the object to complete the top. Erase any excess lines to tidy the drawing.

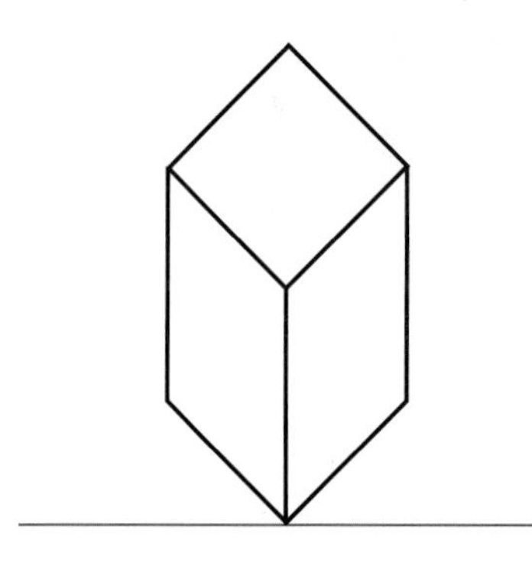

Planometric drawing in environment design

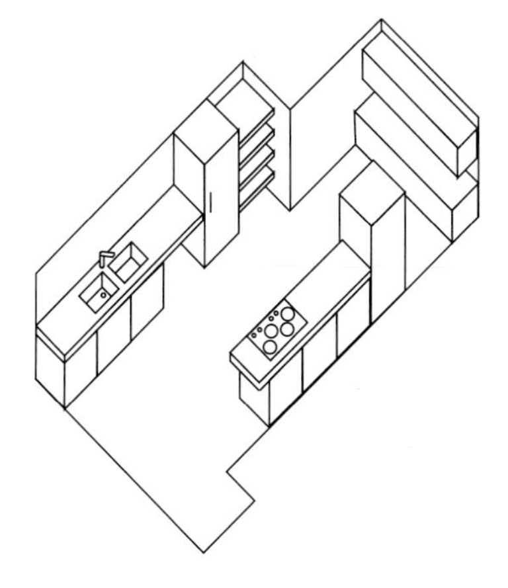

In architectural illustration, planometric drawing is often used to create a three-dimensional representation of a two-dimensional floor plan. When drawing the interior detail of a building or object, the 45° angle allows for an exaggerated viewpoint.

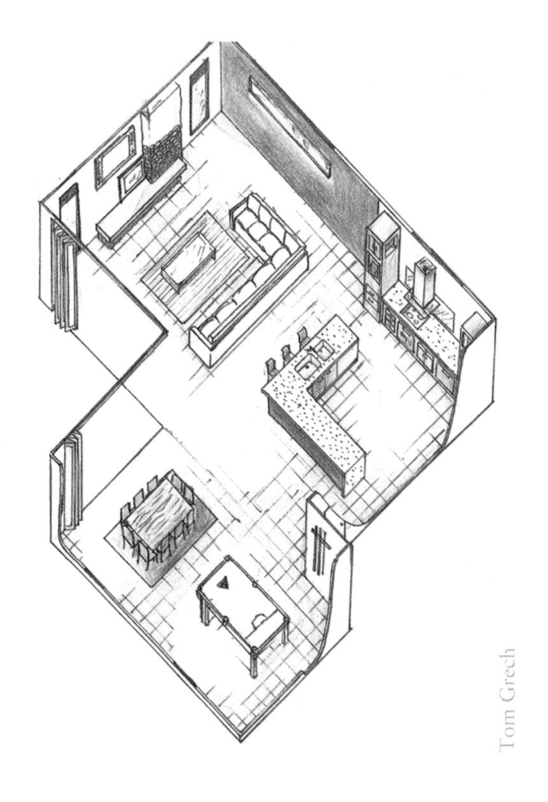

Tom Grech

A STEP-BY-STEP GUIDE TO CREATING A PLANOMETRIC DRAWING FROM A FLOOR PLAN

Step 1: Start with a two-dimensional floor plan.

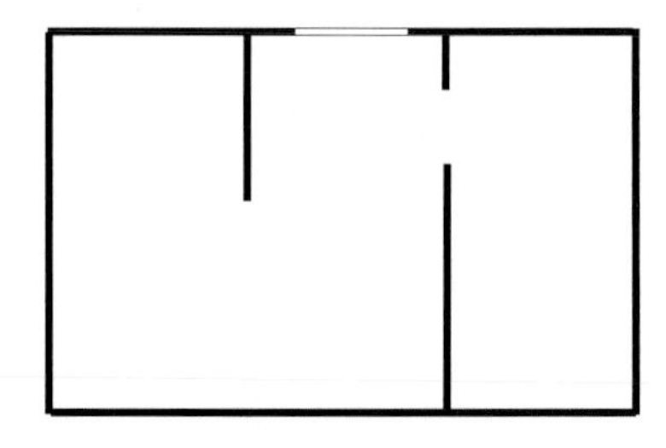

Step 2: Begin by rotating the plan to 45°.

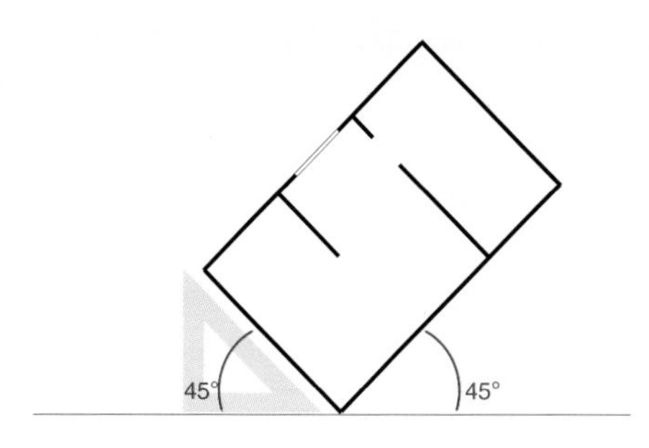

ISBN 9780170349994

Step 3: Project vertical lines from the base of each detail or feature on the plan.

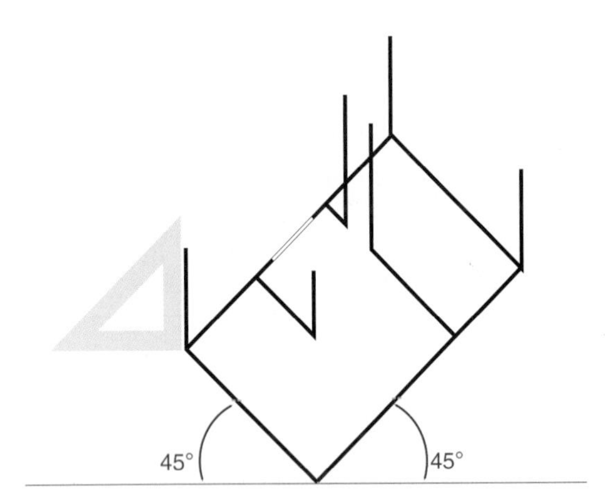

Step 4: Complete the features by drawing 45° lines parallel to the original lines on the plan.

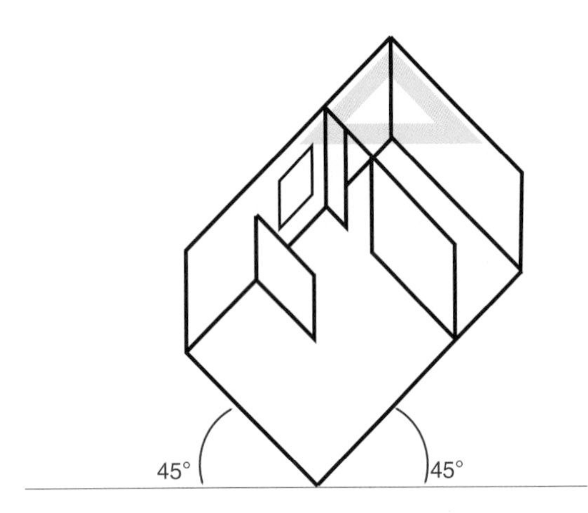

16.2 PERSPECTIVE DRAWING

Perspective drawing closely reflects the way that the human eye perceives an object in space. When we see an object, our brain tells us that the object gets smaller as it recedes into space. We know for a fact that isn't true, but it is what our eye sees. We have an innate understanding of this phenomenon as representing depth and distance.

A long straight railway is a good example of the perspective phenomenon. A railway seems to narrow as it heads towards the horizon. However, if you were to travel along the railway tracks, it would be clear to you that they remain parallel and definitely do not meet.

iStockphoto/inhauscreative

▲ These railways tracks appear to converge as they reach a given point on the horizon.

You might also notice that any houses, trees and power poles along the sides of a railway also appear to diminish in size as they recede into the distance. When drawing in perspective, the principles are the same. A row of objects drawn facing the viewer represents just that – a row. Redraw the same objects so that they appear to diminish in size, approaching a point on the horizon, and you have a composition that implies depth as well as representing the form and detail of the objects themselves.

Interestingly, it is not only the shape and form of objects that appears to change when drawn in perspective. Colour changes also occur when an object appears to recede into the distance. You have probably noticed that when you observe a city skyline from a distance, the buildings seem to be a uniform grey, when in fact they may be a range of colours from dark brown to metallic silver to white.

In a perspective drawing, it is important to remember that bright colours are usually not visible in the far distance and that colours soften as they recede to the horizon. In applying colour to a drawing, it is possible to enhance the concept of distance by developing a contrast between muted colours in the distance and brighter colours in the foreground area. Colour, therefore, can become a helpful tool in your freehand perspective illustrations.

ISBN 9780170349994

iStockphoto/fotoVoyager

▲ Note how the background details are soft and the colours muted in this photograph of a desert road.

freeimages/Thiago Felipe Festa

▲ Note the lack of detail in the far part of this image. Detail diminishes as distance increases. This assists when creating a realistic representation using perspective drawing.

Any details on the surface of an object become less clear as that object recedes into the distance. To enhance the appearance of objects in your perspective drawing, reduce the amount of detail in proportion to the increase in distance; that is, the further away an object appears, the less detail should be identified.

When drawing in perspective it is helpful to visualise what you are planning to draw. It is not always possible to have an object positioned in front of you as you draw, so when drawing images that are less accessible; for example, a car or a building, it can be appropriate to use photographs and images that you have researched.

As with all drawing, don't be afraid to sketch out the concept first – and remember that you can always start again.

ESTABLISHING YOUR POINT OF VIEW

When drawing in perspective, your first task is to visualise what it is you want to represent. Ask yourself: What are the important features of the object I wish to illustrate? What is the key information about this object that I want to convey to the viewer? This will help you to plan the 'point of view' of the object; that is, the position from which you plan to draw it.

In any perspective drawing, the placement of the object in relation to the horizon line will affect the point of view of the depicted object.

The horizon line sits at the level of the viewer's eyes. This is called eye level. An object placed below the horizon line – below eye level – will give more information about the top of the object. Place the object above the horizon line, and then the area underneath the object becomes most obvious. If you place your object directly on the horizon line, the 'point of view' will appear to be quite realistic, as it sits at eye level. Again, it all depends on the effect you wish to create.

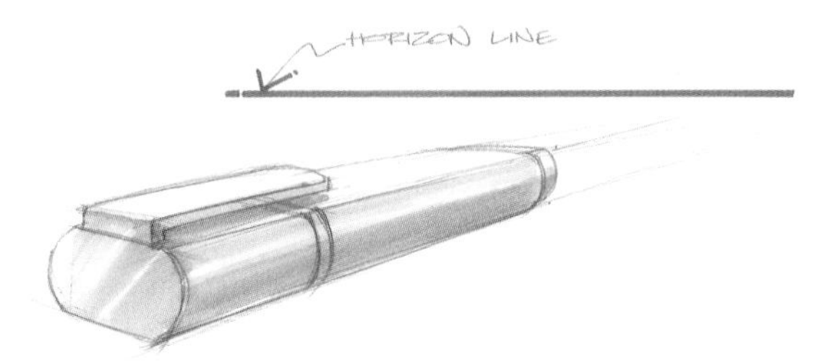

▲ An object that is drawn below the horizon line offers information about the top of the object.

ISBN 9780170349994

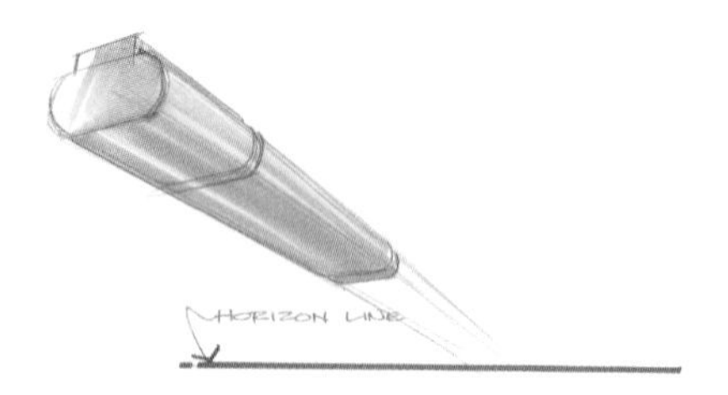

▲ An object that is drawn above the horizon line offers information about the base of the object.

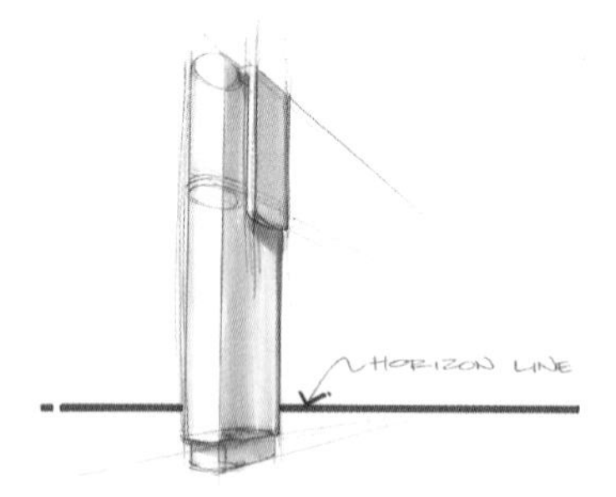

▲ An object that is drawn on the horizon line offers information about the front of the object.

The most common methods of perspective drawing are one-point perspective and two-point perspective. Three-point perspective is sometimes used in illustrations where a dramatic and exaggerated representation is required.

One-point perspective

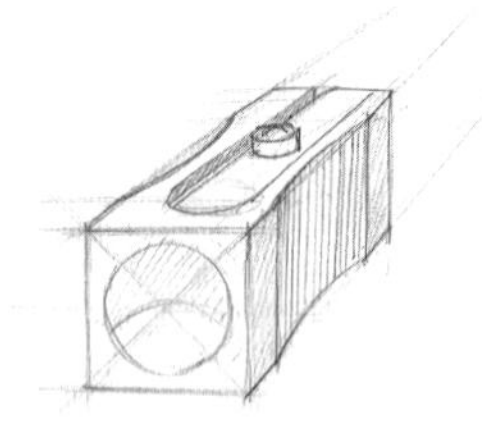

Key concepts to remember when drawing in one-point perspective are:

+ The height and width of the object face the viewer.
+ All depth (or the sides of the object) recedes to one point on the horizon line.

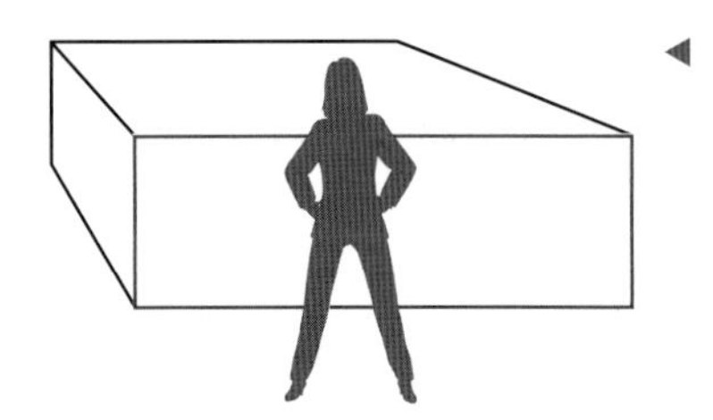

◀ Viewer position: one-point perspective drawing

One-point perspective is sometimes referred to as linear perspective. In one-point perspective an entire plane of an object faces the viewer.

The impression of a road receding to a point on the horizon is one-point perspective. One-point perspective is a method evident in many classic paintings of landscapes and interiors.

A perspective box is a simple way to begin working with this three-dimensional drawing method. Once you can draw a perspective box effectively, you can draw just about anything!

▲ Raphael's School of Athens is a classic representation of one-point perspective.

A STEP-BY-STEP GUIDE TO ONE-POINT PERSPECTIVE

Step 1: Decide on the position of the object in relation to eye level and draw a horizon line.

Horizon line

Step 2: Begin with the height and width of the object facing you. Your page acts as the picture plane (or surface area upon which the object is placed). The height and width of the drawn object should be parallel to the height and width of your page.

Horizon line

ISBN 9780170349994

Step 3: Decide on the angle of view you wish to represent, and place the vanishing point appropriately to the left, right or in the centre. Do you want to depict more of one side? Do you want to depict the front only?

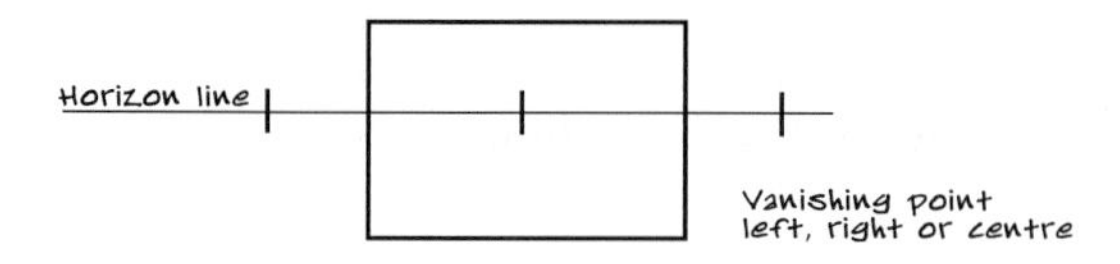

Step 4: Draw light projection lines from the corners of the object to the vanishing point.

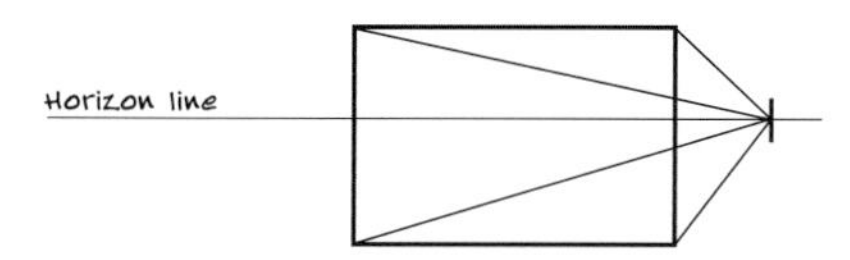

Step 5: Decide on the appropriate depth of the object in perspective, and draw horizontal and vertical lines within the projection lines to complete the back of the box. Making a decision about the size of the object will take some practice and require the application of your visualisation skills. Your aim is to create an image that has realistic proportions. Erase the light projection lines.

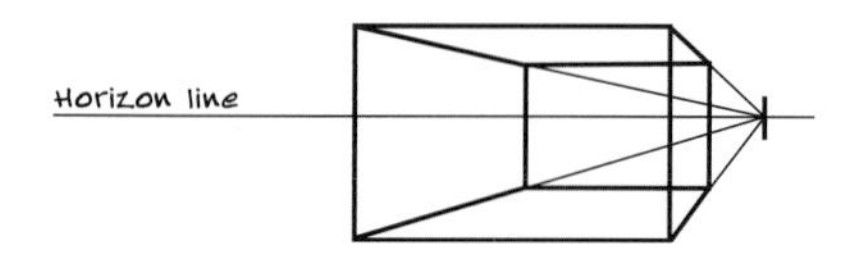

Perspective is often used for the visualisation of concepts and for the representation of architectural ideas in a landscape. More formally, it is used in CADD programs to create an impression of a proposed design. One-point perspective is sometimes used to depict interiors.

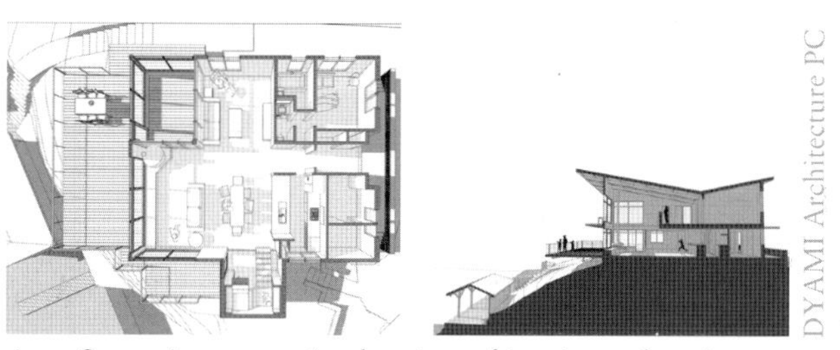

DYAMI Architecture PC

▲ One-point perspective drawings of interior and section views by Dyami Architects

One-point perspective represents the way the human eye sees simple interiors. Once you are able to draw simple geometric shapes in perspective, you can then add details to form highly descriptive illustrations. One-point perspective is particularly useful when illustrating interiors.

iStockphoto/Cimmerian

A STEP-BY-STEP GUIDE TO DRAWING AN INTERIOR

Step 1: Construct a box as illustrated previously – but this time, project lines from the left and right front corners of the box.

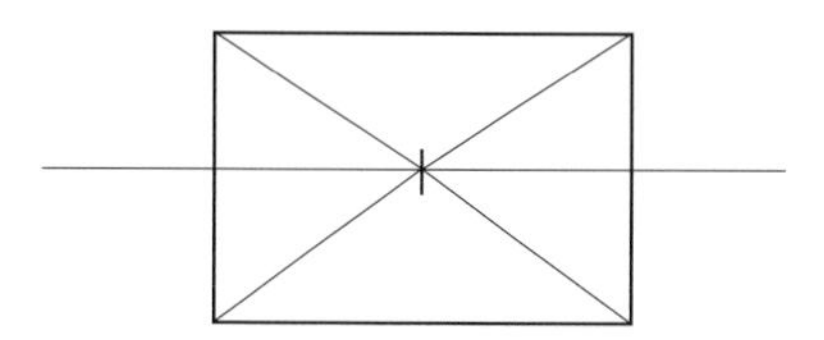

Step 2: Draw vertical lines from the back corners of the box to form the rear wall of the room or object.

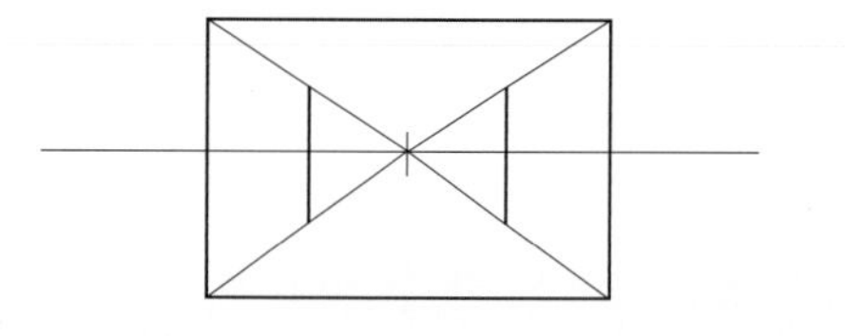

Step 3: Draw a horizontal line to form the base of the back wall.

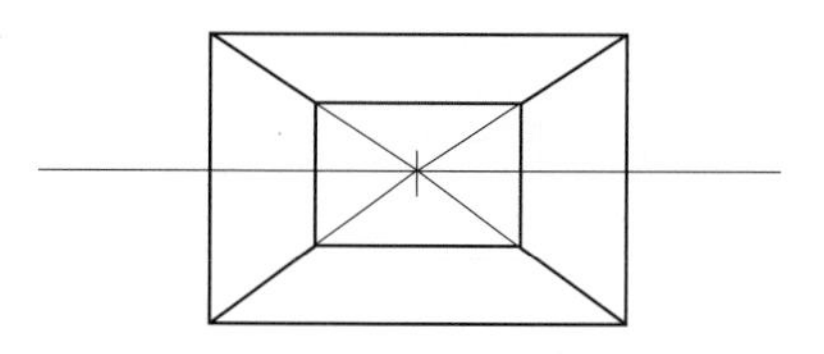

ISBN 9780170349994

Step 4: You now have an empty box to which detail can be added to create any number of possibilities. Any object you create within the interior must recede to the vanishing point.

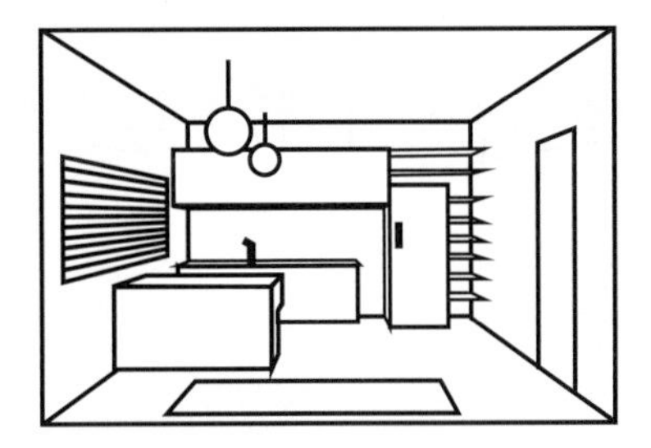

Two-point perspective

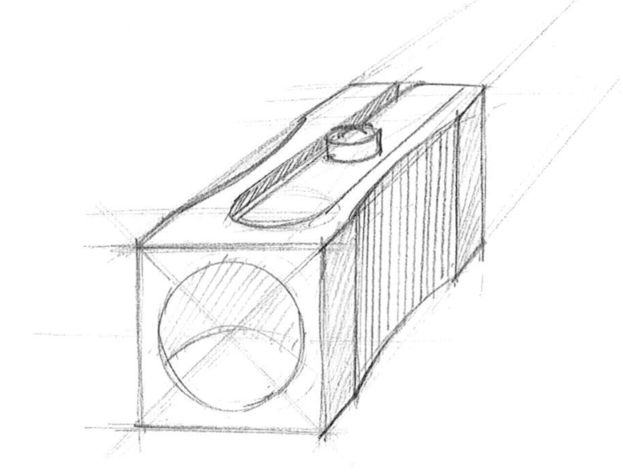

In one-point perspective, the height and width of the object face the viewer. In two-point perspective, only the height faces the viewer, and the depth or sides of the object recede to two vanishing points on the horizon line.

Two-point perspective is sometimes referred to as angular perspective.

Key concepts to remember when drawing in two-point perspective are.

+ The height of the object faces the viewer.
+ All other dimensions recede to two points on the horizon line.

If you stand outside your house, or even outside one of your school buildings, you will become aware that the sides of the buildings recede, ever so slightly, to separate vanishing points.

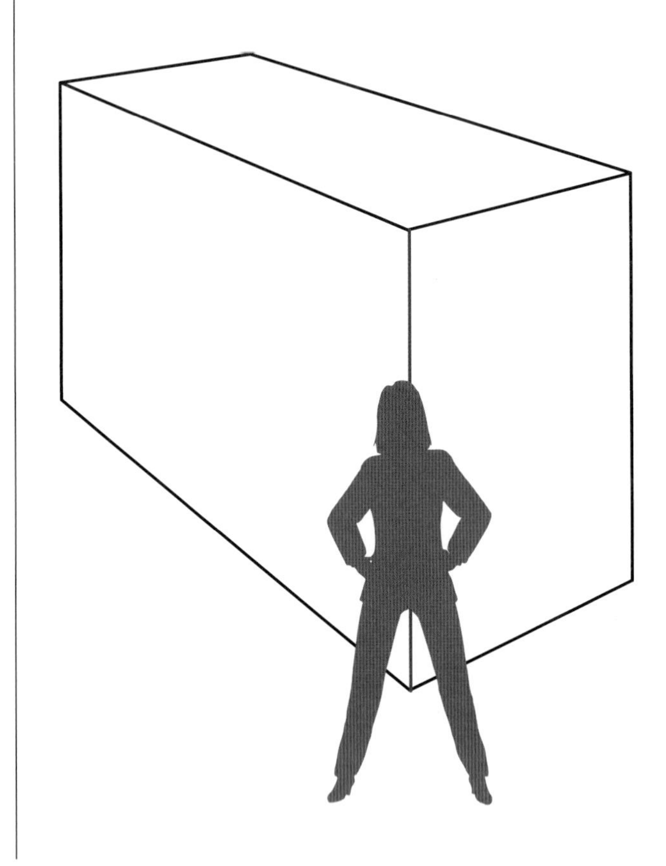

ISBN 9780170349994

A STEP-BY-STEP GUIDE TO TWO-POINT PERSPECTIVE

Step 1: Begin by making a decision about where eye level will be on your drawing. Draw the horizon line.

Horizon line

Step 2: Draw the desired height of the object.

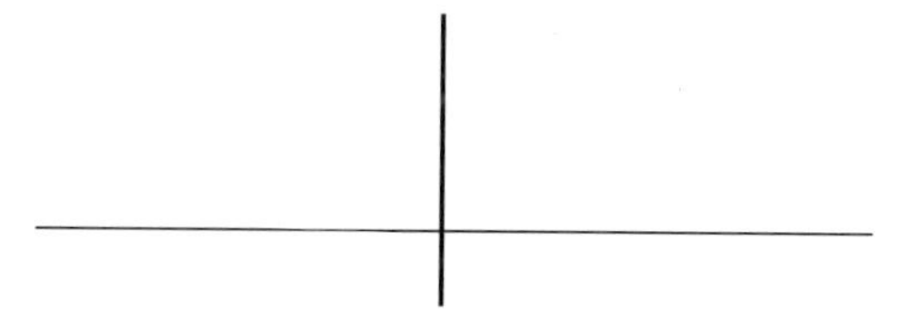

Step 3: Draw vanishing points to the left and right of the height line on the horizon line. Remember, the further apart the vanishing points, the less extreme the perspective will appear to be.

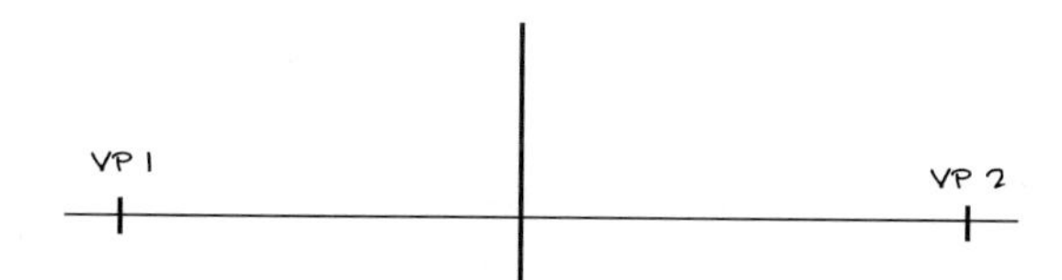

Step 4: Draw light lines from the corners of your height line to the vanishing points.

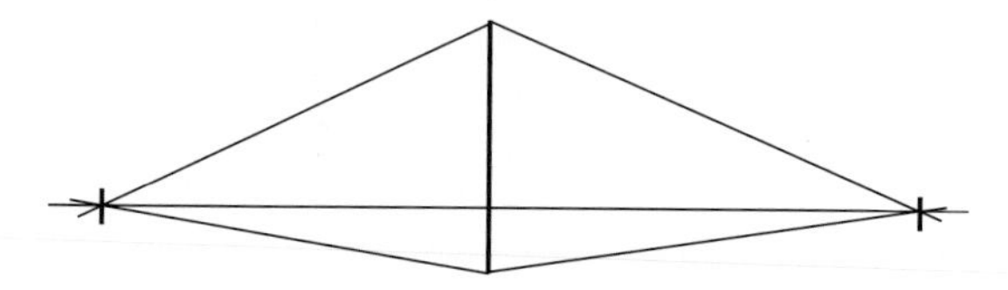

Step 5: Make a decision about the appropriate length for your object and add vertical lines to complete the sides of the object.

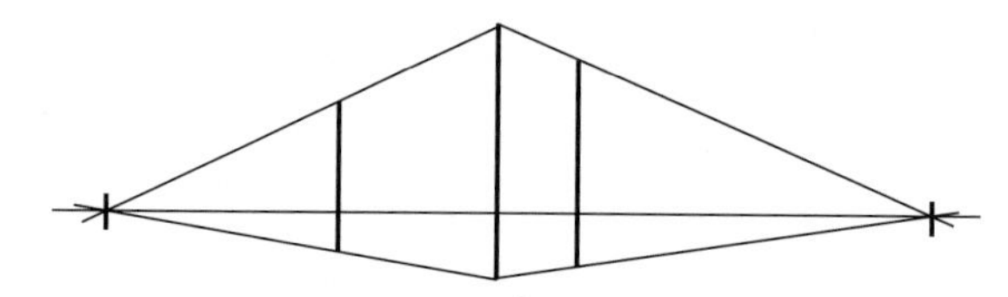

Step 6: Draw the two lines that will complete the top of the object by projecting lines from each of the lines that form the sides.

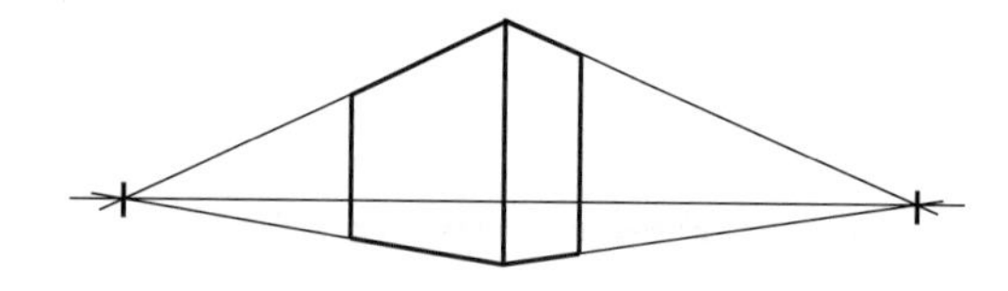

Step 7: Details can be added to a simple two-point perspective drawing to create a more detailed illustration.

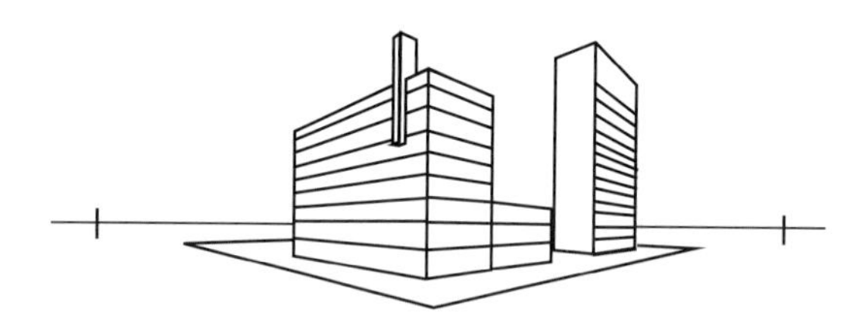

In the area of built environment design, two-point perspective is widely used for both exteriors and interiors.

iStockphoto/SireAnko

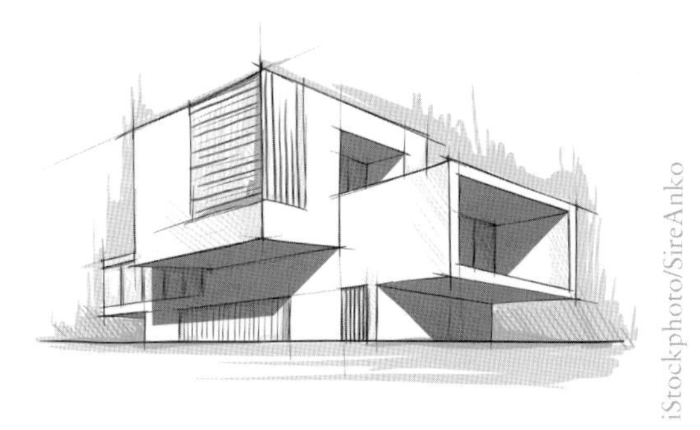

iStockphoto/SireAnko

▲ Two-point perspective drawings of interior and exterior views

ISBN 9780170349994

To create a perspective drawing from a floor plan, the technique of 'plan perspective' can be applied.

A STEP-BY-STEP GUIDE TO DRAWING TWO-POINT PERSPECTIVE FROM A PLAN

Step 1: Draw a horizontal line and label it picture plane. Place the plan view on the picture plane at 45°.

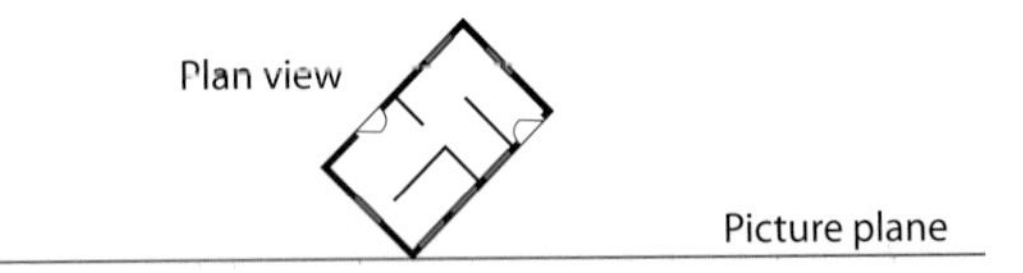

Step 2: Place the elevation view to the right of where the perspective drawing will be and draw a horizontal ground line.

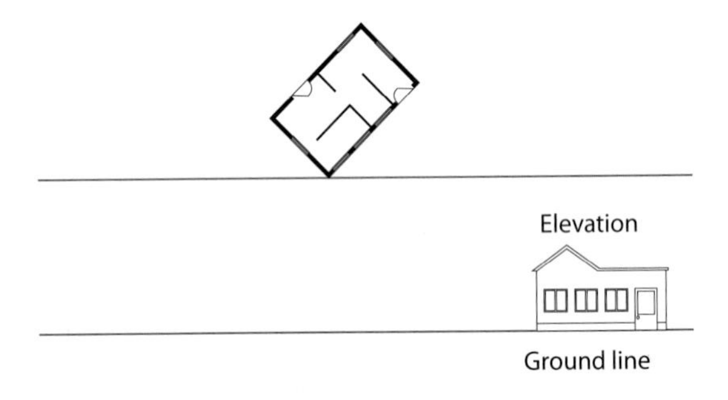

Step 3: Place a viewing point at the most appropriate level to 'view' the drawing. Draw light project lines from each corner of the plan to the viewing point. Don't place the viewing point too close to the ground line or the drawing will become distorted.

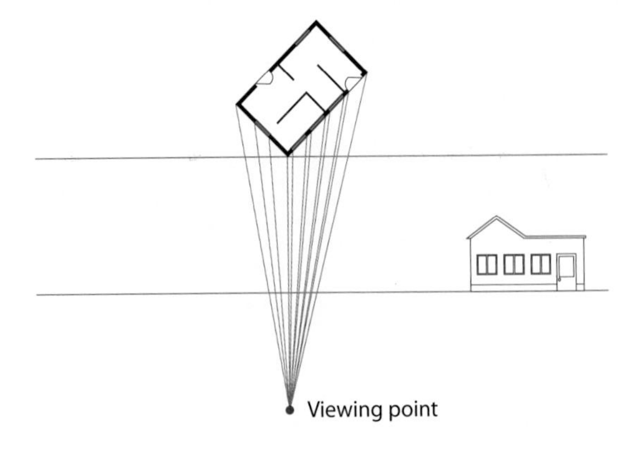

Step 4: Draw a mark where the projection lines intersect with the picture plane (projection lines can now be erased to keep things simple).

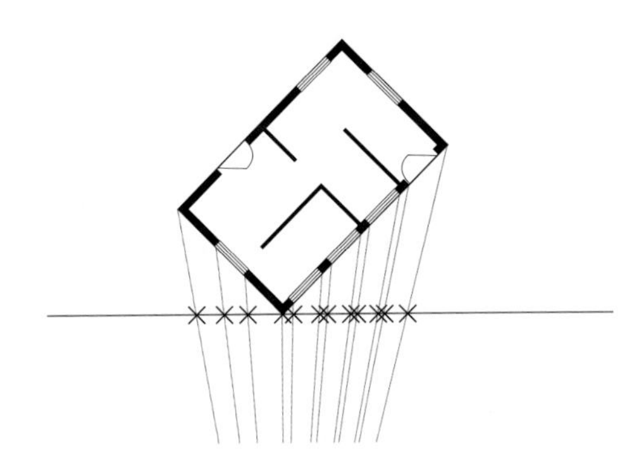

Step 5: Using the marks you just made, draw vertical lines to the ground line to establish the exterior dimensions of the structure.

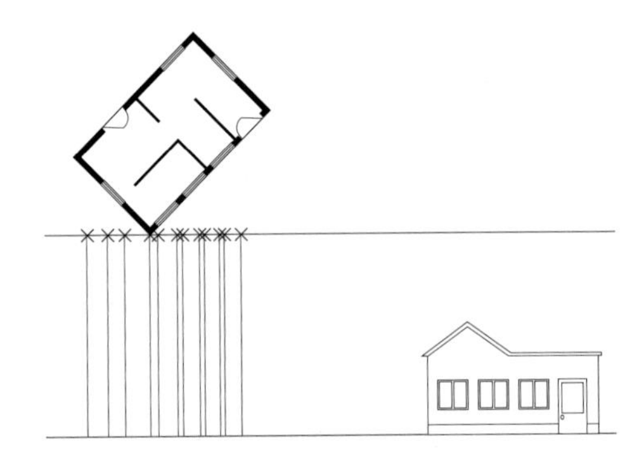

Step 6: Create a horizon line in line with the centre of the elevation. You can vary the position of the horizon line depending on the angle of view you wish to draw.

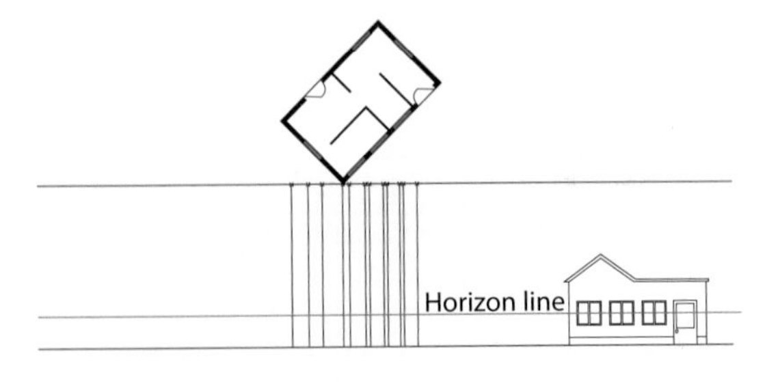

ISBN 9780170349994

Step 7: From the viewing point draw lines that remain parallel to the edges of the plan view.

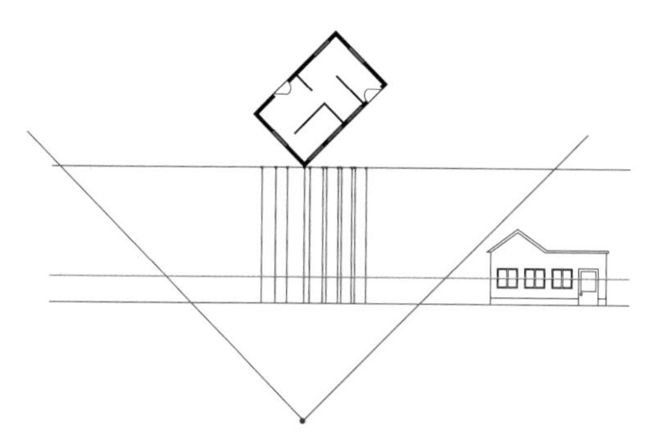

Step 8: Where these lines intersect with the picture plane, draw a vertical line down to meet the horizon line and establish each vanishing point.

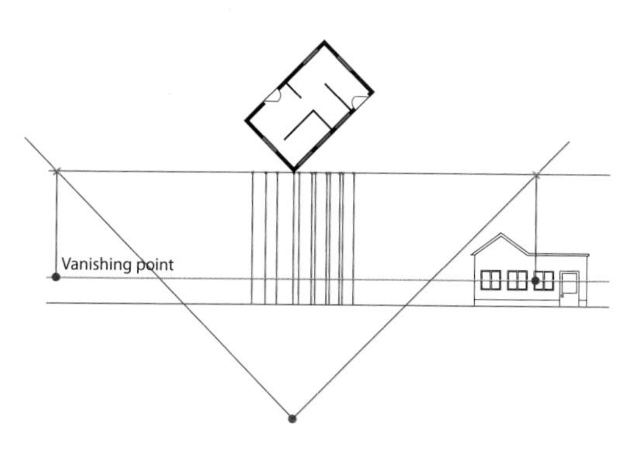

Step 9: Using the elevation to determine the height of features, create vertical lines to represent the wall and roof heights.

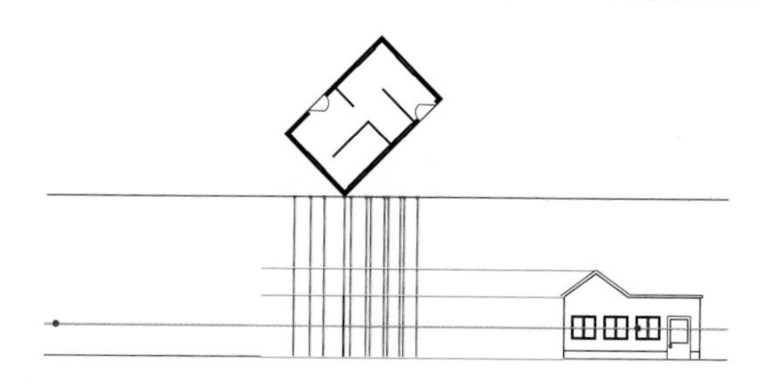

Step 10: Project lines to the vanishing points as per the two-point perspective drawing method to complete the structure.

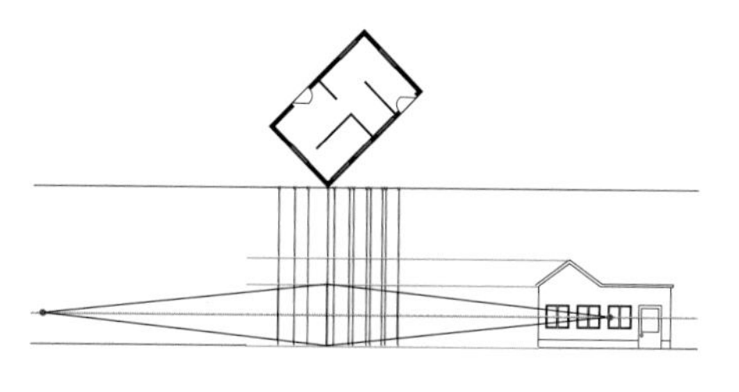

Step 11: Using the projected lines from the plan and elevation views, complete the details such as windows, doors, roofline, etc.

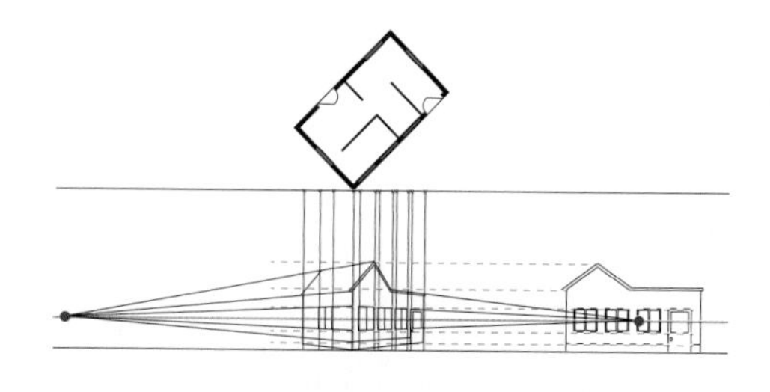

Three-point perspective

Three-point perspective is often used when a dramatic image is required. Three-point perspective can be seen in illustrations used in comics, fantasy and science fiction imagery.

ISBN 9780170349994

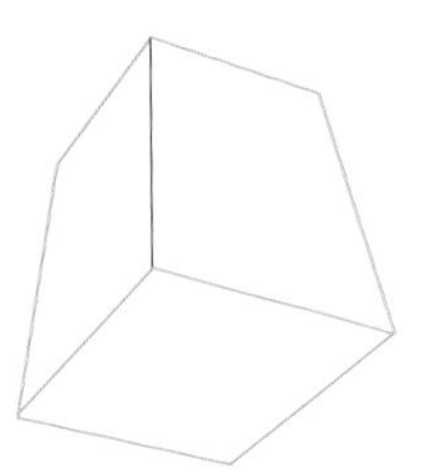

▲ Viewer position: three-point perspective drawing

The nature of three-point perspective creates distortion that can be dramatic and eye-catching; there are similarities between three-point perspective and images that use a fish-eye lens effect.

A STEP-BY-STEP GUIDE TO THREE-POINT PERSPECTIVE

Step 1: Using the same principles of two-point perspective, three-point perspective drawings start with the height of the object and two vanishing points.

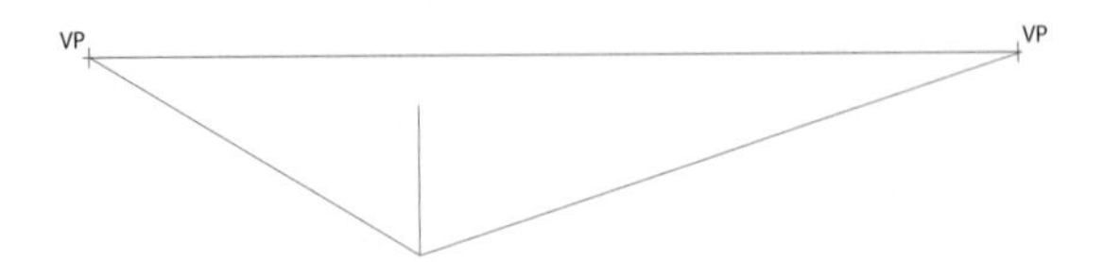

Step 2: Depending on the effect desired, a third vanishing point is added directly above or below the height line.

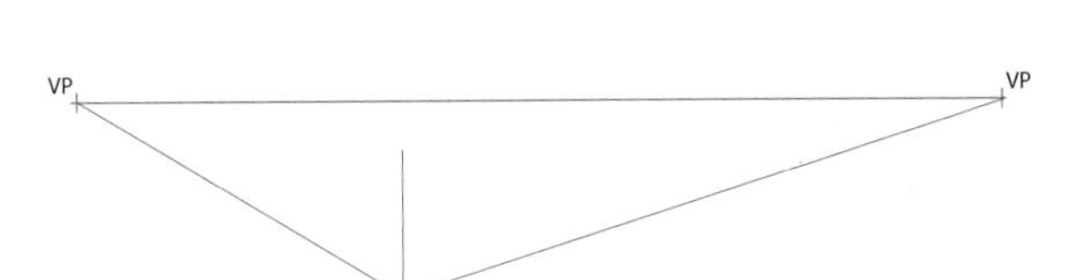

Step 3: Mark the position of the sides of the object.

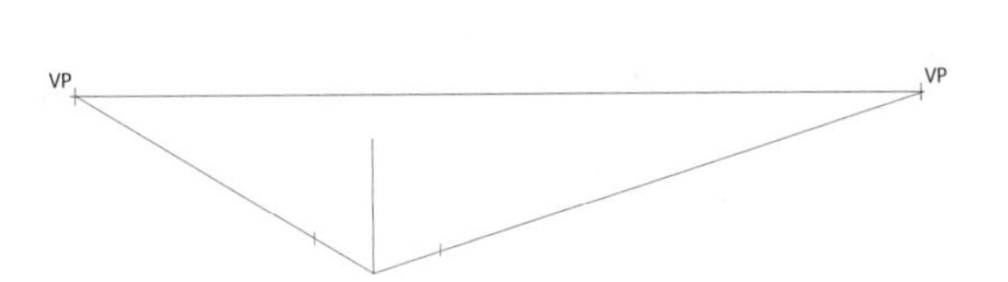

Step 4: Project lines to the third vanishing point to create the sides of the object.

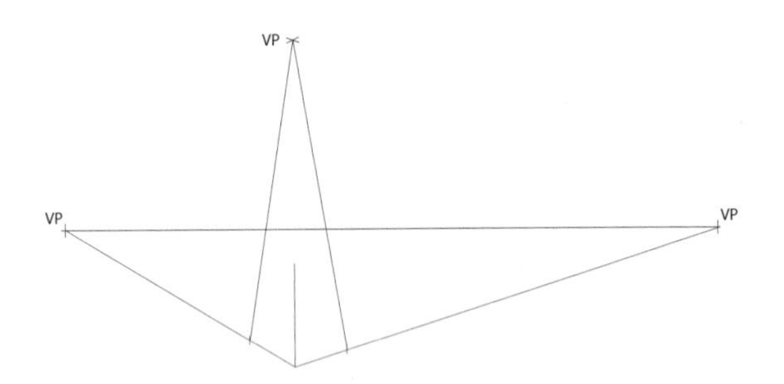

Step 5: Project lines to the original two vanishing points to create the top of the object.

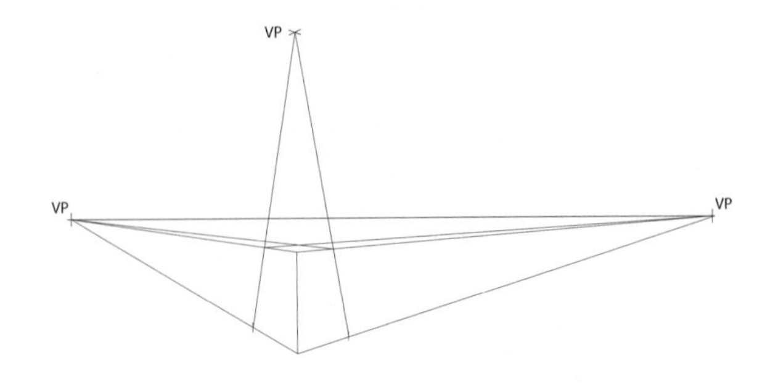

ISBN 9780170349994

Step 6: Remove all working lines and resolve the final object.

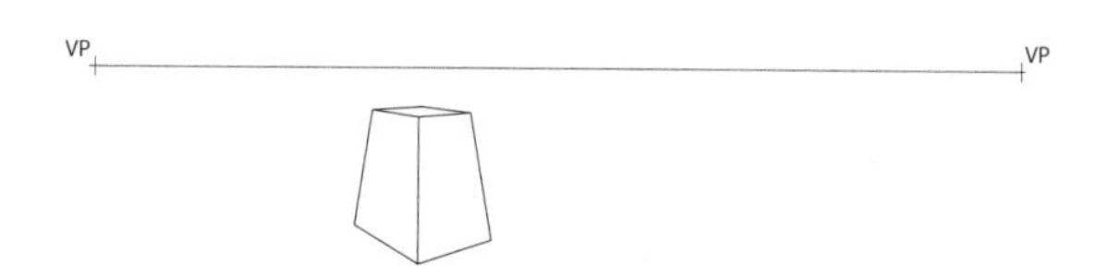

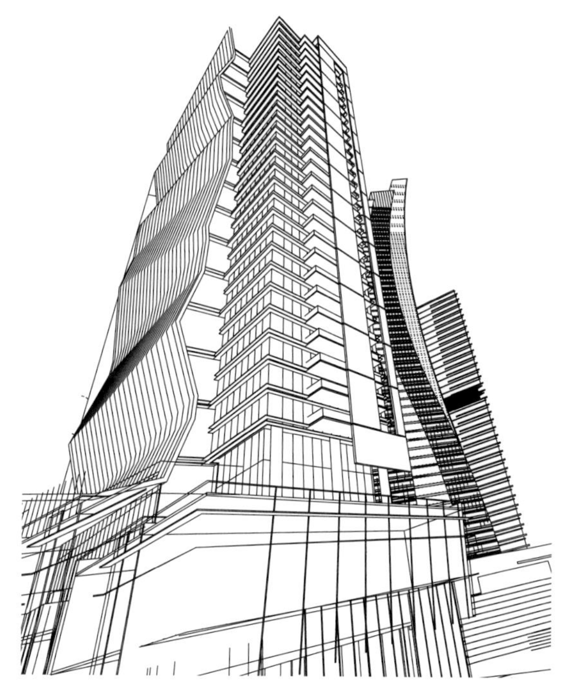

▲ Example of three-point perspective in practice. Wireframe view of multistorey office complex design.

SHADOWS IN PERSPECTIVE

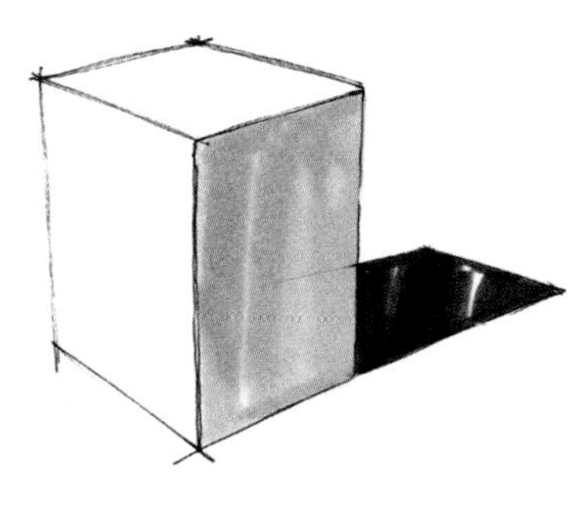

If rendering perspective drawings it is helpful to include elements that create realism; a cast shadow is one such element. The application of a shadow can give an object context and emphasise its scale and volume.

In perspective, constructing a shadow involves projecting a shape onto a surface. The rules of each perspective method apply.

STEP-BY-STEP GUIDE TO SHADOWS IN TWO-POINT PERSPECTIVE

Step 1: Establish a light source. This will direct the type of shadow you wish to apply.

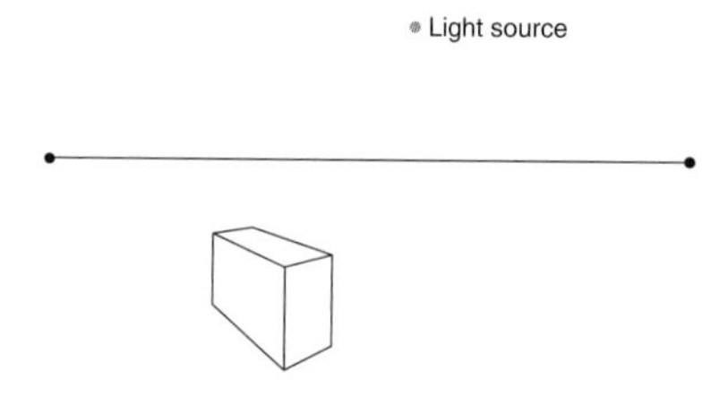

Step 2: Establish a 'shadow vanishing point'. This sits directly below the light source, behind the perspective object and in line with a relevant vanishing point.

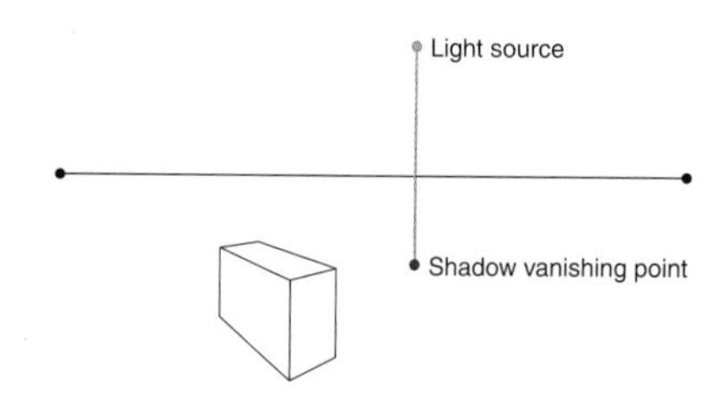

Step 3: Project lines from the shadow vanishing point via the bottom corners of the object where the shadow will be cast.

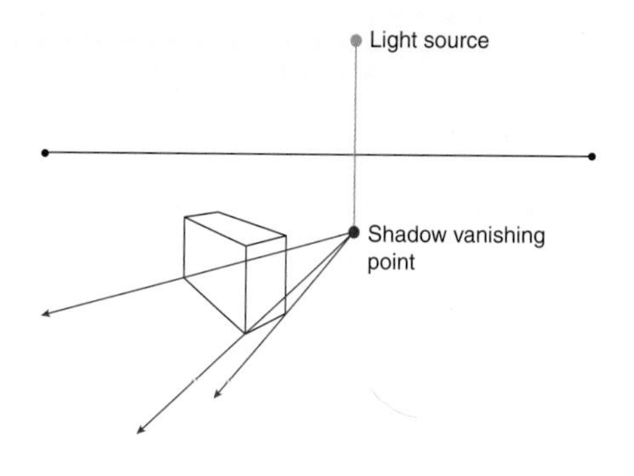

ISBN 9780170349994

Step 4: Project lines from the light source via the top corners of the perspective object.

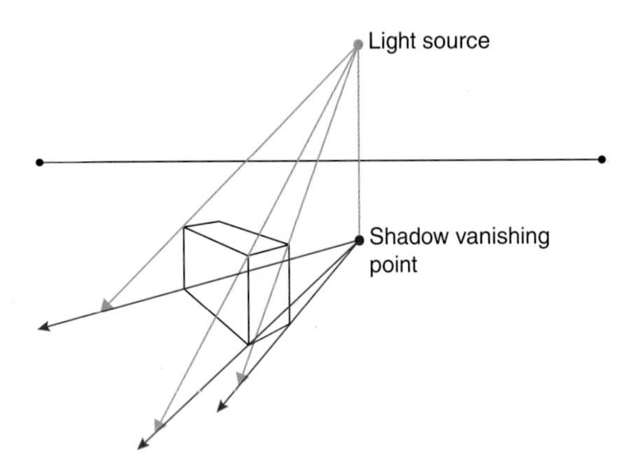

Step 5: Where the light source lines intersect with the shadow vanishing point lines, draw connecting lines.

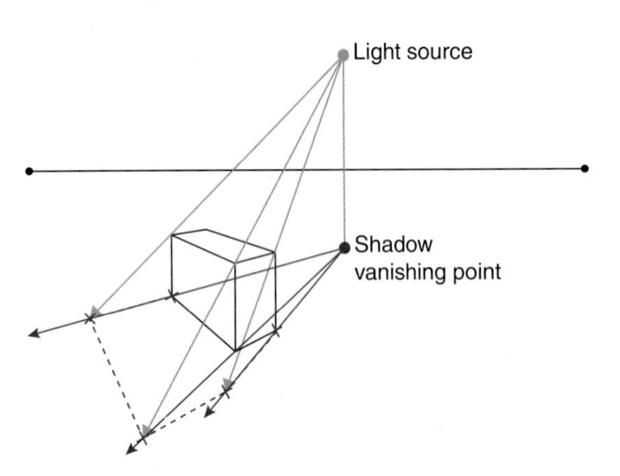

Step 6: The connected points form the shadow. Render as required.

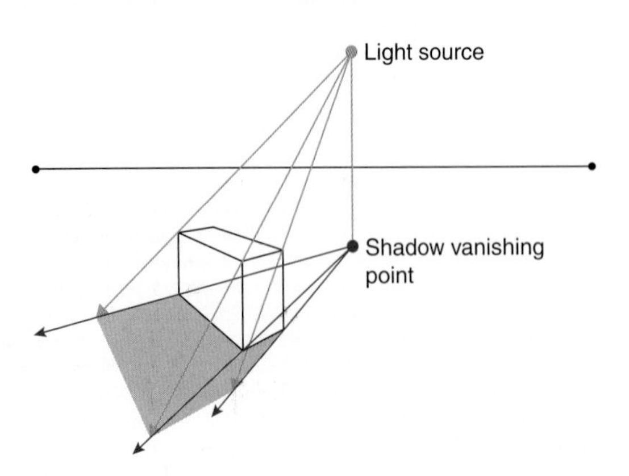

16.3 CIRCLES AND ELLIPSES

When a circle is viewed in perspective or as part of a paraline drawing, it appears as an ellipse. Depicting circular details in three-dimensional drawing can be quite a challenge but, with practice, it will become intuitive.

An ellipse is made up of two axes: a major axis and a minor axis. The major and minor axes are formed when a circle is drawn in perspective or paraline form.

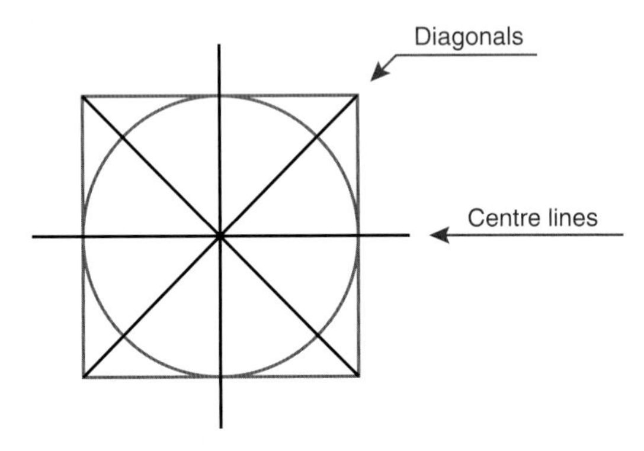

▲ A circle sits within a square at 90°. The circle features centre lines and diagonal lines; these lines assist in creating an ellipse.

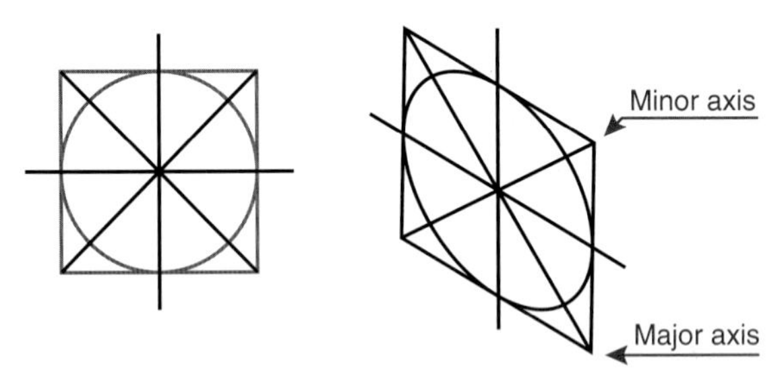

▲ The rotation of the square to 30° (isometric) means that the diagonal lines create a major (long) axis and minor (short) axis. This directs the appearance of the ellipse.

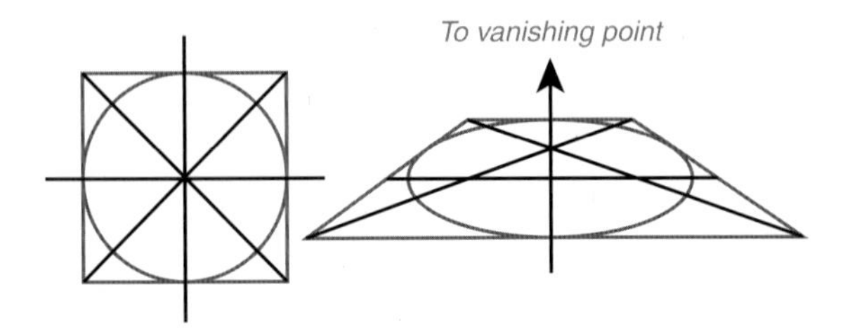

▲ A one-point perspective ellipse. Notice that the centre of the ellipse does not coincide with the intersection of centre lines. This is due to perspective foreshortening.

ISBN 9780170349994

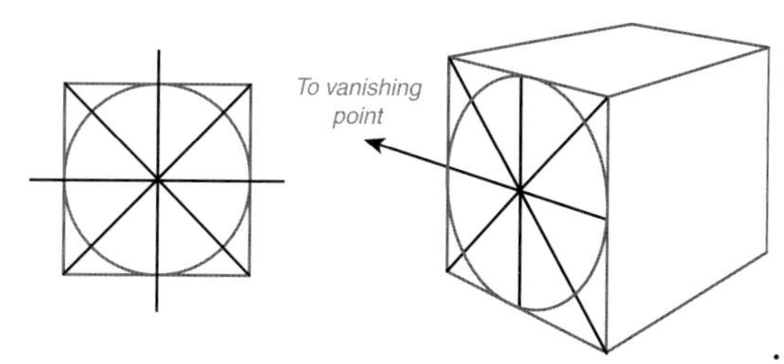

▲ The same principles apply in two-point perspective. The centre lines refer to the vanishing points.

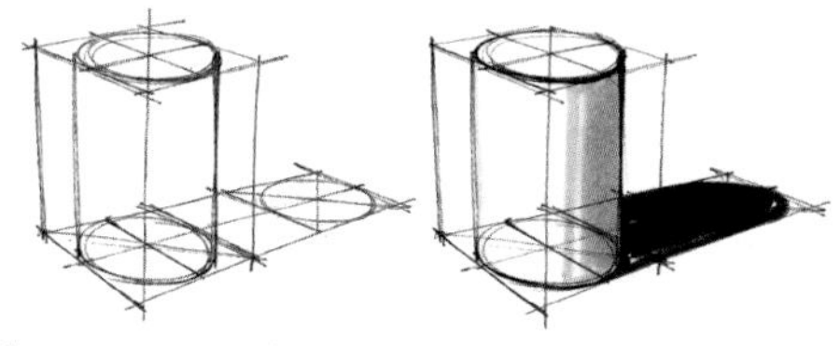

▲ When creating a shadow on a cylindrical or rounded object, use an ellipse to create the rounded edge of the shadow.

ISOMETRIC ELLIPSES

When freehand drawing isometric ellipses, the standard isometric ellipse template is a very helpful piece of equipment. Such a template enables you to draw accurate ellipses. When larger ellipses are required, it is possible to construct them manually using segments of a circle, known as arcs.

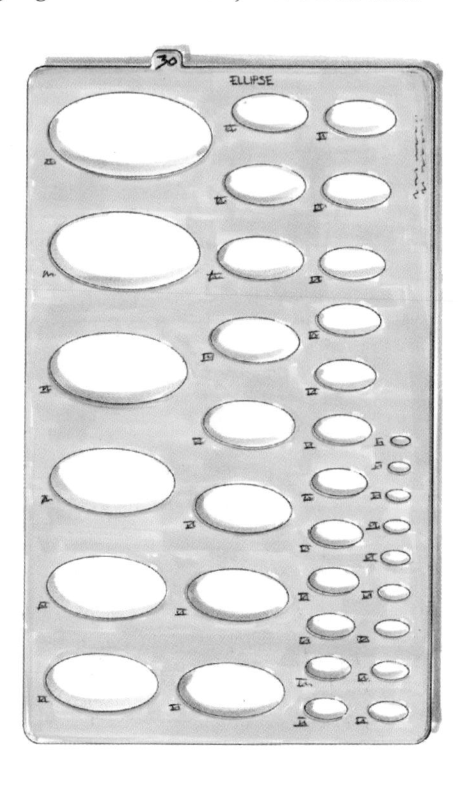

Ellipses are used when developing images such as cylinders, spheres and tubular features. The combination of ellipses creates complex forms.

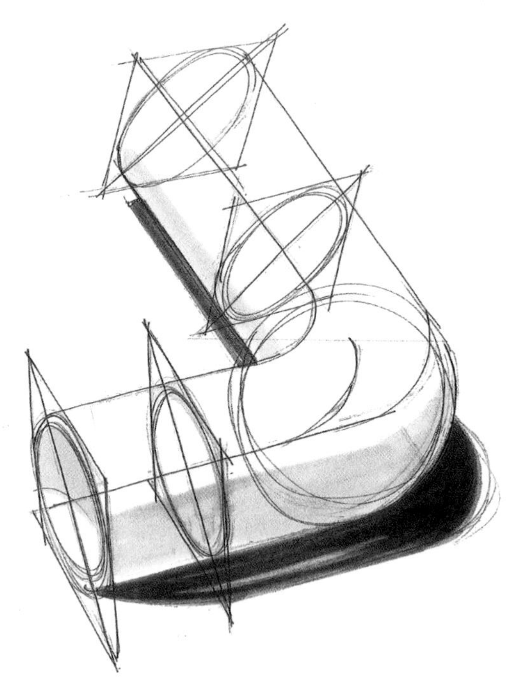

▲ Ellipses can be used in combination to form rounded and cylindrical objects.

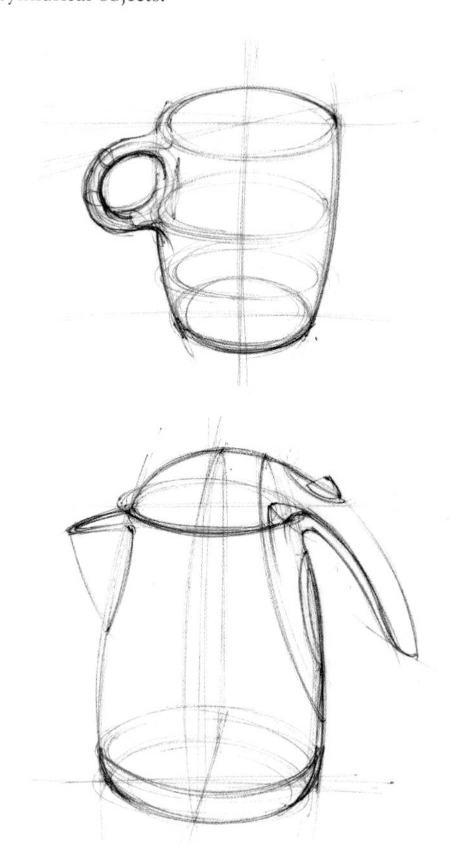

ISBN 9780170349994

16.4 DRAWING COMPLEX OBJECTS

An effective method of drawing complex three-dimensional forms is by using the crating or boxing technique. This technique involves using basic geometric forms as the foundation for constructing complex objects. There are four basic three-dimensional forms: the cone, the sphere, the cube and the cylinder.

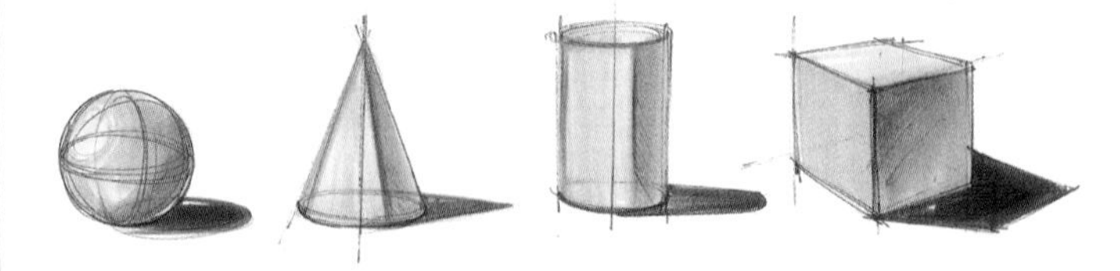

Many objects are made up of variations of these four basic forms. A bottle is a series of cylinders. A wine glass is formed by a partial sphere and cylinder. A compact camera is a combination of cubes and cylinders.

Crating is a drawing method that utilises the basic form of an object as a skeletal structure around which the finished form can be created. When an object is broken down into its most basic shape combinations, realistic proportion and scale can be established.

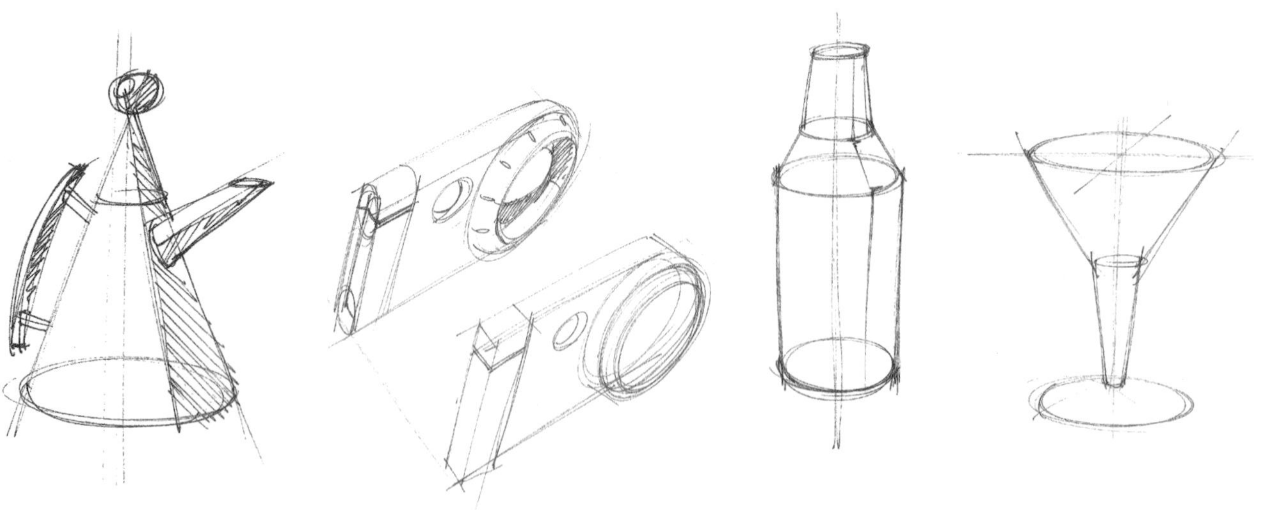

ISBN 9780170349994

USING THE CRATING TECHNIQUE

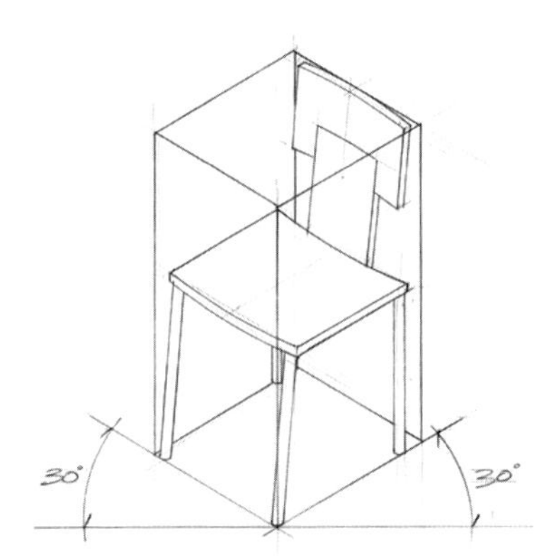

▲ Isometric

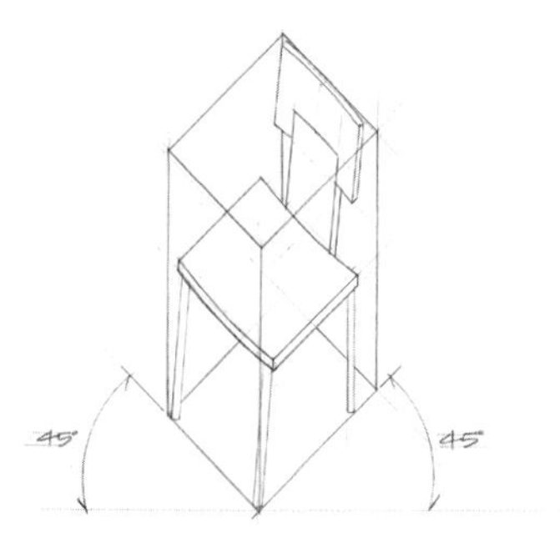

▲ Planometric

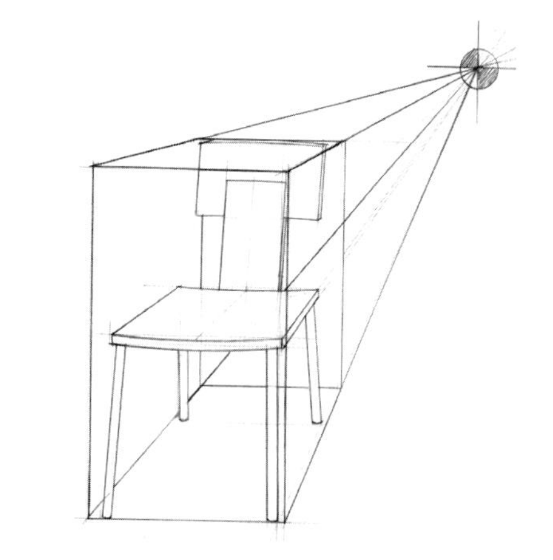

▲ One-point perspective

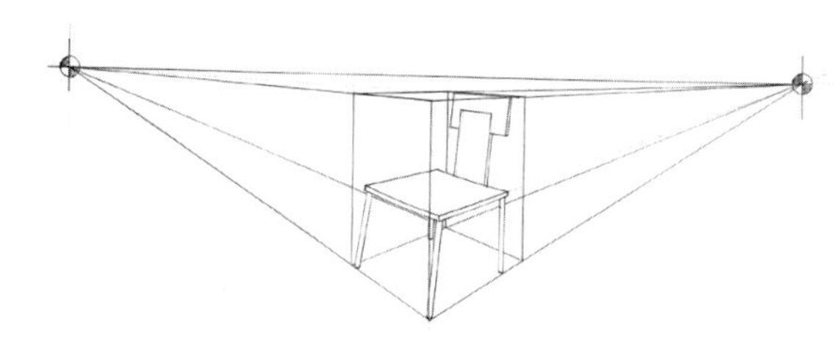

▲ Two-point perspective

DIVIDING PERSPECTIVE OBJECTS

To divide perspective objects, divide the plane of the object by using diagonal lines. Where those lines intersect is the centre of the plane.

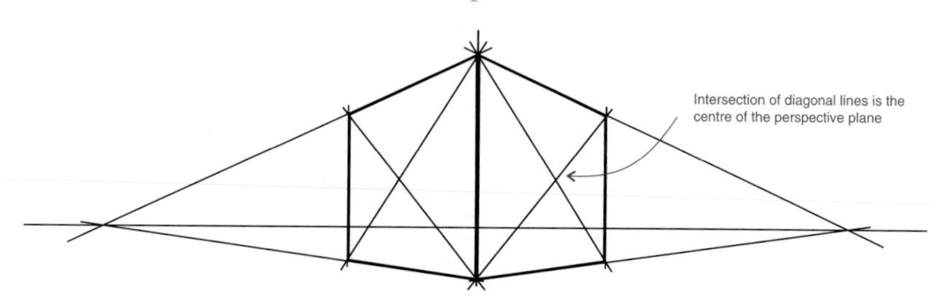

▶ Perspective drawing in context: Ikea uses simple perspective line drawings to display their merchandise. Along with the company's simple and iconic step-by-step visual instructions, these representations of their product line have become a distinctive part of the company brand.

EXPEDIT

KARLSTAD

IKEA/PS

ISBN 9780170349994

APPENDIX
DESIGN PROBLEMS

The design problems in this appendix are provided as starting points for the creation of design briefs (see Chapter 4). They present opportunities for creative problem solving within each of the three design areas defined by the Senior Graphics syllabus (see Part A). The design process is then applied to explore, develop and produce the most effective solution to the initial design problem (see Part B). Design factors are considered, investigated, addressed and applied as appropriate (see Part C). A range of graphical representations is explored with the most appropriate means of presentation applied to depict the design solution (see Part D).

1

Design problem	The need for flexible outdoor furniture suitable for a range of domestic environments but can be neatly collapsed and packed away during wet weather.
Design area	Industrial design.
Client	Innova Furniture.
User/audience	Families in homes with outdoor entertaining space. A range of ages. Interested in entertaining and enjoying the outdoor lifestyle that the Australian climate offers.
Research starters	Research existing furniture designs. Observe and analyse the functionality of objects that fold, pack away and offer flexible design for storage, e.g. flat pack.
Suggested graphical representations	Pictorial illustration of the furniture design/s. Pictorial imagery of the furniture in a likely environment. Creation of a flyer depicting and promoting the new designs to be placed at the point of sale.
Evaluation	Analysis of how effectively the design suits the needs of the target user. Analysis of how efficiently the design functions for wet weather storage.

Add on	Add on a graphic design component by creating pattern designs for application to cushions and tableware; drink ware and flatware (plates and bowls) for use when entertaining outdoors.

Add on	Add on a built environment component by designing an interior retail display for the new furniture designs. Research an existing retail space, e.g. bulky goods store and adapt the interior design to include a high-profile promotional space for the designs.

ISBN 9780170349994

2

Design problem	The design of multipurpose classroom spaces suitable for visual and performing arts education. The requirement is for flexible learning spaces that are engaging and workable for students and staff alike. The areas should address issues of accessibility and sustainability.
Design area	Built environment design.
Client	Queensland School of Visual And Performing Arts.
User/audience	Secondary school students and teachers. Interests include quality teaching and learning outcomes with a particular focus on visual and performing arts. Student ages may range from 12–18 years.
Research starters	Focus carefully on the needs of the user. Observe and existing spaces designed for creativity such as galleries and innovative tertiary or corporate campuses. Research materials, colours, textures and appropriate fixtures.
Suggested graphical representations	Pictorial illustration of the interior designs. Plans and elevations. Three-dimensional rendered simulation.
Evaluation	Analysis of how effectively the design suits the needs of the target user. Analysis of how the spaces could be used to meet the needs of the brief.

Add on	Add on a graphic design component with the design of a corporate identity/logo for the school applicable to a wide range of carriers including uniforms, stationery, web, signage and promotional materials.

Add on	Add on an industrial design component by designing a practical storage and transport solution for student visual art materials. The design should accommodate drawing materials including pencils, markers and instruments. Paints and brushes, pastels and pencils.

ISBN 9780170349994

3

Design problem	The design of the body shape and features of a hybrid sports car with consideration of traditional sports styling. The requirement is for a vehicle that defies the 'conservative' appearance of existing hybrid cars.
Design area	Built environment design.
Client	Xeco Motors.
User/audience	A youthful audience aged 25–30. Males and females whose interest in performance motor vehicles is balanced by consideration of sustainability and environmental responsibility.
Research starters	Research existing designs of motor vehicles and concept cars in particular. Address the physical requirements of hybrid technologies. Investigate current trends in styling, proportions and materials.
Suggested graphical representations	Pictorial illustration of the vehicle design. Orthographic drawing of the vehicle. Three-dimensional rendered image as part of a concept presentation.
Evaluation	Analysis of how effectively the design addressed the balance between style and sustainability. Reflection on the characteristics and features of the design.

Add on	Add on a graphic design component with the design of a corporate identity/logo for the vehicle manufacturer. Create a three-dimensional rendered image of the badge to feature on the vehicle.

Add on	Add on an industrial design component by designing a state of the art electric charging facility for the car. Rendered drawings of the charger in situ could be created.

ISBN 9780170349994

4

Design problem	The design of modular/portable/demountable temporary accommodation for people affected by crisis events, including natural disasters. The need is for accommodation that can be rapidly installed to meet basic human needs of shelter, water and security, under difficult circumstances.
Design area	Built environment design.
Client	Help for Humanity.
User/audience	People in crisis located in any area of the world affected by natural disaster or other crisis. In particular, families and groups who require immediate protection, security and comfort.
Research starters	Research designs of various modular or portable dwellings. Investigate the needs of people in crisis and under extreme stress – identify their most vital priorities.
Suggested graphical representations	Pictorial illustration of the accommodation design. Plans and elevations of the dwelling. Three-dimensional rendered image, including simulated 'fly-through'.
Evaluation	Analysis of how the design could be transported, located and used in times of need. Explanation of the functionality and practical components of the design and how they meet the needs of the users.

Add on	Add on a graphic design component with the design of a corporate identity/logo for Help for Humanity. The design may be applied to help identify the organisation as a non-government organisation. Volunteers and medical staff may wear the logo on a range of clothing options.

Add on	Add on an industrial design component through the design of a water filtration system. Water is usually the first requirement in an emergency. Devise a design that addresses the need for fresh and clean water.

ISBN 9780170349994

5

Design problem	The design of a series of book covers for a popular trilogy or book series. The covers should feature a visual relationship in the use of imagery and typography that identifies them as a series. The books need to stand out in a retail environment.
Design area	Graphic design.
Client	NC Publishers.
User/audience	Young readers aged 12–17 who are interested in imaginative and stimulating plots and characters. They are willing to commit to follow a series and enjoy collecting new issues of their favourite books.
Research starters	Research designs of successful book series and analyse the characteristics of the covers.
Suggested graphical representations	Versions of the covers and/or promotional materials, such as posters for use in store, may be created. Three-dimensional rendered images of the books including animation.
Evaluation	Analysis of how the design will attract the target audience. Evaluation of how effectively the series creates a cohesive and meaningful visual relationship.

Add on	Add on an industrial design component with the design of a point-of-purchase display for the in-store promotion of the new series. The point-of-purchase display should offer an eye-catching and tactile experience for the user/viewer. Two-dimensional technical drawings and renderings of the POP display in the context of the retail space. A model may also be suitable.

Add on	Add on a built environment component by designing an environment for the film adaptation of the series. Using three-dimentional software, create a realistic, textured and colourful space for use in key scenes of the film. Investigate CGI focused film design for inspiration.

ISBN 9780170349994

6

Design problem	The design of a relocatable housing unit or loft-style structure, designed to utilise empty, commercial rooftop spaces in densely populated urban centres. The design should incorporate landscape design considerations. The design should accommodate small households who are interested in sustainable building.
Design area	Built environment design.
Client	Urban Loft Inc.
User/audience	Professionals who live and work in inner urban areas. Young families may also be attracted to the loft designs. Interested in sustainable living, urban design, innovative architecture and landscaping. Located in high density urban areas.
Research starters	Research existing structures and locate an environment for the design. Investigate small space designs, eco design and urban renewal strategies around the world.
Suggested graphical representations	Presentation of plans and rendered elevations. Photo collage of the design placed in an existing urban location. Plan of the landscape design with diagrams indicating how the landscape enhances the sustainable nature of the loft design.
Evaluation	Analysis of how the design will achieve sustainable practices. Evaluation of the relationship between architectural design and landscape design. Analysis of how well the needs of the audience were met.

Add on	Add on a graphic design component with the creation of a multipage editorial spread in a leading architectural magazine, explaining and depicting the features of the housing unit. Three-dimensional rendered images and the photo collage could be used as illustrations. Effective use of design principles, such as hierarchy, balance and proximity along with design elements, such as type and colour, should be considered.

Add on	Add on an industrial design component by designing a recycling, composting or gardening system suitable for rooftop living in an inner-urban environment. Alternatively, design a lightweight, sustainably constructed bicycle that can be used for personal transport within the inner city.

ISBN 9780170349994

7

Design problem	The design of imagery and identity for a musician/ singer-songwriter suitable for both traditional retail and online sales. The identity is to be applied to posters, CDs and other merchandise, including an online social media presence.
Design area	Graphic design.
Client	Your choice.
User/audience	Music lovers and fans of the artist who collect all releases and attend gigs. Aged 18–25, the audience is interested in design, fashion, socialising and attending festivals. They may be students or young professionals living in urban areas.
Research starters	Research existing designs for musicians and bands. Identify the style of music and associate imagery, type and design elements that capture the genre.
Suggested graphical representations	Design of a poster promoting an upcoming gig. Online artwork and CD cover. Merchandise including t-shirts, stickers and media/ press kit about an upcoming album launch.
Evaluation	Analysis of how the design captures the genre/style of the artist. Evaluation of how effective the design is in attracting the target audience.

Add on	Add on an industrial design component with the creation of a custom guitar or other musical instrument for the performer. Create an orthographic representation of the instrument and an exploded view.

Add on	Add on a built environment design component by creating the design of a new music festival. Design the site including stage locations, accommodations, facilities, retail, parking and relaxation areas. Create elevations, landscape designs.

ISBN 9780170349994

8

Design problem	The design of a stimulating educational toy for use by young children in the early years of learning. The toy may be in the form of a puzzle or game and offer educational benefits.
Design area	Industrial design.
Client	Eduplay Toys – educational toy manufacturer.
User/audience	Very young children and toddlers. Parents of young children who are interested in the educational benefits of play.
Research starters	Examine research of developmental stages of children. Research existing products that offer educational/play combinations. Observe or investigate the reactions of children to various elements of design, e.g. colour, texture, form.
Suggested graphical representations	Presentation of orthographic and pictorial views. Rendered images of the design. Animation of how the design works. Diagram of educational benefits.
Evaluation	Analysis of how children interact with the design. Evaluation of responses from both children and parents – examination of how well the design meets the user needs.

Add on	Add on a graphic design component with packaging design for the promotion of the toy. Designed for display in a retail environment, the packaging should be appealing to children and parents alike. A scale model of the packaging with surface graphics applied.

Add on	Add on a built environment design component by creating a design for a day care centre where the educational benefits of play is the priority. Design interactive and stimulating interior and exteriors spaces for children to sue and explore. Elevations, plans and a simulation are possible presentations formats.

9

Design problem	Exhibition design for an upcoming exhibition about a significant historical event or issue in Australian history. Considerations of movement and presentation of artifacts in a contemporary and engaging manner is required.
Design area	Built environment design.
Client	Museum Australis.
User/audience	Families and individuals interested in Australian history and culture. Students and school groups who are interested in the educational value of the exhibition experience.
Research starters	Identification of theme for the exhibition. Research of effective exhibition design and the use of public spaces that allow viewing, engagement and interaction while protecting precious artifacts.
Suggested graphical representations	Plans, elevations and rendered concept drawings of the spaces within the exhibition. Map/plan of the event using three-dimensional images to identify each section of the exhibition.
Evaluation	Analysis of how effectively the design allows movement and engagement within the exhibition. Evaluation of the successful use of spaces.

Add on	Add on a graphic design component in the design of a poster and billboard designs for the promotion of the exhibition to the wider public. Digital images may be used to show the designs in contexts such as existing billboards, urban areas.

Add on	Add on an industrial design component by creating a design for a collectible souvenir available for purchase at the museum shop. The souvenir should capture the theme of the exhibition and have strong appeal to the target audience. A three-dimensional render or scale model would be appropriate.

ISBN 9780170349994

10

Design problem	The design of an innovative motorised watercraft that can be used above and below the surface of waterways. The craft may also offer watertight storage for snorkelling and fishing gear.
Design area	Industrial design.
Client	Aquatica – maritime manufacturer.
User/audience	Young, active users aged 18–30. Males and females interested in water activities and an outdoor lifestyle. The users may be very interested in marine diversity, recreational fishing or conservation. The vehicle may also be used by government and not for profit organisations to observe and maintain marine life and reef ecosystems for research.
Research starters	Research of forms and shapes suitable for aquadynamic transportation. Research of existing craft both land and water-based for inspiration. Observation of marine mammals for inspiration.
Suggested graphical representations	Orthographic drawing of the design. Pictorial drawing including exploded view of the design.
Evaluation	Analysis of how effectively the design meets the needs of the users. Evaluation of safety and suitability of the design for the environments in which it will be used.

Add on	Add on a graphic design component in the design of a website to promote the watercraft. Design should include the corporate identity of the design and diagrams that explain features and functions. Animations may show how the watercraft works.

Add on	Add on a built environment design component by designing a boathouse designed to accommodate the watercraft, diving/snorkelling accessories and a trailer. The boathouse could also provide bathing facilities and a small kitchenette.

ISBN 9780170349994

11

Design problem	The design of a secure and safe overnight camping shelter for hikers in a national park. The shelter should be in keeping with the protected natural environment.
Design area	Built environment design.
Client	Department of Natural Resources.
User/audience	Active individuals who are passionate about hiking and spending time in native bushland. Users enjoy the natural environment and the isolation that hiking can provide. Users may range from experienced hikers to first-time campers in small groups.
Research starters	Research of the environment – selection of an area for the focus of the design task. Research of similar dwellings or other structures of a similar size but different purpose. Consideration of climatic and environmental factors e.g. heat, wind, weather variation etc.
Suggested graphical representations	Plans and elevations of the dwelling. Interior images of the design including features and fixtures.
Evaluation	Analysis of how effectively the design meets the needs of the users. Evaluation of suitability of the design for the environment in which it will be used.

Add on	Add on an industrial design component in the design of a weatherproof ‘survival pack’ for hikers. Contents might include products to assist in addressing first aid needs, hunger and communication.

Add on	Add on a graphic design component by designing a safety/survival brochure or app to be provided for hikers venturing into remote areas. Information should include simple diagrams to assist in emergency situations.

ISBN 9780170349994

12

Design problem	The design of a new brand and uniform for a roller derby team based in a Queensland city, which will participate in a national competition series. The uniform should address safety concerns while retaining the style and aesthetic of this contact sport.
Design area	Graphic design.
Client	Sunshine Skate.
User/audience	Young and physically active females aged 18–35. Interests include technology, music, fashion, tattoos and socialising. They are passionate about the sport and aim to raise awareness of it using social media.
Research starters	Research of the sport itself including rules and strategies. Research of the design of garments for sporting enterprises – focus on materials.
Suggested graphical representations	Rendered drawing of the uniforms.
Evaluation	Analysis of how effectively the design meets the needs of the users. Evaluation of suitability of the design for the nature of the sport.

Add on	Add on an industrial design component in the design of a new rollerskate design. The skate should offer aerodynamic qualities, be ergonomic and use innovative materials. The design should be durable and flexible.

Add on	Add on a built environment design component by designing a clubhouse for the roller derby team. The design should include areas for fitness training and socialising. Murals and other visual features should be incorporated into the design.

ISBN 9780170349994

13

Design problem	Design of the global relay torch for the 2024 Olympic Games. Runners hold the torch for short periods during a marathon relay that takes place in the year prior to the Games.
Design area	Industrial design.
Client	2024 Olympic Games.
User/audience	A broad audience of marathon participants including adults and children selected to run. The torch should be comfortable to hold and be lightweight enough for the broad range of runners. Safety is a concern.
Research starters	Research of the history of the torch relay and the different designs that have been used in the past. Research ergonomics to ensure the design is suitable for the purpose.
Suggested graphical representations	Rendered drawing of the torch showing materials. Orthographic drawing of the design. Exploded isometric view of the torch indicating parts.
Evaluation	Evaluation of how the design suits the varied needs of the users. Summary of how the torch functions in a safe but effective and memorable manner.

Add on	Add on a built environment design component through the design of a 'recovery pod' for athletes to use between events. Designed for sleeping, oxygen recovery or the cooling of muscles, each pod enables athletes to prepare for their next event.

Add on	Add on a graphic design component by creating a logo design and signage system for the Games. Signage to represent key sporting events, their locations. The logo should symbolise the Olympic spirit. The task may also feature a mascot or character design to promote the games.

ISBN 9780170349994

14

Design problem	Redesign one or more buildings within your school environment to make them more accessible for people with a range of disabilities.
Design area	Built environment design.
Client	Your school.
User/audience	School users, both staff and students who live with a disability and require the use of accessible spaces and functions.
Research starters	Investigate and test the suitability of the selected building for different abilities. For example, walk blindfolded around your administration building to understand the challenges faced by vision-impaired employees.
Suggested graphical representations	Plans and elevations of the new design. Diagram that explains the accessible features and how they are utilised. Site plan of the school.
Evaluation	Evaluation of how the design suits the accessibility needs of the users. Evaluation of user experience.

Add on	Add on a graphic design component through the design of a way finding system for your school. Create a useful map or app and link its features to the signage system for ease of navigation. Consider accessibility issues to enhance the user-centred design of the map.

Add on	Add on an industrial design component by creating furniture for the reception area of the school. The new designs should be ergonomic and reflect contemporary design aesthetics. The use of sustainable or recycled materials is encouraged. Furniture may include the desk, seating, information or display areas and shelving. Orthographic and isometric views required.

ISBN 9780170349994

15

Design problem	The design of a comfortable and mixed purpose transportation shelter for use at an international airport. The shelter may offer interactive information panels or kiosks to assist travellers.
Design area	Built environment design.
Client	Brisbane Airport Corporation.
User/audience	Travellers, commuters and visitors to Brisbane airport. Staff may also use the shelter for transportation to staff car parks.
Research starters	Investigate the appearance and functionality of similar structures. Research the ideal functions and design of a transportation hub.
Suggested graphical representations	Plans and elevations of the new design. Site plan of the airport including the location of the new shelter.
Evaluation	Evaluation of how the design suits the airport environment. Summary of the functionality of the structure with clear explanations of each feature.

Add on	Add on a graphic design component through the design of maps, diagrams and information graphics to assist travellers to locate and make use of the shelter/terminal.

Add on	Add on an industrial design component by creating design concepts for furniture and lighting within the shelter. Ergonomics, accessibility and sustainability factors should be accounted for.

ISBN 9780170349994

16

Design problem	Typeface and poster design for the promotion of a new dance event to tour throughout Australia.
Design area	Graphic design.
Client	Contemporary Dance Company.
User/audience	Adults aged 30–50 years who are interested in cultural events and the performing arts. Located in urban centres throughout Australia. Other interests might include music, eating out and contemporary art.
Research starters	Research promotional materials for varying arts events and festivals, including the use of typography. Establish theme and focus of the event -investigate related imagery.
Suggested graphical representations	Final poster and related free postcards. Website for promotion of the event.
Evaluation	Evaluation of how the design suits the interests of the users. Summary of how the design captures the theme of the event.

Add on	Add on a built environment design component in the set design for the dance production that reflects the desired theme of the production. Renderings, elevations and plans required.

Add on	Add on an industrial design component by creating a series of props that are to be used as part of the production (defined by theme). Orthographic and exploded views required.

ISBN 9780170349994

17

Design problem	The design of a temporary retail kiosk for a shopping centre. The kiosk may be used to sell seasonal products such as Christmas gifts or decorations, Valentine's Day products etc.
Design area	Built environment design.
Client	Suburban retail centre.
User/audience	Consumers aged 16–50 who celebrate significant events and enjoy giving and receiving special occasion gifts. Interests include shopping, collecting and socialising.
Research starters	Research existing small retail spaces or 'pop-up' stores. Research imagery and colour appropriate to the event that is being promoted.
Suggested graphical representations	Plans and elevations are required. A scale model of the kiosk may be suitable. Perspective rendering of the kiosk in context.
Evaluation	Evaluation of how the design suits the environment and promotes the required events. Summary of how the design captures the attention of the target audience.

Add on	Add on a graphic design component in the design of pattern and imagery to be applied to the kiosk in the form of banners, digital imagery or other promotional media.

Add on	Add on an industrial design component by designing a set of headphones that can be altered with collectible inserts or changes in materials/features to suit special occasions, clothing colours or the mood of the user. To be sold in a retail stores and online. Visual explanations of how the product works are required including exploded diagrams and orthographic drawings.

ISBN 9780170349994

18

Design problem	The interior design of a beachside cafe with outdoor seating area on the sand. The café may have a sustainable focus and use recycled materials in its design.
Design area	Built environment design.
Client	Beach café.
User/audience	Consumers aged 18–40 who enjoy dining in the outdoors, like to indulge in healthy food and who make time to socialise.
Research starters	Research appropriate sustainable materials for a beachside location. Investigation of current trends in restaurant and café design.
Suggested graphical representations	Plans and elevations. A scale model or three-dimensional rendered model of the cafe. Perspective rendering of the interior.
Evaluation	Evaluation of how the design sits within the environment and makes use of recycled materials in its construction. Summary of the design features.

Add on	Add on a graphic design component in the design of an identity for the cafe including signage, menus and a web page.

Add on	Add on an industrial design component by designing a lighting feature for the café. Using the recycling theme, the lighting design should emphasise the environment/location of the café.

ISBN 9780170349994

19

Design problem	The design of a storage device to contain cables and charging devices for laptops, tablets and smartphones. The device should protect loose cables but be easy to use and accessible. It may have a retractable function.
Design area	Industrial design.
Client	NexTech products.
User/audience	Tech-savvy consumers and people working in businesses where the compact packing/storage of charging packs and cables are important. Business travellers and home users who prefer a tidy workplace.
Research starters	Research storage systems across a range of products. Analysis of functions and needs of the target users.
Suggested graphical representations	Exploded drawing. Orthographic drawing. Product rendering using perspective and illustrating the product in context.
Evaluation	Evaluation of how the design meets the needs of the user. Explanation of the functionality of the product.

Add on	Add on a graphic design component in the design of a web page or magazine advertisement targeting business users. A brand identity and packaging may also be required.

Add on	Add on built environment design component by an innovative office space for the manufacturing company. Investigate engaging business campuses, such as Facebook and Google, to research cutting edge corporate environment designs.

ISBN 9780170349994

20

Design problem	The design of an information pack or smartphone/tablet app for young people travelling overseas for the first time. The pack may include safety and security information, contact details and advice for inexperienced travellers. The content must be engaging and memorable.
Design area	Graphic design.
Client	SmartTraveller.gov.au.
User/audience	Young people aged 17–20 years who are travelling overseas for the first time. Interests include parties, sightseeing, sports and experiencing new cultures.
Research starters	Research the key information needed by first time travellers. Identify risks and research effective information designs to help convey information in an engaging and memorable way.
Suggested graphical representations	Production of the pack. Diagrams, maps, graphs and symbols.
Evaluation	Evaluation of how the design meets the needs of the user and conveys the key information effectively. Explanation of how the information might be accessed or used.

Add on	Add on an industrial design component in the design of a pickpocket-proof wallet or extra-secure daypack/backpack for travellers to use while away.

Add on	Add on built environment design component by designing the 'ideal' youth hostel to be constructed in a popular resort area. Elements might include safety considerations; secure areas and zones for different interests, ages and experiences. Rooms and whole floors may be designed and decorated in distinctive and creative styles.

21

Design problem	The design of a food truck showcasing contemporary and innovative new cuisines to be used at festivals and markets. The truck is a mobile vendor that attracts passers-by and provides a culinary experience for diners. It may be part of a fleet of trucks servicing the coast.
Design area	Industrial design.
Client	Your choice of cuisine.
User/audience	Festival-goers and market attendees who are interested in trying new food and beverages. The audience may range from families with young children to young people and couples. Additional audience interests may include music, shopping, dining out and the outdoors.
Research starters	Research the requirements of a food truck, including functionality and required fixtures. Investigate other mobile vehicles that offer multiple functions, e.g. caravans, mobile homes.
Suggested graphical representations	Orthographic representation of the van. Interior sketches or renderings. Plans indicating flow and function.
Evaluation	Evaluation of how the design functions and provides a service to users. Explanation of how the design addresses the requirements of the design brief.

Add on	Add on a graphic design component in the design of an identity for the truck. The identity will be applied to an app (so users can find the truck location), signage, decals, packaging and staff uniforms.

Add on	Add on built environment design component by designing a restaurant or café that is the central home of the food truck fleet. Consider the link between the cuisine and the interior/exterior spaces. Produce plans, elevations and interior renderings of the spaces.

50problems50days.com
UK designer, Peter Smart travelled through Europe and identified 50 design problems on each day of his travels. His interactive website illustrates his proposed solutions to each problems as well as providing insights into his design process.

ISBN 9780170349994

GLOSSARY

abstract Imagery that does not realistically represent life

accessibility Consideration of a range of abilities and disabilities in design and adjustments made to maximise the use of a space, product or graphic

aesthetic Considerations of appearances that are attractive or in good taste

alignment The position of text or images in a composition in relation to a grid or axis

analysis The detailed examination of the composition, elements, principles and components of a design

animation A series of images arranged in a timed sequence suggesting continuous movement

Art Deco Design style characterised by the use of geometric shapes and forms

art director An individual with responsibility for managing the creative and production process usually within a design studio or advertising agency

Art Nouveau Design movement characterised by decorative, organic forms and inspired by Asian art

ascender In typography, the parts of lowercase letters that rise above the x-height, e.g. b, d, f, h, k, i, and t

axis An imaginary straight line around which compositional elements are grouped

axonometric drawing Sometimes referred to as paraline drawing. Constructed of lines that remain parallel and do not converge at any given point

baseline In typography, the horizontal line upon which the main body of the type sits. Rounded letters actually dip slightly below the baseline to give optical balance

body text Term used to describe type used for long passages of text, such as articles in a newspaper or magazine or chapters in a book

brainstorming Visual, verbal and written techniques designed to rapidly generate creative ideas and solutions

CADD Computer-aided design and drafting

client The individual or organisation for whom designs are created. Clients usually provide payment for design services

CMYK The four process colour inks: Cyan, Magenta, Yellow and Black

collage Method of pasting shapes cut from materials including paper and newsprint onto a surface

composition The arrangement of design elements and visual information on a surface

concept/conceptual Concerned with ideas rather than the tangible, real outcome or product

constraints Creative restrictions placed upon a designer, usually outlined in the design brief

contemporary Belonging to the current era

context The circumstances surrounding a visual communication, i.e. its physical location

contour A line that traces the outer surfaces and form of an object

contrast Application of opposing elements for visual effect

copyright Legal protection against copying or misuse provided to films, images, music, broadcasts, artistic works and theatrical products. Copyright is automatically applied under Australian law

cropping The removal of visual material to enhance visual impact

crosshatching Rendering technique that uses overlapping diagonal lines to suggest tone

Dada A group of reactionary artists, poets and writers that originated in Zurich, Switzerland

design brief Written or verbal instruction to a designer outlining a design task. It features information including the client need, design constraints, audience, purpose and context

design elements The building blocks that we use to construct a composition; the fundamental components of a composition. There are eight design elements: colour, form, line, point, shape, texture, tone and type

design factors Design considerations used to inspire, inform and assess successful design outcomes: including user-centred design, elements and principles of design, design technologies, legal responsibilities, project management, sustainability and materials

design principles Principles that direct how we use design elements to develop a composition

design process The cyclical process involved in generating, exploring, developing and producing design solutions

design technologies The range of tools, processes and skills needed to realise graphical solutions

digital imagery Images created electronically

dimensions Written measurements placed on a technical drawing

dot rendering Rendering technique that uses uniform dots of colour to create tone

dpi Dots per inch. Used in inkjet printing. Refers to the number of dots per square inch of image

elements of design The building blocks that are used to construct a composition; the fundamental components of a composition. They include colour, space, form, line, point, shape, texture, tone and type

ergonomics The study of human factors, such as comfort and usability, in product and interface design

evaluate To assess effectiveness

foreshortening Drawing technique that visually indicates objects that are closer to the viewer

golden section Mathematical calculation also known as golden mean. Refers to the height to width ratio between elements of a composition

graphical conventions The conventions, rules, standards or requirements that are applied in the production of graphical representations. Conventions are applicable to particular design areas and will change according to circumstance and audience. Conventions include the Australian Standards

graphical representations (sometimes called design products) The range of graphical products that demonstrate both the development of ideas and design solutions

icon Symbolic or sacred imagery that represents a concept

ideation The process of generating new concepts

intellectual property Legal protection for registered products such as new inventions, brands, designs or artistic creations. Intellectual property rights are not automatically applied under Australian law

isometric drawing Paraline drawing method where the length and width are drawn at 30°

kerning The manipulation of space between individual letters of a typeface

layout The arrangement of visual elements, usually in a two-dimensional context

leading The distance between two lines of type

legibility The visual clarity of text

Minimalism Style of design where decoration and detail are minimal. Shapes, spaces and forms are 'clean' and uncluttered

mixed media The application of different materials and methods in the production of an illustration

modernism A design aesthetic characterised by the use of modern materials (such as steel and glass), the application of abstract forms, the manipulation of space and a conservative colour palette

multimedia The use of varied software packages to create digital products using sound and vision

opacity The density of a colour or tonal value

opaque Not transparent

organic Irregular shapes based on and inspired by natural shapes and form

orthographic drawing Sometimes referred to as multiview drawing. A series of drawings – known as 'views' – are drawn to show every part of the object clearly

paraline drawing Drawing methods where all lines remain parallel

perspective drawing Drawing method in which objects appear to recede to given points in the distance

pixel A very small unit of visual information in digital form

planometric drawing Paraline drawing method where the length and width are drawn at 45°

postmodernism A complex term used to describe the progressive architecture, design, literature, visual communications, music, sociology and film that has evolved since the 1960s. The postmodernism movement was a reaction against the perceived structural constraints of modernism, and is characterised by decoration, ornamentation and experimental approaches to design

ppi Pixels per inch. Used in software. Refers to the number of pixels per square inch of image

principles of design Principles that direct how we use design elements to develop a composition. They include alignment, balance, hierarchy, contrast, unity, proximity, repetition, proportion, scale, consistency and cropping

proportion The relative scale of objects in relation to each other

ISBN 9780170349994

prototype Experimental model designed to test and evaluate a design

proximity Placement of elements close together, creating a visual relationship

raster Images made of pixels, e.g. scanned images and digital photographs. Raster software includes Adobe Photoshop

rendering The use of tone and colour in drawing to create texture, surface detail and form

resolution The quality of a digital image, determined by the number of pixels per square inch (ppi)

RGB Colour system for capturing images on a digital screen: Red, Green, Blue

sans serif Typefaces that do not have serif

serif A counterstroke on letterforms, projecting from the ends of the main strokes

stylised Visual communications that appear to be structured around a set of compositional rules

subjective Not objective. Displaying personal judgements and feelings

sustainability Involves the connection and interaction between social, ethical, economic and environmental systems to ensure sustainable outcomes

symmetry Visual elements mirrored on either side of an invisible axis

taper An object that gradually becomes smaller towards one end

thumbnails Small sketches used to generate ideas quickly

tonal scale The range of grey variations from white to black

user-centred design Designs in which the needs of the user are prioritised

vanishing point Point on the horizon where lines appear to converge

vector Images made from paths: lines and shapes. Created in programs such as Adobe Illustrator

views Representations of surfaces in orthogonal drawing. Common views in third-angle projection are front, top, right- and left-hand sides

visual diary Bound sketchbook used as a workbook for developmental drawings

visual weight 'Heaviness' or 'lightness' of elements (including white space) within a composition

wayfinding The design and use of maps and signage

white space Areas of a composition without visual material. Can create balance and visual weight

INDEX

ISBN 9780170349994

ISBN 9780170349994

ISBN 9780170349994